3rd International Conference on the Application of Superabsorbent Polymers (SAP) and Other New Admixtures Towards Smart Concrete

RILEM BOOKSERIES
Volume 24

RILEM, The International Union of Laboratories and Experts in Construction Materials, Systems and Structures, founded in 1947, is a non-governmental scientific association whose goal is to contribute to progress in the construction sciences, techniques and industries, essentially by means of the communication it fosters between research and practice. RILEM's focus is on construction materials and their use in building and civil engineering structures, covering all phases of the building process from manufacture to use and recycling of materials. More information on RILEM and its previous publications can be found on www.RILEM.net. Indexed in SCOPUS, Google Scholar and SpringerLink.

More information about this series at http://www.springer.com/series/8781

William P. Boshoff · Riaan Combrinck ·
Viktor Mechtcherine · Mateusz Wyrzykowski
Editors

3rd International Conference on the Application of Superabsorbent Polymers (SAP) and Other New Admixtures Towards Smart Concrete

Springer

Editors
William P. Boshoff
EBIT
University of Pretoria
Hatfield, South Africa

Viktor Mechtcherine
Fakultät Bauingenieurwesen
TU Dresden
Dresden, Germany

Riaan Combrinck
Stellenbosch University
Stellenbosch, South Africa

Mateusz Wyrzykowski
Empa
Dübendorf, Switzerland

ISSN 2211-0844 ISSN 2211-0852 (electronic)
RILEM Bookseries
ISBN 978-3-030-33341-6 ISBN 978-3-030-33342-3 (eBook)
https://doi.org/10.1007/978-3-030-33342-3

This Springer imprint is published by the registered company Springer Nature Switzerland AG
The registered company address is: Gewerbestrasse 11, 6330 Cham, Switzerland

Preface

In classical handbooks on concrete technology, concrete is usually described as a composite made of cement, aggregates, water, and admixtures. The chemical admixtures, even though contributing to a usually negligible part of the volume of a concrete mix, play a central role in modern concrete technology. They allow to modify such key aspects as rheological behavior of concrete (flow and setting times), air content, density, shrinkage, and others. In particular, the development of the science of chemical admixtures enables (and is at the same time promoted by) the technology of high-performance concrete, in particular concrete based on alternative, supplementary cementitious materials.

One of the recent advancements in the technology and science of concrete admixtures regards the application of superabsorbent polymers (SAP). Numerous studies have proven the multi-purpose applicability of the SAP in modern concrete technology, in particular for the reduction of early age (autogenous and plastic) shrinkage and the modification of pore structure (improved freeze thaw resistance). While the SAP are being currently introduced in practice, there is a need for further scientific studies on the properties of the SAP themselves and their effect on the concrete properties.

This aspect was the main motivation for the International Conference on the Application of Superabsorbent Polymers (SAP) and Other New Admixtures Towards Smart Concrete (November 25–27, Skukuza, South Africa). This conference is a third in a series (the first took place in Lyngby, Denmark, on August 15–18, 2010, and the second in Dresden, Germany, on September 14–17, 2014). The third conference builds on the experience and summarizes the activities of the RILEM Technical Committee 260-RSC "Recommendations for Use of Superabsorbent Polymers in Concrete Construction" and scientifically confronts them with other prominent research activities in the field of concrete admixtures.

The main goal of the RILEM TC 260-RSC was to facilitate the application of SAP in concrete practice by delivering a set of practically oriented RILEM recommendations (three recommendations published by the committee in the journal Materials and Structures). These recommendations stem from the number of reviews and interlaboratory studies obtained by both this committee and the predecessor RILEM TC 225-SAP. We believe that the international conference will further contribute to this important goal.

The presented conference proceedings gather the papers devoted in major part to the studies of the behavior of SAP in cement-based materials, mainly concrete (in particular the absorption behavior) as well as the effect of the SAP on fresh and hardened concrete properties. Further topics include other modern admixtures, in particular rheology-modifying admixtures, including the recently emerging field of bio- or waste-derived admixtures. We believe that the proceedings offer the most recent overview of the topics of superabsorbent polymers and other modern admixtures in concrete technology from the leading worldwide experts in the field. All papers presented in these proceedings were reviewed by an international scientific committee.

The scientific committee of the conference stems from the RILEM TC 260-RSC. We would like to thank for the informational support of the conference provided by RILEM - International Union of Laboratories and Experts in Construction Materials, Systems and Structures and by *fib* - International Federation of Structural Concrete. We would also like to thank the Local Organizing Committee from the University of Pretoria and the Stellenbosch University.

William P. Boshoff
Editor and Conference Chair

Viktor Mechtcherine
Mateusz Wyrzykowski
Editors and Conference Co-chairs

Riaan Combrinck
Editor

Contents

Mechanical Properties and Durability

Shrinkage and Associated Cracking

Rheology and Admixtures

Applications and Innovations

RILEM Publications

The following list is presenting the global offer of RILEM Publications, sorted by series. Each publication is available in printed version and/or in online version.

RILEM Proceedings (PRO)

PRO 1: Durability of High Performance Concrete (ISBN: 2-912143-03-9; e-ISBN: 2-351580-12-5; e-ISBN: 2351580125); *Ed. H. Sommer*

PRO 2: Chloride Penetration into Concrete (ISBN: 2-912143-00-04; e-ISBN: 2912143454); *Eds. L.-O. Nilsson and J.-P. Ollivier*

PRO 3: Evaluation and Strengthening of Existing Masonry Structures (ISBN: 2-912143-02-0; e-ISBN: 2351580141); *Eds. L. Binda and C. Modena*

PRO 4: Concrete: From Material to Structure (ISBN: 2-912143-04-7; e-ISBN: 2351580206); *Eds. J.-P. Bournazel and Y. Malier*

PRO 5: The Role of Admixtures in High Performance Concrete (ISBN: 2-912143-05-5; e-ISBN: 2351580214); *Eds. J. G. Cabrera and R. Rivera-Villarreal*

PRO 6: High Performance Fiber Reinforced Cement Composites - HPFRCC 3 (ISBN: 2-912143-06-3; e-ISBN: 2351580222); *Eds. H. W. Reinhardt and A. E. Naaman*

PRO 7: 1st International RILEM Symposium on Self-Compacting Concrete (ISBN: 2-912143-09-8; e-ISBN: 2912143721); *Eds. Å. Skarendahl and Ö. Petersson*

PRO 8: International RILEM Symposium on Timber Engineering (ISBN: 2-912143-10-1; e-ISBN: 2351580230); *Ed. L. Boström*

PRO 9: 2nd International RILEM Symposium on Adhesion between Polymers and Concrete ISAP '99 (ISBN: 2-912143-11-X; e-ISBN: 2351580249); *Eds. Y. Ohama and M. Puterman*

PRO 10: 3rd International RILEM Symposium on Durability of Building and Construction Sealants (ISBN: 2-912143-13-6; e-ISBN: 2351580257); *Eds. A. T. Wolf*

PRO 11: 4th International RILEM Conference on Reflective Cracking in Pavements (ISBN: 2-912143-14-4; e-ISBN: 2351580265); *Eds. A. O. Abd El Halim, D. A. Taylor and El H. H. Mohamed*

PRO 12: International RILEM Workshop on Historic Mortars: Characteristics and Tests (ISBN: 2-912143-15-2; e-ISBN: 2351580273); *Eds. P. Bartos, C. Groot and J. J. Hughes*

PRO 13: 2nd International RILEM Symposium on Hydration and Setting (ISBN: 2-912143-16-0; e-ISBN: 2351580281); *Ed. A. Nonat*

PRO 14: Integrated Life-Cycle Design of Materials and Structures - ILCDES 2000 (ISBN: 951-758-408-3; e-ISBN: 235158029X); (ISSN: 0356-9403); *Ed. S. Sarja*

PRO 15: Fifth RILEM Symposium on Fibre-Reinforced Concretes (FRC) - BEFIB'2000 (ISBN: 2-912143-18-7; e-ISBN: 291214373X); *Eds. P. Rossi and G. Chanvillard*

PRO 16: Life Prediction and Management of Concrete Structures (ISBN: 2-912143-19-5; e-ISBN: 2351580303); *Ed. D. Naus*

PRO 17: Shrinkage of Concrete – Shrinkage 2000 (ISBN: 2-912143-20-9; e-ISBN: 2351580311); *Eds. V. Baroghel-Bouny and P.-C. Aïtcin*

PRO 18: Measurement and Interpretation of the On-Site Corrosion Rate (ISBN: 2-912143-21-7; e-ISBN: 235158032X); *Eds. C. Andrade, C. Alonso, J. Fullea, J. Polimon and J. Rodriguez*

PRO 19: Testing and Modelling the Chloride Ingress into Concrete (ISBN: 2-912143-22-5; e-ISBN: 2351580338); *Eds. C. Andrade and J. Kropp*

PRO 20: 1st International RILEM Workshop on Microbial Impacts on Building Materials (CD 02) (e-ISBN 978-2-35158-013-4); *Ed. M. Ribas Silva*

PRO 21: International RILEM Symposium on Connections between Steel and Concrete (ISBN: 2-912143-25-X; e-ISBN: 2351580346); *Ed. R. Eligehausen*

PRO 22: International RILEM Symposium on Joints in Timber Structures (ISBN: 2-912143-28-4; e-ISBN: 2351580354); *Eds. S. Aicher and H.-W. Reinhardt*

PRO 23: International RILEM Conference on Early Age Cracking in Cementitious Systems (ISBN: 2-912143-29-2; e-ISBN: 2351580362); *Eds. K. Kovler and A. Bentur*

PRO 24: 2nd International RILEM Workshop on Frost Resistance of Concrete (ISBN: 2-912143-30-6; e-ISBN: 2351580370); *Eds. M. J. Setzer, R. Auberg and H.-J. Keck*

PRO 25: International RILEM Workshop on Frost Damage in Concrete (ISBN: 2-912143-31-4; e-ISBN: 2351580389); *Eds. D. J. Janssen, M. J. Setzer and M. B. Snyder*

PRO 26: International RILEM Workshop on On-Site Control and Evaluation of Masonry Structures (ISBN: 2-912143-34-9; e-ISBN: 2351580141); *Eds. L. Binda and R. C. de Vekey*

PRO 27: International RILEM Symposium on Building Joint Sealants (CD03; e-ISBN: 235158015X); *Ed. A. T. Wolf*

PRO 28: 6th International RILEM Symposium on Performance Testing and Evaluation of Bituminous Materials - PTEBM'03 (ISBN: 2-912143-35-7; e-ISBN: 978-2-912143-77-8); *Ed. M. N. Partl*

PRO 29: 2nd International RILEM Workshop on Life Prediction and Ageing Management of Concrete Structures (ISBN: 2-912143-36-5; e-ISBN: 2912143780); *Ed. D. J. Naus*

PRO 30: 4th International RILEM Workshop on High Performance Fiber Reinforced Cement Composites - HPFRCC 4 (ISBN: 2-912143-37-3; e-ISBN: 2912143799); *Eds. A. E. Naaman and H. W. Reinhardt*

PRO 31: International RILEM Workshop on Test and Design Methods for Steel Fibre Reinforced Concrete: Background and Experiences (ISBN: 2-912143-38-1; e-ISBN: 2351580168); *Eds. B. Schnütgen and L. Vandewalle*

PRO 32: International Conference on Advances in Concrete and Structures 2 vol. (ISBN (set): 2-912143-41-1; e-ISBN: 2351580176); *Eds. Ying-shu Yuan, Surendra P. Shah and Heng-lin Lü*

PRO 33: 3rd International Symposium on Self-Compacting Concrete (ISBN: 2-912143-42-X; e-ISBN: 2912143713); *Eds. Ó. Wallevik and I. Níelsson*

PRO 34: International RILEM Conference on Microbial Impact on Building Materials (ISBN: 2-912143-43-8; e-ISBN: 2351580184); *Ed. M. Ribas Silva*

PRO 35: International RILEM TC 186-ISA on Internal Sulfate Attack and Delayed Ettringite Formation (ISBN: 2-912143-44-6; e-ISBN: 2912143802); *Eds. K. Scrivener and J. Skalny*

PRO 36: International RILEM Symposium on Concrete Science and Engineering – A Tribute to Arnon Bentur (ISBN: 2-912143-46-2; e-ISBN: 2912143586); *Eds. K. Kovler, J. Marchand, S. Mindess and J. Weiss*

PRO 37: 5th International RILEM Conference on Cracking in Pavements – Mitigation, Risk Assessment and Prevention (ISBN: 2-912143-47-0; e-ISBN: 2912143764); *Eds. C. Petit, I. Al-Qadi and A. Millien*

PRO 38: 3rd International RILEM Workshop on Testing and Modelling the Chloride Ingress into Concrete (ISBN: 2-912143-48-9; e-ISBN: 2912143578); *Eds. C. Andrade and J. Kropp*

PRO 39: 6th International RILEM Symposium on Fibre-Reinforced Concretes - BEFIB 2004 (ISBN: 2-912143-51-9; e-ISBN: 2912143748); *Eds. M. Di Prisco, R. Felicetti and G. A. Plizzari*

PRO 40: International RILEM Conference on the Use of Recycled Materials in Buildings and Structures (ISBN: 2-912143-52-7; e-ISBN: 2912143756); *Eds. E. Vázquez, Ch. F. Hendriks and G. M. T. Janssen*

PRO 41: RILEM International Symposium on Environment-Conscious Materials and Systems for Sustainable Development (ISBN: 2-912143-55-1; e-ISBN: 2912143640); *Eds. N. Kashino and Y. Ohama*

PRO 42: SCC'2005 - China: 1st International Symposium on Design, Performance and Use of Self-Consolidating Concrete (ISBN: 2-912143-61-6; e-ISBN: 2912143624); *Eds. Zhiwu Yu, Caijun Shi, Kamal Henri Khayat and Youjun Xie*

PRO 43: International RILEM Workshop on Bonded Concrete Overlays (e-ISBN: 2-912143-83-7); *Eds. J. L. Granju and J. Silfwerbrand*

PRO 44: 2nd International RILEM Workshop on Microbial Impacts on Building Materials (CD11) (e-ISBN: 2-912143-84-5); *Ed. M. Ribas Silva*

PRO 45: 2nd International Symposium on Nanotechnology in Construction, Bilbao (ISBN: 2-912143-87-X; e-ISBN: 2912143888); *Eds. Peter J. M. Bartos, Yolanda de Miguel and Antonio Porro*

PRO 46: ConcreteLife'06 - International RILEM-JCI Seminar on Concrete Durability and Service Life Planning: Curing, Crack Control, Performance in Harsh Environments (ISBN: 2-912143-89-6; e-ISBN: 291214390X); *Ed. K. Kovler*

PRO 47: International RILEM Workshop on Performance Based Evaluation and Indicators for Concrete Durability (ISBN: 978-2-912143-95-2; e-ISBN: 9782912143969); *Eds. V. Baroghel-Bouny, C. Andrade, R. Torrent and K. Scrivener*

PRO 48: 1st International RILEM Symposium on Advances in Concrete through Science and Engineering (e-ISBN: 2-912143-92-6); *Eds. J. Weiss, K. Kovler, J. Marchand, and S. Mindess*

PRO 49: International RILEM Workshop on High Performance Fiber Reinforced Cementitious Composites in Structural Applications (ISBN: 2-912143-93-4; e-ISBN: 2912143942); *Eds. G. Fischer and V. C. Li*

PRO 50: 1st International RILEM Symposium on Textile Reinforced Concrete (ISBN: 2-912143-97-7; e-ISBN: 2351580087); *Eds. Josef Hegger, Wolfgang Brameshuber and Norbert Will*

PRO 51: 2nd International Symposium on Advances in Concrete through Science and Engineering (ISBN: 2-35158-003-6; e-ISBN: 2-35158-002-8); *Eds. J. Marchand, B. Bissonnette, R. Gagné, M. Jolin and F. Paradis*

PRO 52: Volume Changes of Hardening Concrete: Testing and Mitigation (ISBN: 2-35158-004-4; e-ISBN: 2-35158-005-2); *Eds. O. M. Jensen, P. Lura and K. Kovler*

PRO 53: High Performance Fiber Reinforced Cement Composites - HPFRCC5 (ISBN: 978-2-35158-046-2; e-ISBN: 978-2-35158-089-9); *Eds. H. W. Reinhardt and A. E. Naaman*

PRO 54: 5th International RILEM Symposium on Self-Compacting Concrete (ISBN: 978-2-35158-047-9; e-ISBN: 978-2-35158-088-2); *Eds. G. De Schutter and V. Boel*

PRO 55: International RILEM Symposium Photocatalysis, Environment and Construction Materials (ISBN: 978-2-35158-056-1; e-ISBN: 978-2-35158-057-8); *Eds. P. Baglioni and L. Cassar*

PRO 56: International RILEM Workshop on Integral Service Life Modelling of Concrete Structures (ISBN 978-2-35158-058-5; e-ISBN: 978-2-35158-090-5); *Eds. R. M. Ferreira, J. Gulikers and C. Andrade*

PRO 57: RILEM Workshop on Performance of cement-based materials in aggressive aqueous environments (e-ISBN: 978-2-35158-059-2); *Ed. N. De Belie*

PRO 58: International RILEM Symposium on Concrete Modelling - CONMOD'08 (ISBN: 978-2-35158-060-8; e-ISBN: 978-2-35158-076-9); *Eds. E. Schlangen and G. De Schutter*

PRO 59: International RILEM Conference on On Site Assessment of Concrete, Masonry and Timber Structures - SACoMaTiS 2008 (ISBN set: 978-2-35158-061-5; e-ISBN: 978-2-35158-075-2); *Eds. L. Binda, M. di Prisco and R. Felicetti*

PRO 60: Seventh RILEM International Symposium on Fibre Reinforced Concrete: Design and Applications - BEFIB 2008 (ISBN: 978-2-35158-064-6; e-ISBN: 978-2-35158-086-8); *Ed. R. Gettu*

PRO 61: 1st International Conference on Microstructure Related Durability of Cementitious Composites 2 vol., (ISBN: 978-2-35158-065-3; e-ISBN: 978-2-35158-084-4); *Eds. W. Sun, K. van Breugel, C. Miao, G. Ye and H. Chen*

PRO 62: NSF/ RILEM Workshop: In-situ Evaluation of Historic Wood and Masonry Structures (e-ISBN: 978-2-35158-068-4); *Eds. B. Kasal, R. Anthony and M. Drdácký*

PRO 63: Concrete in Aggressive Aqueous Environments: Performance, Testing and Modelling, 2 vol., (ISBN: 978-2-35158-071-4; e-ISBN: 978-2-35158-082-0); *Eds. M. G. Alexander and A. Bertron*

PRO 64: Long Term Performance of Cementitious Barriers and Reinforced Concrete in Nuclear Power Plants and Waste Management - NUCPERF 2009 (ISBN: 978-2-35158-072-1; e-ISBN: 978-2-35158-087-5); *Eds. V. L'Hostis, R. Gens, C. Gallé*

PRO 65: Design Performance and Use of Self-consolidating Concrete - SCC'2009 (ISBN: 978-2-35158-073-8; e-ISBN: 978-2-35158-093-6); *Eds. C. Shi, Z. Yu, K. H. Khayat and P. Yan*

PRO 66: 2nd International RILEM Workshop on Concrete Durability and Service Life Planning - ConcreteLife'09 (ISBN: 978-2-35158-074-5; ISBN: 978-2-35158-074-5); *Ed. K. Kovler*

PRO 67: Repairs Mortars for Historic Masonry (e-ISBN: 978-2-35158-083-7); *Ed. C. Groot*

PRO 68: Proceedings of the 3rd International RILEM Symposium on 'Rheology of Cement Suspensions such as Fresh Concrete (ISBN 978-2-35158-091-2; e-ISBN: 978-2-35158-092-9); *Eds. O. H. Wallevik, S. Kubens and S. Oesterheld*

PRO 69: 3rd International PhD Student Workshop on 'Modelling the Durability of Reinforced Concrete (ISBN: 978-2-35158-095-0); *Eds. R. M. Ferreira, J. Gulikers and C. Andrade*

PRO 70: 2nd International Conference on 'Service Life Design for Infrastructure' (ISBN set: 978-2-35158-096-7, e-ISBN: 978-2-35158-097-4); *Ed. K. van Breugel, G. Ye and Y. Yuan*

PRO 71: Advances in Civil Engineering Materials - The 50-year Teaching Anniversary of Prof. Sun Wei' (ISBN: 978-2-35158-098-1; e-ISBN: 978-2-35158-099-8); *Eds. C. Miao, G. Ye, and H. Chen*

PRO 72: First International Conference on 'Advances in Chemically-Activated Materials – CAM'2010' (2010), 264 pp, ISBN: 978-2-35158-101-8; e-ISBN: 978-2-35158-115-5, *Eds. Caijun Shi and Xiaodong Shen*

PRO 73: 2nd International Conference on 'Waste Engineering and Management - ICWEM 2010' (2010), 894 pp, ISBN: 978-2-35158-102-5; e-ISBN: 978-2-35158-103-2, *Eds. J. Zh. Xiao, Y. Zhang, M. S. Cheung and R. Chu*

PRO 74: International RILEM Conference on 'Use of Superabsorsorbent Polymers and Other New Addditives in Concrete' (2010) 374 pp., ISBN: 978-2-35158-104-9; e-ISBN: 978-2-35158-105-6; *Eds. O. M. Jensen, M. T. Hasholt, and S. Laustsen*

PRO 75: International Conference on 'Material Science - 2nd ICTRC - Textile Reinforced Concrete - Theme 1' (2010) 436 pp., ISBN: 978-2-35158-106-3; e-ISBN: 978-2-35158-107-0; *Ed. W. Brameshuber*

PRO 76: International Conference on 'Material Science - HetMat - Modelling of Heterogeneous Materials - Theme 2' (2010) 255 pp., ISBN: 978-2-35158-108-7; e-ISBN: 978-2-35158-109-4; *Ed. W. Brameshuber*

PRO 77: International Conference on 'Material Science - AdIPoC - Additions Improving Properties of Concrete - Theme 3' (2010) 459 pp., ISBN: 978-2-35158-110-0; e-ISBN: 978-2-35158-111-7; *Ed. W. Brameshuber*

PRO 78: 2nd Historic Mortars Conference and RILEM TC 203-RHM Final Workshop – HMC2010 (2010) 1416 pp., e-ISBN: 978-2-35158-112-4; *Eds. J. Válek, C. Groot, and J. J. Hughes*

PRO 79: International RILEM Conference on Advances in Construction Materials Through Science and Engineering (2011) 213 pp., ISBN: 978-2-35158-116-2, e-ISBN: 978-2-35158-117-9; *Eds. Christopher Leung and K. T. Wan*

PRO 80: 2nd International RILEM Conference on Concrete Spalling due to Fire Exposure (2011) 453 pp., ISBN: 978-2-35158-118-6, e-ISBN: 978-2-35158-119-3; *Eds. E. A. B. Koenders and F. Dehn*

PRO 81: 2nd International RILEM Conference on Strain Hardening Cementitious Composites (SHCC2-Rio) (2011) 451 pp., ISBN: 978-2-35158-120-9, e-ISBN: 978-2-35158-121-6; *Eds. R. D. Toledo Filho, F. A. Silva, E. A. B. Koenders and E. M. R. Fairbairn*

PRO 82: 2nd International RILEM Conference on Progress of Recycling in the Built Environment (2011) 507 pp., e-ISBN: 978-2-35158-122-3; *Eds. V. M. John, E. Vazquez, S. C. Angulo and C. Ulsen*

PRO 83: 2nd International Conference on Microstructural-related Durability of Cementitious Composites (2012) 250 pp., ISBN: 978-2-35158-129-2; e-ISBN: 978-2-35158-123-0; *Eds. G. Ye, K. van Breugel, W. Sun and C. Miao*

PRO 84: CONSEC13 - Seventh International Conference on Concrete under Severe Conditions – Environment and Loading (2013) 1930 pp., ISBN: 978-2-35158-124-7; e-ISBN: 978-2- 35158-134-6; *Eds. Z. J. Li, W. Sun, C. W. Miao, K. Sakai, O. E. Gjorv & N. Banthia*

PRO 85: RILEM-JCI International Workshop on Crack Control of Mass Concrete and Related issues concerning Early-Age of Concrete Structures – ConCrack 3 – Control of Cracking in Concrete Structures 3 (2012) 237 pp., ISBN: 978-2-35158-125-4; e-ISBN: 978-2-35158-126-1; *Eds. F. Toutlemonde and J.-M. Torrenti*

PRO 86: International Symposium on Life Cycle Assessment and Construction (2012) 414 pp., ISBN: 978-2-35158-127-8, e-ISBN: 978-2-35158-128-5; *Eds. A. Ventura and C. de la Roche*

PRO 87: UHPFRC 2013 – RILEM-fib-AFGC International Symposium on Ultra-High Performance Fibre-Reinforced Concrete (2013), ISBN: 978-2-35158-130-8, e-ISBN: 978-2-35158-131-5; *Eds. F. Toutlemonde*

PRO 88: 8th RILEM International Symposium on Fibre Reinforced Concrete (2012) 344 pp., ISBN: 978-2-35158-132-2, e-ISBN: 978-2-35158-133-9; *Eds. Joaquim A. O. Barros*

PRO 89: RILEM International workshop on performance-based specification and control of concrete durability (2014) 678 pp, ISBN: 978-2-35158-135-3, e-ISBN: 978-2-35158-136-0; *Eds. D. Bjegović, H. Beushausen and M. Serdar*

PRO 90: 7th RILEM International Conference on Self-Compacting Concrete and of the 1st RILEM International Conference on Rheology and Processing of Construction Materials (2013) 396 pp, ISBN: 978-2-35158-137-7, e-ISBN: 978-2-35158-138-4; *Eds. Nicolas Roussel and Hela Bessaies-Bey*

PRO 91: CONMOD 2014 - RILEM International Symposium on Concrete Modelling (2014), ISBN: 978-2-35158-139-1; e-ISBN: 978-2-35158-140-7; *Eds. Kefei Li, Peiyu Yan and Rongwei Yang*

PRO 92: CAM 2014 - 2nd International Conference on advances in chemically-activated materials (2014) 392 pp., ISBN: 978-2-35158-141-4; e-ISBN: 978-2-35158-142-1; *Eds. Caijun Shi and Xiadong Shen*

PRO 93: SCC 2014 - 3rd International Symposium on Design, Performance and Use of Self-Consolidating Concrete (2014) 438 pp., ISBN: 978-2-35158-143-8; e-ISBN: 978-2-35158-144-5; *Eds. Caijun Shi, Zhihua Ou, Kamal H. Khayat*

PRO 94 (online version): HPFRCC-7 - 7th RILEM conference on High performance fiber reinforced cement composites (2015), e-ISBN: 978-2-35158-146-9; *Eds. H. W. Reinhardt, G. J. Parra-Montesinos, H. Garrecht*

PRO 95: International RILEM Conference on Application of superabsorbent polymers and other new admixtures in concrete construction (2014), ISBN: 978-2-35158-147-6; e-ISBN: 978-2-35158-148-3; *Eds. Viktor Mechtcherine, Christof Schroefl*

PRO 96 (online version): XIII DBMC: XIII International Conference on Durability of Building Materials and Components (2015), e-ISBN: 978-2-35158-149-0; *Eds. M. Quattrone, V. M. John*

PRO 97: SHCC3 – 3rd International RILEM Conference on Strain Hardening Cementitious Composites (2014), ISBN: 978-2-35158-150-6; e-ISBN: 978-2-35158-151-3; *Eds. E. Schlangen, M. G. Sierra Beltran, M. Lukovic, G. Ye*

PRO 98: FERRO-11 – 11th International Symposium on Ferrocement and 3rd ICTRC - International Conference on Textile Reinforced Concrete (2015), ISBN: 978-2-35158-152-0; e-ISBN: 978-2-35158-153-7; *Ed. W. Brameshuber*

PRO 99 (online version): ICBBM 2015 - 1st International Conference on Bio-Based Building Materials (2015), e-ISBN: 978-2-35158-154-4; *Eds. S. Amziane, M. Sonebi*

PRO 100: SCC16 - RILEM Self-Consolidating Concrete Conference (2016), ISBN: 978-2-35158-156-8; e-ISBN: 978-2-35158-157-5; *Ed. Kamal H. Kayat*

PRO 101 (online version): III Progress of Recycling in the Built Environment (2015), e-ISBN: 978-2-35158-158-2; *Eds. I. Martins, C. Ulsen and S. C. Angulo*

PRO 102 (online version): RILEM Conference on Microorganisms-Cementitious Materials Interactions (2016), e-ISBN: 978-2-35158-160-5; *Eds. Alexandra Bertron, Henk Jonkers, Virginie Wiktor*

PRO 103 (online version): ACESC'16 - Advances in Civil Engineering and Sustainable Construction (2016), e-ISBN: 978-2-35158-161-2; *Eds. T. Ch. Madhavi, G. Prabhakar, Santhosh Ram and P. M. Rameshwaran*

PRO 104 (online version): SSCS'2015 - Numerical Modeling - Strategies for Sustainable Concrete Structures (2015), e-ISBN: 978-2-35158-162-9

PRO 105: 1st International Conference on UHPC Materials and Structures (2016), ISBN: 978-2-35158-164-3, e-ISBN: 978-2-35158-165-0

PRO 106: AFGC-ACI-fib-RILEM International Conference on Ultra-High-Performance Fibre-Reinforced Concrete – UHPFRC 2017 (2017), ISBN: 978-2-35158-166-7, e-ISBN: 978-2-35158-167-4; *Eds. François Toutlemonde & Jacques Resplendino*

PRO 107 (online version): XIV DBMC – 14th International Conference on Durability of Building Materials and Components (2017), e-ISBN: 978-2-35158-159-9; *Eds. Geert De Schutter, Nele De Belie, Arnold Janssens, Nathan Van Den Bossche*

PRO 108: MSSCE 2016 - Innovation of Teaching in Materials and Structures (2016), ISBN: 978-2-35158-178-0, e-ISBN: 978-2-35158-179-7; *Ed. Per Goltermann*

PRO 109 (2 volumes): MSSCE 2016 - Service Life of Cement-Based Materials and Structures (2016), ISBN Vol. 1: 978-2-35158-170-4, Vol. 2: 978-2-35158-171-4, Set Vol. 1&2: 978-2-35158-172-8, e-ISBN : 978-2-35158-173-5; *Eds. Miguel Azenha, Ivan Gabrijel, Dirk Schlicke, Terje Kanstad and Ole Mejlhede Jensen*

PRO 110: MSSCE 2016 - Historical Masonry (2016), ISBN: 978-2-35158-178-0, e-ISBN: 978-2-35158-179-7; *Eds. Inge Rörig-Dalgaard and Ioannis Ioannou*

PRO 111: MSSCE 2016 - Electrochemistry in Civil Engineering (2016), ISBN: 978-2-35158-176-6, e-ISBN: 978-2-35158-177-3; *Ed. Lisbeth M. Ottosen*

PRO 112: MSSCE 2016 - Moisture in Materials and Structures (2016), ISBN: 978-2-35158-178-0, e-ISBN: 978-2-35158-179-7; *Eds. Kurt Kielsgaard Hansen, Carsten Rode and Lars-Olof Nilsson*

PRO 113: MSSCE 2016 - Concrete with Supplementary Cementitious Materials (2016), ISBN: 978-2-35158-178-0, e-ISBN: 978-2-35158-179-7; *Eds. Ole Mejlhede Jensen, Konstantin Kovler and Nele De Belie*

PRO 114: MSSCE 2016 - Frost Action in Concrete (2016), ISBN: 978-2-35158-182-7, e-ISBN: 978-2-35158-183-4; *Eds. Marianne Tange Hasholt, Katja Fridh and R. Doug Hooton*

PRO 115: MSSCE 2016 - Fresh Concrete (2016), ISBN: 978-2-35158-184-1, e-ISBN: 978-2-35158-185-8; *Eds. Lars N. Thrane, Claus Pade, Oldrich Svec and Nicolas Roussel*

PRO 116: BEFIB 2016 – 9th RILEM International Symposium on Fiber Reinforced Concrete (2016), ISBN: 978-2-35158-187-2, e-ISBN: 978-2-35158-186-5; *Eds. N. Banthia, M. di Prisco and S. Soleimani-Dashtaki*

PRO 117: 3rd International RILEM Conference on Microstructure Related Durability of Cementitious Composites (2016), ISBN: 978-2-35158-188-9, e-ISBN: 978-2-35158-189-6; *Eds. Changwen Miao, Wei Sun, Jiaping Liu, Huisu Chen, Guang Ye and Klaas van Breugel*

PRO 118 (4 volumes): International Conference on Advances in Construction Materials and Systems (2017), ISBN Set: 978-2-35158-190-2, Vol. 1: 978-2-35158-193-3, Vol. 2: 978-2-35158-194-0, Vol. 3: ISBN:978-2-35158-195-7, Vol. 4: ISBN:978-2-35158-196-4, e-ISBN: 978-2-35158-191-9; *Eds. Manu Santhanam, Ravindra Gettu, Radhakrishna G. Pillai and Sunitha K. Nayar*

PRO 119 (online version): ICBBM 2017 - Second International RILEM Conference on Bio-based Building Materials, (2017), e-ISBN: 978-2-35158-192-6; *Eds. Sofiane Amziane, Mohammed Sonebi*

PRO 120 (2 volumes): EAC-02 - 2nd International RILEM/COST Conference on Early Age Cracking and Serviceability in Cement-based Materials and Structures, (2017), Vol. 1: 978-2-35158-199-5, Vol. 2: 978-2-35158-200-8, Set: 978-2-35158-197-1, e-ISBN: 978-2-35158-198-8; *Eds. Stéphanie Staquet and Dimitrios Aggelis*

PRO 121 (2 volumes): SynerCrete18: Interdisciplinary Approaches for Cement-based Materials and Structural Concrete: Synergizing Expertise and Bridging Scales of Space and Time, (2018), Set: 978-2-35158-202-2, Vol.1: 978-2-35158-211-4, Vol.2: 978-2-35158-212-1, e-ISBN: 978-2-35158-203-9; *Eds. Miguel Azenha, Dirk Schlicke, Farid Benboudjema, Agnieszka Knoppik*

PRO 122: SCC'2018 China - Fourth International Symposium on Design, Performance and Use of Self-Consolidating Concrete, (2018), ISBN: 978-2-35158-204-6, e-ISBN: 978-2-35158-205-3; *Eds. C. Shi, Z. Zhang, K. H. Khayat*

PRO 123: Final Conference of RILEM TC 253-MCI: Microorganisms-Cementitious Materials Interactions (2018), Set: 978-2-35158-207-7, Vol.1: 978-2-35158-209-1, Vol.2: 978-2-35158-210-7, e-ISBN: 978-2-35158-206-0; *Ed. Alexandra Bertron*

PRO 124 (online version): Fourth International Conference Progress of Recycling in the Built Environment (2018), e-ISBN: 978-2-35158-208-4; *Eds. Isabel M. Martins, Carina Ulsen, Yury Villagran*

PRO 125 (online version): SLD4 - 4th International Conference on Service Life Design for Infrastructures (2018), e-ISBN: 978-2-35158-213-8; *Eds. Guang Ye, Yong Yuan, Claudia Romero Rodriguez, Hongzhi Zhang, Branko Savija*

PRO 126: Workshop on Concrete Modelling and Material Behaviour in honor of Professor Klaas van Breugel (2018), ISBN: 978-2-35158-214-5, e-ISBN: 978-2-35158-215-2; *Ed. Guang Ye*

PRO 127 (online version): CONMOD2018 - Symposium on Concrete Modelling (2018), e-ISBN: 978-2-35158-216-9; *Eds. Erik Schlangen, Geert de Schutter, Branko Savija, Hongzhi Zhang, Claudia Romero Rodriguez*

PRO 128: SMSS2019 - International Conference on Sustainable Materials, Systems and Structures (2019), ISBN: 978-2-35158-217-6, e-ISBN: 978-2-35158-218-3

PRO 129: 2nd International Conference on UHPC Materials and Structures (UHPC2018-China), ISBN: 978-2-35158-219-0, e-ISBN: 978-2-35158-220-6;

PRO 130: 5th Historic Mortars Conference (2019), ISBN: 978-2-35158-221-3, e-ISBN: 978-2-35158-222-0; *Eds. José Ignacio Álvarez, José María Fernández, Íñigo Navarro, Adrián Durán, Rafael Sirera*

PRO 131 (online version): 3rd International Conference on Bio-Based Building Materials (ICBBM2019), e-ISBN: 978-2-35158-229-9; *Eds. Mohammed Sonebi, Sofiane Amziane, Jonathan Page*

PRO 132: IRWRMC'18 - International RILEM Workshop on Rheological Measurements of Cement-based Materials (2018), ISBN: 978-2-35158-230-5, e-ISBN: 978-2-35158-231-2; *Eds. Chafika Djelal, Yannick Vanhove*

PRO 133 (online version): CO2STO2019 - International Workshop CO2 Storage in Concrete (2019), e-ISBN: 978-2-35158-232-9; *Eds. Assia Djerbi, Othman Omikrine-Metalssi, Teddy Fen-Chong*

RILEM Reports (REP)

Report 19: Considerations for Use in Managing the Aging of Nuclear Power Plant Concrete Structures (ISBN: 2 912143 07 1); *Ed. D. J. Naus*

Report 20: Engineering and Transport Properties of the Interfacial Transition Zone in Cementitious Composites (ISBN: 2-912143-08-X); *Eds. M. G. Alexander, G. Arliguie, G. Ballivy, A. Bentur and J. Marchand*

Report 21: Durability of Building Sealants (ISBN: 2-912143-12-8); *Ed. A. T. Wolf*

Report 22: Sustainable Raw Materials - Construction and Demolition Waste (ISBN: 2-912143-17-9); *Eds. C. F. Hendriks and H. S. Pietersen*

Report 23: Self-Compacting Concrete state-of-the-art report (ISBN: 2-912143-23-3); *Eds. Å. Skarendahl and Ö. Petersson*

Report 24: Workability and Rheology of Fresh Concrete: Compendium of Tests (ISBN: 2-912143-32-2); *Eds. P. J. M. Bartos, M. Sonebi and A. K. Tamimi*

Report 25: Early Age Cracking in Cementitious Systems (ISBN: 2-912143-33-0); *Ed. A. Bentur*

Report 26: Towards Sustainable Roofing (Joint Committee CIB/RILEM) (CD 07) (e-ISBN 978-2-912143-65-5); *Eds. Thomas W. Hutchinson and Keith Roberts*

Report 27: Condition Assessment of Roofs (Joint Committee CIB/RILEM) (CD 08) (e-ISBN 978-2-912143-66-2); *Ed. CIB W 83/RILEM TC166-RMS*

Report 28: Final report of RILEM TC 167-COM 'Characterisation of Old Mortars with Respect to Their Repair (ISBN: 978-2-912143-56-3); *Eds. C. Groot, G. Ashall and J. Hughes*

Report 29: Pavement Performance Prediction and Evaluation (PPPE): Interlaboratory Tests (e-ISBN: 2-912143-68-3); *Eds. M. Partl and H. Piber*

Report 30: Final Report of RILEM TC 198-URM 'Use of Recycled Materials' (ISBN: 2-912143-82-9; e-ISBN: 2-912143-69-1); *Eds. Ch. F. Hendriks, G. M. T. Janssen and E. Vázquez*

Report 31: Final Report of RILEM TC 185-ATC 'Advanced testing of cement-based materials during setting and hardening' (ISBN: 2-912143-81-0; e-ISBN: 2-912143-70-5); *Eds. H. W. Reinhardt and C. U. Grosse*

Report 32: Probabilistic Assessment of Existing Structures. A JCSS publication (ISBN 2-912143-24-1); *Ed. D. Diamantidis*

Report 33: State-of-the-Art Report of RILEM Technical Committee TC 184-IFE 'Industrial Floors' (ISBN 2-35158-006-0); *Ed. P. Seidler*

Report 34: Report of RILEM Technical Committee TC 147-FMB 'Fracture mechanics applications to anchorage and bond' Tension of Reinforced Concrete Prisms – Round Robin Analysis and Tests on Bond (e-ISBN 2-912143-91-8); *Eds. L. Elfgren and K. Noghabai*

Report 35: Final Report of RILEM Technical Committee TC 188-CSC 'Casting of Self Compacting Concrete' (ISBN 2-35158-001-X; e-ISBN: 2-912143-98-5); *Eds. Å. Skarendahl and P. Billberg*

Report 36: State-of-the-Art Report of RILEM Technical Committee TC 201-TRC 'Textile Reinforced Concrete' (ISBN 2-912143-99-3); *Ed. W. Brameshuber*

Report 37: State-of-the-Art Report of RILEM Technical Committee TC 192-ECM 'Environment-conscious construction materials and systems' (ISBN: 978-2-35158-053-0); *Eds. N. Kashino, D. Van Gemert and K. Imamoto*

Report 38: State-of-the-Art Report of RILEM Technical Committee TC 205-DSC 'Durability of Self-Compacting Concrete' (ISBN: 978-2-35158-048-6); *Eds. G. De Schutter and K. Audenaert*

Report 39: Final Report of RILEM Technical Committee TC 187-SOC 'Experimental determination of the stress-crack opening curve for concrete in tension' (ISBN 978-2-35158-049-3); *Ed. J. Planas*

Report 40: State-of-the-Art Report of RILEM Technical Committee TC 189-NEC 'Non-Destructive Evaluation of the Penetrability and Thickness of the Concrete Cover' (ISBN 978-2-35158-054-7); *Eds. R. Torrent and L. Fernández Luco*

Report 41: State-of-the-Art Report of RILEM Technical Committee TC 196-ICC 'Internal Curing of Concrete' (ISBN 978-2-35158-009-7); *Eds. K. Kovler and O. M. Jensen*

Report 42: 'Acoustic Emission and Related Non-destructive Evaluation Techniques for Crack Detection and Damage Evaluation in Concrete' - Final Report of RILEM Technical Committee 212-ACD (e-ISBN: 978-2-35158-100-1); *Ed. M. Ohtsu*

Report 45: Repair Mortars for Historic Masonry - State-of-the-Art Report of RILEM Technical Committee TC 203-RHM (e-ISBN: 978-2-35158-163-6); *Eds. Paul Maurenbrecher and Caspar Groot*

Report 46: *S*urface delamination of concrete industrial floors and other durability related aspects guide - Report of RILEM Technical Committee TC 268-SIF (e-ISBN: 978-2-35158-201-5); *Ed. Valerie Pollet*

Microstructure, Hydration and Absorption

Calorimetry Analysis of High Strength Cement Pastes Containing Superabsorbent Polymer (SAP)

Livia B. Agostinho[1(✉)], Thyala A. Cunha[1], Daiane V. M. R. Cupertino[2], and Eugênia F. Silva[3]

[1] Structures and Civil Construction, University of Brasília, Brasília, Brazil
liviaborbagostinho@gmail.com
[2] Civil Engineering, Eletrobras Furnas Hydroelectric Company, Goiânia, Brazil
[3] Civil Engineering Department, University of Brasília, Brasília, Federal District, Brazil

Abstract. With the application of isothermal calorimetry technique, the speed of the hydration reactions can be evaluated in a simplified and efficient way, over time, by the heat evolution curves. This technique can become a suitable tool to understand the process of absorption and desorption of the SAPs in the cement matrix, a fundamental parameter that determines the efficiency of the polymers as mitigating agents of autogenous shrinkage and their behavior during the fresh state. It is also possible to identify the water retention or early release during the fluid period of the pastes. This methodology also permits a better understanding of the participation of the water incorporated by the polymer, in the kinetics of hydration of the cement, over time. It was possible to observe that the presence of the SAP slightly alters the reaction kinetics of the cement as it reduces its rate of acceleration of the curve of the second calorimetric peak. The higher the SAP content, the lower the acceleration rate of the curves. The addition of the SAP generated a lightly delay of the second peak time of the pastes and a deceleration of the kinetics reaction of the cement, as compared to the reference mixture. This behavior seems to be related to the desorption kinetics of the polymers. It was also possible to conclude that the higher the SAP amount, the higher the total quantity of accumulated heat in the end of 3 days of test, that is, a higher volume of hydration reactions.

Keywords: Superabsorbent Polymers (SAP) · Isothermal calorimetry · Cement pastes · Hydration

1 Introduction

Isothermal calorimetry is a methodology that measures thermal power (heat production rate) during the cement hydration by means of monitoring the heat flow of the mass, when they are in isothermal conditions, without acceleration caused by the heat released (Quarcioni 2008). Using this technique, the speed of the hydration reactions can be evaluated in a simplified and efficient way, over time, by the heat evolution curves.

W. P. Boshoff et al. (Eds.): SAP 2019, RILEM Bookseries 24, pp. 3–10, 2020.
https://doi.org/10.1007/978-3-030-33342-3_1

The isothermal conduction calorimetry technique allows to follow the hydration kinetics of the cement and this process can be divided in several stages.

Stage I - initial dissolution, occurs in the first minutes immediately after the addition of water, when the cement grains begin to react and the readily water soluble components such as alkalis, calcium sulphate begin to dissolve. In this phase, the first peak of the heat release of hydration is presented. From the rheological point of view, it is the most important stage, since the initial dissolution rate of tricalcium aluminate (C_3A) can be increased by the dissolved alkalis. The main responsible reaction of this peak is the dissolution of the C_3A and of the calcium sulfate ($CaSO_4$), forming ettringite (AFt). After a few minutes, an amorphous gel rich in alumina and silicate (also with a high content of calcium and sulfate) is formed around the surface of the cement grains.

This gel causes period II - induction, also called dormancy period, has a few hours duration and low rate of hydration that can vary according to temperature, cement composition, water/cement ratio, chemical additives, among others. It is in this period that the concrete must be transported and applied, since it is still workable (Bianchi 2014).

Stage III - acceleration, is responsible for the second exothermic peak. It is caused by ionic supersaturation of the aqueous phase (dissolution and precipitation mechanism). At this stage, the main constituents of the cement, tricalcium silicate (C_3S) and dicalcium silicate (C_2S), react with water and quickly form amorphous hydrated calcium silicate (C-S-H) and calcium hydroxide ($Ca(OH)_2$). There is also a decrease in the sulfate and calcium ions due to the formation of ettringite. In this stage occurs the loss of workability, that is, the setting time of the cement admixture.

The period IV - deceleration, begins with the reduction of the heat rate, due to the low concentration of the ions in the solution. Hydration is now controlled by the topochemical mechanism. Finally, the V stage- slow reaction stage occurs, associated with the transformation of AFt into Monosulphatus (AFm).

Few studies have been found in the literature using the isothermal calorimetry technique in cementitious materials modified with SAPs. The works of: Zhutovsky and Kovler (2013), Justs et al. (2014), Justs et al. (2015), Wang et al. (2016) and Manzano (2016).

Zhutovsky and Kovler (2013) evaluated the efficiency of different cure types by varying the w/c ratio of pastes (0.21, 0.25, 0.33 and 0.45) and using SAP as an internal curing agent. The presence of SAP and internal curing water shifted the curves a little to the right, indicating a slight delay and a small reduction of the main peak of heat release. They concluded that the lower the w/c ratio, the greater the efficiency of the different types of cure, including internal cure (SAP).

Justs et al. (2014) evaluated the effect of SAP addition on pastes with low w/c ratio (0.20 and 0.30). The authors concluded that the principal hydration peaks of polymer-containing pastes are lower and appear earlier compared to blends without SAP having the same total w/c ratio. This behavior is caused by the gradual release of the internal curing water from the SAP. On the other hand, in admixtures with the same basic w/c ratio, those with SAP were slightly retarded compared to the reference mixture. Two possible explanations are: the leaching of the soluble fraction of SAP that can interfere on the hydration process, and the absorption of alkaline ions by SAP.

Manzano (2016), carried out calorimetric tests on pastes with basic w/c ratio of 0.3 and varied the type of SAP also adding internal curing water (extra). Even with the delay caused by the superplasticizer additive, it was possible to evaluate that the addition of the polymers changed the profile of the calorimetric curves. It was observed, as in other works, the right shift of the second peak of heat release, that is, a slight delay in the peak time for SAP-containing pastes.

The use of the isothermal calorimetry technique can become an adequate tool to understand the process of absorption and desorption of SAP in the cement matrix. That is a fundamental parameter that determines the efficiency of polymers as mitigating agents of autogenous shrinkage, and their behavior during the fresh state, being able to identify the water retention or early release during the fluid period of the pastes. This methodology also allows to give subsidies on the participation of the water incorporated by the polymer, in the kinetics of hydration of the cement, over time.

2 Materials and Methods

The CPV-ARI cement was chosen due to the quantity of clinker and because it is frequently used in high strength concretes (HSC) in Brazil and Table 1 shows the chemical composition of Portland cement used in this research.

Table 1. Chemical composition of Portland cement and silica fume.

Material	Chemical Composition (%)									
	SO_3	MgO	SiO_2	Fe_2O_3	Al_2O_3	CaO	Free CaO	$CaSO_4$	Na_2O	K_2O
Cement	3.28	4.36	24.41	3.02	7.09	53.44	2.16	2.16	0.29	0.77

The values of the chemical and physical properties were within the limits established by the Brazilian specific standards. The insoluble residue content was low with no evidence of pozzolanic material. The levels of free calcium oxide and magnesium oxide were at acceptable levels (2.16% and 4.6%, respectively) and, probably, did not cause expansive cement reaction due to the later hydration of these compounds (CaO and MgO).

The Glenium superplasticizer additive was used and it is a white viscous liquid with ether carboxylate as chemical basis, specific mass between 1.067 and 1.107 g/cm^3 and 30% of solid contents.

The SAP used in this study was developed at the Denmark Technical University (DTU), and donated by Prof. Ole Mejlhede Jensen. Chemically, this polymer is a covalent cross-linked acrylic acid/acrylamide produced by the reverse suspension polymerization technique. The SAP was specially developed for using in high alkaline environment, such as the cementitious suspension.

Manzano (2016) characterized the SAP used in this research and the results are shown in Table 2. The author reported a SAP absorption of 15 g/g, that is, each gram of dry polymer absorbs 15 g of water in a cement mixture. This data is especially important for determining the internal curing water (w/ic) to be added to the mixture.

Table 2. Characteristics of the superabsorbent polymer

Characteristic	Result	Method
Specific mass (g/cm^3)	1.456	Pycnometer
Diameter of dry particle (μm)	66,3	SEM
Absorption in water (g/g)	95.8	Graduate test tube
	80.3	Optical microscope
Absorption in cement (g/g)	15.0	Slump flow

The experimental program consisted on three pastes, one as a reference and other two with SAP content varied between 0.15% and 0.30% while all other components of the cement mixtures remained constant. All pastes contained 1.2% of superplasticizer additive, and the w/c ratio was also constant and equal to 0.4. In the mixtures containing SAP, extra water was added, in order to avoid significantly changes on rheological behavior on the fresh state. The total w/c ratio of mixtures SAP15 and SAP30 were respectively 0,43 and 0,45.

The isothermal conduction calorimetry tests were performed at FURNAS Centrais Elétricas SA - Aparecida de Goiânia Laboratory using the TAM AIR isothermal conduction calorimeter (Fig. 1) composed of eight channels with temperature control, manufactured by TA Instruments with the characteristics presented in Table 3 and with data acquisition made by PicoLog software.

Table 3. Specifications of the TAM AIR equipment used in the isothermal calorimetry tests

Temperature amplitude	5 °C a 90 °C
Thermostat type	Air
Thermostat stability	±0,02 °C
Maximum sample capacity	20 ml
Detection limit	4 μW
Detection accuracy	±20 μW

The objective of this study was to evaluate the release of heat from the chemical reaction of cement mixture, making it possible to analyze the influence of SAP on the hydration reactions of the studied cement pastes.

The preparation of the samples for all calorimetry tests consisted of weighing the materials in precision scale to two decimal places, where the cement and liquid materials were measured separately. The mixtures were prepared manually using a spatula for 1 min, followed by mechanical mixing using a digital IKA RW 20 mixer at a rotational speed of about 2500 rpm, for three minutes.

The external blending methodology, does not allow the acquisition of the complete heat release data since the mixture is introduced into the equipment in a period in which the hydration reactions of the cement are already occurring, thus losing data of the first release peak of heat, but it guarantee a repeatability on subsequent measurements and negligible error.

Fig. 1. Overview of the isothermal calorimeter.

On the Fig. 2 there are some steps mentioned above for the preparation and execution of the essay.

(a) (b) (c)

Fig. 2. Procedure for preparation of the samples and execution of the external test: (a) weighing of the ampoule; (b) introducing the paste into the ampoule; (c) ampoule seal

3 Analisys and Results

The data presented here were obtained from the average of three essays of each paste proposed. The use of superplasticizer Glenium retarded the reaction kinetics of cement in all pastes (delayed the setting time in about 12 h from previous results of rheological essays), especially during the dormancy and induction period. This may have happened because of the expiration date of the superplasticizer (expired about 2 months before the date of execution of the essay) that could interfere on the compatibility of the elements of the superplasticizer and the mixture components. The calorimetry curve of the three pastes studied on this research are shown in Fig. 3.

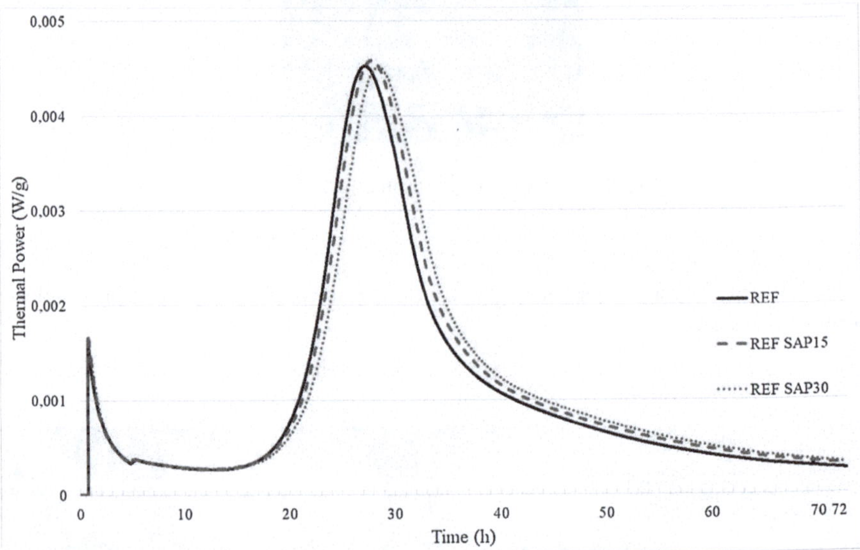

Fig. 3. Curves of thermal power versus time of the three pastes studied.

Analyzing the effect of SAP on the heat flow curves, it generates a slight increase in the time of occurrence of the calorimetric peak (about 10%), when compared to the pastes without SAP causing a displacement of the curves to the right.

The acceleration rate was calculated from a linear regression of the specific segment of the heat flow curve as a function of time in hours (slope of the acceleration curve), and the results are shown in Table 4.

Table 4. Summary of the calculated parameters of the isothermal calorimetry assay of the pastes.

	REF	SAP15	SAP30
Accumulate Heat (J/g)	268,5	281,5	284,4
Aceleration rate (%$_o$)	0,78	0,70	0,69

The presence of the SAP slightly alters the reaction kinetics of the cement, as they reduce their rate of acceleration of the curve of the second calorimetric peak. The higher the SAP content, the lower the rate of acceleration of the curves.

The presence of SAP generated a delay of the peak time of the pastes and a deceleration of the reaction kinetics of the cement, when compared to the reference mixture with the same basic w/c ratio. This behavior seems to be related to the desorption kinetics of the polymers, the leaching of some non-crosslinked SAP particles and the absorption of alkaline ions into the polymers (Justs et al. 2014)

Analyzing the accumulated heat during the 72 h of test, the data obtained are in Table 4 and represented in Fig. 4. The highest total amount of heat released by the mixtures with SAP are due to the internal curing water participating in the hydration reactions. Also, the higher the amount of SAP, the more total heat is released, that is because an additional water is added to the mixtures and allows more cement getting hydrated.

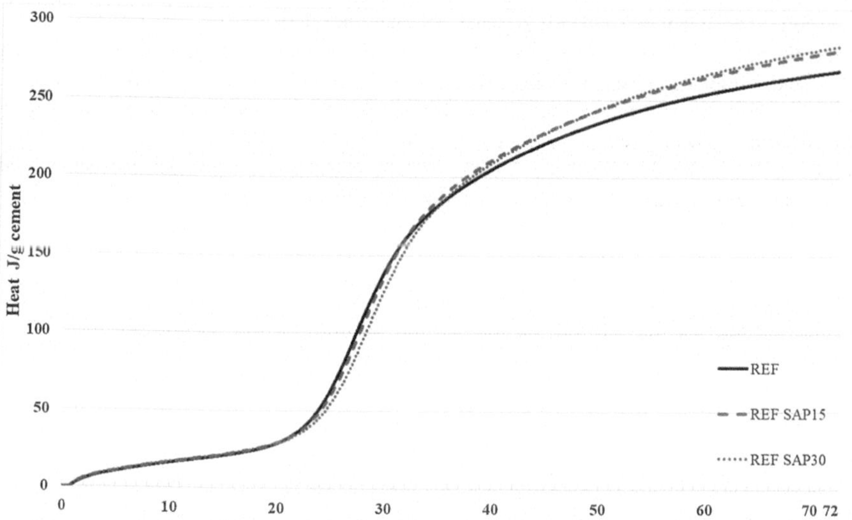

Fig. 4. Accumulated heat curves as a function of time.

4 Conclusions

The presence of SAP changes slightly the kinetics of cement hydration, as lowering the acceleration rates, and increasing the amounts of total heat released. These changes may be due to the presence of extra water added to be the internal curing water. Even if the water was supposed to be in the mixture to fill up the capillarities, they participate to the hydration process, altering the power heat curves and promoting more hydration products.

Acknowledgements. The authors thank University of Brasilia and Furnas Central Hydropower S. A. for the partnership in this research. Thanks are also given to The National Council for Scientific and Technological Development (CNPq) for the financial support and Professor Jensen and DTU for providing the studied SAP.

References

Bianchi, G.Q.: Application of nano-silica in concrete. Ph.D. thesis. Technische Universiteit Eindhoven, Eindhoven, the Netherlands (2014)

Justs, J., Wyrzykowski, M., Winnefeld, F., Bajare, D., Lura, P.: Influence of superabsorbent polymers on hydration of cement pastes with low water-to-binder ratio: a calorimetry study. J. Therm. Anal. Calorim. **115**(1), 425–432 (2014)

Justs, J., Wyrzykowski, M., Bajare, D., Lura, P.: Internal curing by superabsorbent polymers in ultra-high performance concrete. Cem. Concr. Res. **76**, 82–90 (2015)

Manzano, R., Alejandro, M.: Estudo Experimental de Materiais Cimentícios de Alta Resistência modificados com Polímeros Superabsorventes (PSAs) como Agentes de Cura Interna. Doctoral thesis. Universidade De Brasília, Brasília-DF (2016)

Quarcioni, V.A.: Influência da cal hidratada nas idades iniciais da hidratação do cimento Portland: estudo em pasta. Technical Bulletin, p. 172. USP, São Paulo, Brazil (2008)

Wang, F., Yang, J., Hu, S., Li, X., Cheng, H.: Influence of superabsorbent polymers on the surrounding cement paste. Cem. Concr. Res. **81**, 112–121 (2016)

Zhutovsky, S., Kovler, K.: Hydration kinetics of high-performance cementitious systems under different curing conditions. Mater. Struct. **46**(10), 1599–1611 (2013)

Comparison of Sorption Assessment Methods for Superabsorbent Polymers

João Custódio[1(✉)], Paulo Francinete[2], António Ribeiro[1], Ana Gonzalez[1], and Eugênia Silva[3]

[1] LNEC – National Laboratory for Civil Engineering, Avenida do Brasil 101, 1700-066 Lisbon, Portugal
jcustodio@lnec.pt
[2] IFG – Federal Institute of Goiás, Rua 75, 46, Centro, Goiânia 74055-110, Brazil
[3] UnB – Brasília University, Darcy Ribeiro Campus, Brasília 70910-900, Brazil

Abstract. Currently, superabsorbent polymers (SAP) constitute a promising class of chemical admixtures for concrete. However, since their sorption capability and kinetics can differ significantly depending on chemical composition and grading, when choosing a SAP for a particular application it should be pretested with respect to its sorption capacity before actual implementation in concrete. Due to the influence that the test liquid ionic composition has on SAP sorption behaviour, they should be tested not only for free water sorption but also for sorption of solutions with chemical compositions as close as possible of those of concrete or mortar pore solutions; furthermore, this should be complemented with the evaluation of sorption behaviour in an environment similar to the one the SAP will experience when in a mortar or concrete. This article focuses two free sorption assessment testing methods, teabag and filtration methods, and two methods that indirectly provide the SAP sorption capacity, slump flow consistence method, using a cementitious mortar, and a new method, plunger penetration consistence method, using a cement paste. It was found that a good agreement exists between the results obtained with the two indirect methods, and that both the direct and indirect methods produced similar sorption trends. Hence, from the results obtained so far, it appears that the plunger penetration method could be a useful complement to the teabag and filtration methods, by providing a simple and quick way of estimating SAP sorption capacity in an environment more similar to that found by the SAP in concrete.

Keywords: Superabsorbent polymers · Sorption behaviour · Assessment methods · Deionised water · Cement filtrate solution

1 Introduction

Superabsorbent polymers (SAP) are polymeric materials with great ability to absorb liquid from the environment and to retain this liquid without dissolving. These polymers constitute a promising class of chemical admixtures for concrete, because they can, for instance, mitigate autogenous shrinkage of concrete with low water-to-cement ratio (Jensen and Hansen 2001), enhance freeze-thaw resistance of concrete

W. P. Boshoff et al. (Eds.): SAP 2019, RILEM Bookseries 24, pp. 11–19, 2020.
https://doi.org/10.1007/978-3-030-33342-3_2

(Jensen and Hansen 2002), mitigate crack formation due to plastic shrinkage in concrete (Mechtcherine and Dudziak 2012), and induce self-sealing of concrete (Lee et al. 2010).

The most important property for the use of SAP in cementitious materials is their absorption capacity, defined as the ratio of the mass of the liquid absorbed until equilibrium is met and the initial mass of dry polymer. SAP sorption capacity, in general terms, depends upon the polymer properties and the exposure environment. The main SAP characteristics that may influence its sorption capacity are the polymer production process, the density of anionic groups, the crosslinking degree and the particles size distribution. Regarding the exposure environment, the main parameters are liquid ionic composition, system temperature and spatial confinement.

Existing standardized methods for determining SAP sorption capability are generally performed in water or saline solutions, which do not replicate the actual chemical conditions occurring in the cementitious material. Since sorption capability changes as a function of the liquid ionic composition, it is essential that SAP properties are also evaluated with test liquids similar to the concrete pore solution. SAP sorption capability determination in a cementitious environment is a subject still under investigation.

In the present study, a new methodology for the determination of SAP sorption capacity in cementitious environment (plunger penetration consistence test method) is presented, and it is compared with the modified slump flow consistence test method and the teabag and the filtration test methods. In the latter two methods, the sorption capacity was determined using deionized water and a filtered solution of cement slurry. The tests were performed using three superabsorbent polymers with different absorption capability. The main objective of the study was to verify if the proposed new methodology can be used to determine the SAP sorption capacity.

2 Materials and Methods

The test campaign, devised to determine SAP absorption capacity and to link that with the effects SAP produce when used as an admixture in cementitious materials, is presented in Table 1. It comprises two free sorption assessment testing methods, teabag and filtration methods (Snoeck et al. 2018), and two methods that indirectly provide SAP absorption capacity, the plunger penetration consistence method, using a cement paste, and the slump flow consistence method, using a cementitious mortar. In the latter two methods, SAP absorption capacity is determined by comparing different consistence or slump flow measurements of reference mixtures with varying water content with that of the mixture containing the SAP. The direct sorption tests were carried out with deionised water and a cement filtrate solution. The latter was prepared by filtrating the solution obtained after 24 h of continuous automated stirring of a mixture composed of a CEM I 42.5 R cement (CIMPOR Alhandra) and deionised water, and having a water to cement ratio of 5. Three superabsorbent polymers were evaluated (SAP 1, SAP 2, SAP 3). SAP 1 is a cross-linked poly(acrylate-co-acrylamide) with qualitatively intermediate crosslinking and charge density. SAP 2 is a cross-linked polyacrylate with qualitatively intermediate crosslinking density and charge density. SAP 3 is a covalently cross-linked acrylamide/acrylic acid copolymer. All SAP were studied in their original grading. SAP 1 and SAP 2 have a particle size distribution in the range of 0–1500 μm, whilst SAP 3 has a finer particle size distribution (0–100 μm). SAP 1[1] and SAP 2[2] were used in several previous studies. Further details on SAP 3 can be found in (Silva Júnior 2017).

Table 1. Test campaign.

Test method	Superabsorbent polymers	Test medium
Teabag (Snoeck et al. 2018)	SAP 1, SAP 2, SAP 3	Deionised water
Filtration (Snoeck et al. 2018)		Cement filtrate solution
Plunger penetration consistence		Cement paste
Slump flow consistence		Cementitious mortar

Slump Flow Consistence Test Method. In this method, SAP sorption capacity is determined by comparing different slump flow measurements of reference mixtures with varying water content with that of the mixture containing the SAP. The method concept used in this study follows that of (Mönnig 2005), but the test duration is increased from 15 to 56 min. The consistence of the freshly mixed mortars is determined according to EN 1015-3; the only change made is that jolting is not applied to the flow table and the mixing procedure utilised to prepare the mortars is the one used in (Mechtcherine et al. 2014). Table 2 shows the composition of the reference mortars assessed. All mortars produced with SAP comprised 245 g of deionised water and 0.30% of SAP (the weight percentages refer to the mass of cement). Slump flow measurements were made at regular time intervals (8, 14, 20, 26, 32, 38, 44, 50 and 56 min). Time 0 min corresponds to the moment when the cement first contacted with water. The mortar is remixed between measurements at high speed for 2 min.

Table 2. Composition of the reference mortar

Constituents	Description	Mass (g)
Cement	CEM I 42.5 R (CIMPOR Alhandra)	700
Sand	River sand 0–4 mm, graded as in (Mechtcherine et al. 2014)	1340
Water	Deionised	196 g, 210 g, 224 g, 227.5 g, 231 g, 238 g
Superplasticiser	BASF Master Glenium ACE 426	10.5
Silica fume	BASF Mayco MS610	70

[1] SAP 1 – SAP 1 native in (Mechtcherine et al. 2018), SAP used in (Secrieru et al. 2016), SAP 2 in (Schröfl et al. 2015), SAP-DN in (Mechtcherine et al. 2015), and SAP D in (Schröfl et al. 2012).

[2] SAP 2 – SAP 2 native in (Mechtcherine et al. 2018), SAP 1 in (Mechtcherine et al. 2014) and (Mechtcherine et al. 2016), SAP 1 in (Schröfl et al. 2015), SAP B in (Mechtcherine et al. 2015), and SAP B in (Schröfl et al. 2012).

Plunger Penetration Consistence Test Method. With this test method, SAP sorption capacity is determined by comparing different consistence measurements of reference mixtures with varying water content (125, 127.5, 132, 135, 137, 139, 141, 143 and 145 g) with the one of the mixture containing the SAP. The consistence test itself is performed according to EN 196-3. The cement pastes were produced with deionised water and the same cement as that used to prepare the cement filtrate solution for the teabag and filtration tests. All pastes produced with SAP comprised 143 g of deionised water (the amount that produced a distance between the bottom face of the plunger and the base-plate of 2 mm in the reference paste). Three cementitious pastes were evaluated per SAP, namely, one with 0.06%, another one with 0.15% and a third one with 0.20% of SAP (the weight percentages refer to the mass of cement). These percentages produced a distance between the bottom face of the plunger and the base-plate ranging from 5 to 34 mm, thus covering adequately the plunger penetration range.

3 Results and Discussion

The results obtained for the SAP sorption capacity, determined with the four methods described in the previous section, are summarized in Table 3.

The absorption capacity of all SAP samples, determined with teabag and filtration methods, was approximately one order of magnitude larger in deionised water compared to that in cement filtrate solution. This effect was more pronounced for SAP 2 and less intense for SAP 3. The absorption capacity of SAP 1, SAP 2 and SAP 3 determined using a cement filtrate solution was, respectively, 16%, 12% and 20% of that obtained in deionised water. It has been suggested that this difference is due to the fact that the anionic functional groups of the polymers can strongly interact with ions in the solution (*e.g.*, Ca^{2+}). Since the anionic functional groups of the polymer may complex those ions in a very stable way, the effective anionic charge density in the SAP diminishes and, as a consequence, the polymer chains experience a lower osmotic pressure than the more highly charged uncomplexed chains; thus, resulting in a diminished absorption capacity (Plank and Sachsenhauser 2009; Schröfl et al. 2012).

All SAP samples exhibited an increasing absorption trend in deionised water from 1 min up to 1440 min of contact time in the teabag and filtration methods, the increase being greater for SAP 1 (73%) and SAP 2 (70%) than for SAP 3 (30%). However, SAP 3 had the highest relative sorption capacity at 1 min in the teabag and filtration methods (SAP 1 = 27%; SAP 2 = 29%; SAP 3 = 70%). More than 80% of the SAP water absorption has occurred in the first 30 min of testing (SAP 1 = 89%; SAP 2 = 87%; SAP 3 = 82%). SAP 2 registered the highest and SAP 3 the lowest deionised water absorption capacity of the three assessed SAP at 1440 min.

All SAP displayed stable swelling properties in deionized water, when monitoring the swelling capacity in time with teabag and filtration methods. However, this behaviour was not observed for the cement filtrate solution, where all SAP samples reached a maximum absorption capacity after about 5–180 min of contact with the test fluid (SAP 1 – 180 min; SAP 2 – 30 min; SAP 3 – 5 min), and then experienced a gradual loss of test fluid. SAP 2 and SAP 3 were more prone to this effect. The extent of test fluid desorption was more pronounced in the filtration method in comparison with the teabag method for SAP 1 (Teabag = 0%; Filtration = −15%), the opposite trend was observed

Table 3. SAP sorption capacity

SAP	Time (min)	*SC* TB_{H2O} (g_{H2O}/g_{SAP})	*SC* TB_{CS} (g_{H2O}/g_{SAP})	*SC* F_{H2O} (g_{H2O}/g_{SAP})	*SC* F_{CS} (g_{H2O}/g_{SAP})	*SC PPC* (g_{H2O}/g_{SAP})	*SC SFC* (g_{H2O}/g_{SAP})
1	1	28	11	123	29	19	22
	5	100	25	182	36		
	10	140	31	217	39		
	30	183	34	282	40		
	60	189	31	297	41		
	180	190	36	284	43		
	1440	219	38	301	37		
2	1	15	13	209	35	7	8
	5	59	33	267	35		
	10	103	38	312	34		
	30	227	48	383	29		
	60	252	42	398	27		
	180	254	29	388	23		
	1440	294	18	385	11		
3	1	69	21	114	26	13	14
	5	73	23	114	27		
	10	79	22	118	26		
	30	95	21	118	26		
	60	114	21	116	26		
	180	125	19	126	28		
	1440	127	17	132	25		

Notation: *SC* TB_{H2O} – SAP sorption capacity determined with teabag method and deionised water. *SC* TB_{CS} – SAP sorption capacity determined with teabag method and cement slurry. *SC* F_{H2O} – SAP sorption capacity determined with filtration method and deionised water. *SC* F_{CS} – SAP sorption capacity determined with filtration method and cement slurry. *SC PPC* – SAP sorption capacity determined with plunger penetration consistence method; the values presented consist of the arithmetic mean of the values obtained with cement pastes with three different SAP dosages; the plunger penetration measurements were made roughly 4 min after the cement and SAP contacted with deionised water. *SC SFC* – SAP sorption capacity determined with slump flow consistence method; the values presented were calculated with the slump flow measurements performed 56 min after the cement and SAP contacted with the deionised water.

for SAP 3 (Teabag = −23%; Filtration = −12%), and no significant difference was verified for SAP 2 (Teabag = −64%; Filtration = −67%). SAP 2 registered the highest overall desorption (SAP 1 = −8%; SAP 2 = −65%; SAP 3 = −18%). SAP 1 registered the highest and SAP 2 the lowest cement slurry filtrate absorption capacity of the three assessed SAP at 1440 min. From the three SAP assessed, SAP 2 is the only one of the polyacrylate type, which is known to be very sensitive to test liquids with multivalent cations, as is the case of the cement slurry filtrate; the other two SAP have acrylamide in their composition, which normally results in a better stability in multivalent-cation rich environments.

The 24 h sorption values in deionised water for all SAP were systematically higher for the filtration method than for the teabag method (sorption of SAP 1, SAP 2 and SAP 3 determined with the teabag method was, respectively, 73%, 76% and 97% of that obtained with the filtration method). The 24 h sorption results in cement filtrate solution for all SAP were higher for the filtration method than for the teabag method only for SAP 3, for SAP 1 it was practically the same in both methods.

In terms of results dispersion, a higher standard deviation was observed when deionised water was used as test fluid, in comparison to the cement filtrate solution.

Looking now into the results obtained indirectly (*i.e.* with the plunger penetration and the slump flow consistence methods) and directly (*i.e.* with teabag and filtration methods, using the cement filtrate solution), it is readily observable that the sorption capacities determined indirectly are lower than those obtained with the direct methods. This behaviour could be expected because in cement pastes and cementitious mortars the SAP cannot expand freely, as is under the influence of external forces and the ionic composition of the test fluid is not exactly the same as that of the cement filtrate solution; furthermore, a part of the measured mass gain in the teabag and filtration methods is due to liquid held by capillary action in-between the SAP particles. Nevertheless, there was a good agreement between the results obtained with the two indirect methods for the three SAP assessed, even though the consistence measurements, used to calculate the sorption capacity, were made roughly 4 min and 56 min after the cement and SAP entered in contact with the water, respectively for the plunger penetration and slump flow methods. This may, in part, be due to the different level of restraint to SAP swelling and chemical environment existing in the paste and mortar.

Comparing the SAP absorption capacity values obtained with the plunger penetration consistence method with those of teabag and filtration methods at 5 min of SAP contact time with the cement filtrate solution (Table 4), it is seen that SAP 1 and SAP 3 absorption capacities determined by the consistence method are roughly half of the ones obtained with the filtration method, and about two thirds of those obtained with the teabag method; for SAP 2, this difference was even more significant (as seen above, from all assessed polymers, SAP 2 was the one experiencing the greatest desorption of cement slurry filtrate). The differences obtained between the plunger penetration consistence method and the teabag and filtration methods, might actually be smaller, because the plunger penetration measurements were made four minutes after the cement and SAP entered in contact with water and, as already mentioned, some of the measured mass gain in the teabag and filtration methods is due to liquid held by capillary action in-between the SAP particles.

Table 4. SAP sorption capacity at 5 min of testing

SAP	Time (min)	*SC TB$_{CS}$* (g_{H2O}/g_{SAP})	*SC F$_{CS}$* (g_{H2O}/g_{SAP})	*SC PPC* (g_{H2O}/g_{SAP})	*SC TB$_{CS}$* (%)	*SC F$_{CS}$* (%)	*SC PPC* (%)
1	5	25	36	19	70	100	52
2	5	33	35	7	93	100	21
3	5	23	27	13	86	100	51

Note: The penetration measurements were made roughly 4 min after the cement and SAP contacted with deionised water.

Comparing the SAP absorption capacity values obtained with the plunger penetration consistence method with those of teabag and filtration methods at 10, 30 and 60 min of SAP contact time with the cement filtrate solution (Table 5), it is seen that the absorption capacities determined by the slump flow method and the filtration method follow the same trend observed for the plunger penetration measurements, *i.e.*, the ones determined with the plunger penetration consistence method are lower than those obtained of teabag and filtration methods. From the results it is also seen that SAP 1 experienced a gradual increase in sorption capacity throughout the plunger penetration test duration (a behaviour similar to that observed in the full test period of the teabag and filtration methods), SAP 2 a slight decrease in its sorption capacity with the test progression (a more significant decrease was observed in the full test period of the teabag and filtration methods), and SAP 3 exhibited a gradual loss of test fluid with the increase of testing time (similarly to the behaviour shown in the full test period of the teabag and filtration methods).

Table 5. SAP sorption capacity at 10, 30 and 60 min of testing

SAP	Time (min)	*SC TB*$_{CS}$ (g_{H2O}/g_{SAP})	*SC F*$_{CS}$ (g_{H2O}/g_{SAP})	*SC SFC* (g_{H2O}/g_{SAP})	*SC TB*$_{CS}$ (%)	*SC F*$_{CS}$ (%)	*SC SFC* (%)
1	10	31	39	14	80	100	36
2	10	38	34	9	100	91	23
3	10	22	26	22	88	100	86
1	30	34	40	18	85	100	45
2	30	48	29	8	100	61	17
3	30	21	26	19	82	100	75
1	60	31	41	22	76	100	53
2	60	42	27	8	100	64	18
3	60	21	26	14	78	100	54

Note: Slump flow measurements performed 8, 32 and 56 min after the cement and SAP contacted with deionised water.

4 Conclusions

This work showed that SAP sorption kinetics and absorption capacity depend on polymer type and on surrounding medium. So, whenever SAP are to be used in cement-based materials, they should always be pre-tested with respect to their absorption capacity and kinetics of absorption and desorption. This characterisation may involve water sorption, but preferably the determination of sorption behaviour in extracted or synthetic pore solutions, as test liquids ionic composition strongly influences their sorption response. Currently, methods of free sorption testing in cementitious solutions exist; however, since they cannot truly mimic chemically the solution existing in the cement based material, nor the forces, temperature and other factors to which they are subject to when present in a cementitious medium, it is important to complement SAP characteristics assessed via free sorption tests with tests performed in cement pastes or mortars, prior to their implementation in concrete.

The main goal of this study was to contribute to the current discussion on the testing of SAP in an environment as close as possible to that observed in concrete, but that could be used in routine tests to assess potential SAP candidates for specific intended uses. Its main findings can be summarized as follows: (i) free sorption capacity of all SAP was approximately one order of magnitude larger in deionised water than in cement filtrate solution; (ii) all SAP exhibited an increasing absorption trend in water from 1 min up to 1440 min of contact time; (iii) all SAP exhibited stable swelling properties in deionised water (however, for the cement filtrate, all SAP reached a maximum sorption capacity after about 5–180 min of contact, and then experienced a gradual loss of test fluid); (iv) the 24 h sorption values in water for all SAP were systematically higher for the filtration method than for the teabag method; (v) the 24 h sorption results in cement filtrate solution for all SAP were higher for the filtration method than for the teabag method only for SAP 3, for SAP 1 it was practically the same in both methods; (vi) SAP sorption capacities determined indirectly were lower than those obtained with the direct methods; (vii) there was a good agreement between the results obtained with the two indirect methods for the three SAP assessed; (viii) similar sorption trends were observed with the direct and indirect methods for the three SAP assessed. Hence, from the results obtained so far, it appears that the plunger penetration consistence method provides a simple and quick way of estimating the SAP sorption capacity in an environment more similar to that found by the SAP in concrete.

Acknowledgments. Authors wish to acknowledge the support given by CIMPOR and BASF GmbH to the research study. João Custódio carried out the work within the scope of the FCT Investigator Project IF/00595/2015, financed by the Fundação para a Ciência e a Tecnologia, FCT (Portugal); the author wishes to acknowledge this financial support.

References

Jensen, O.M., Hansen, P.F.: Water-entrained cement-based materials: I. Principles and theoretical background. Cem. Concr. Res. **31**(4), 647–654 (2001)

Jensen, O.M., Hansen, P.F.: Water-entrained cement-based materials: II. Experimental observations. Cem. Concr. Res. **32**(6), 973–978 (2002)

Lee, H.X.D., Wong, H.S., et al.: Self-sealing cement-based materials using superabsorbent polymers. In: Jensen, O.M., Hasholt, M.T., Laustsen, S. (eds.) Proceedings pro074: International RILEM Conference on Use of Superabsorbent Polymers and Other New Additives in Concrete. RILEM Publications (2010)

Mechtcherine, V., Dudziak, L.: Effects of superabsorbent polymers on shrinkage of concrete: plastic, autogenous, drying. In: Mechtcherine, V., Reinhardt, H.-W. (eds.) Application of Super Absorbent Polymers (SAP) in Concrete Construction, pp. 63–98. Springer, Dordrecht (2012)

Mechtcherine, V., Gorges, M., et al.: Effect of internal curing by using superabsorbent polymers (SAP) on autogenous shrinkage and other properties of a high-performance fine-grained concrete: results of a RILEM round-robin test. Mater. Struct. **47**(3), 541–562 (2014)

Mechtcherine, V., Schröfl, C., et al.: Effect of superabsorbent polymers (SAP) on the freeze–thaw resistance of concrete: results of a RILEM interlaboratory study. Mater. Struct. **50**(1), 14 (2016)

Mechtcherine, V., Secrieru, E., et al.: Effect of superabsorbent polymers (SAPs) on rheological properties of fresh cement-based mortars - development of yield stress and plastic viscosity over time. Cem. Concr. Res. **67**, 52–65 (2015)

Mechtcherine, V., Snoeck, D., et al.: Testing superabsorbent polymer (SAP) sorption properties prior to implementation in concrete: results of a RILEM Round-Robin Test. Mater. Struct. **51** (1), 28 (2018)

Mönnig, S.: Water saturated super-absorbent polymers used in high strength concrete. Otto Graf J. **16**, 193–202 (2005)

Plank, J., Sachsenhauser, B.: Experimental determination of the effective anionic charge density of polycarboxylate superplasticizers in cement pore solution. Cem. Concr. Res. **39**(1), 1–5 (2009)

Schröfl, C., Mechtcherine, V., et al.: Relation between the molecular structure and the efficiency of superabsorbent polymers (SAP) as concrete admixture to mitigate autogenous shrinkage. Cem. Concr. Res. **42**(6), 865–873 (2012)

Schröfl, C., Mechtcherine, V., et al.: Sorption kinetics of superabsorbent polymers (SAPs) in fresh Portland cement-based pastes visualized and quantified by neutron radiography and correlated to the progress of cement hydration. Cem. Concr. Res. **75**, 1–13 (2015)

Secrieru, E., Mechtcherine, V., et al.: Rheological characterisation and prediction of pumpability of strain-hardening cement-based-composites (SHCC) with and without addition of superabsorbent polymers (SAP) at various temperatures. Constr. Build. Mater. **112**, 581–594 (2016)

Silva Júnior, P.F.: Estudo da estabilidade dimensional de concretos de alta resistência com adição de polímero superabsorvente e nanopartículas de sílica. Ph.D. thesis, University of Brasília, Brasília, Brazil (2017)

Snoeck, D., Schröfl, C., et al.: Recommendation of RILEM TC 260-RSC: testing sorption by superabsorbent polymers (SAP) prior to implementation in cement-based materials. Mater. Struct. **51**(5), 116 (2018)

Altering the Crosslinking Density of Polyacrylamide Hydrogels to Increase Swelling Capacity and Promote Calcium Hydroxide Growth in Cement Voids

Cole R. Davis, Baishakhi Bose, Alejandro M. Alcaraz, Carlos J. Martinez, and Kendra A. Erk(✉)

School of Materials Engineering, Purdue University, 701 West Stadium Avenue, West Lafayette, IN 47907, USA
erk@purdue.edu

Abstract. Polyacrylamide (PAM) superabsorbent polymer microspheres were synthesized for use as internal curing agents in low water-to-cement ratio mixtures. An increase in PAM swelling capacity in pore solutions was achieved by reducing the crosslinking density during synthesis: 5.1 g/g swelling was observed for 2 wt.% crosslinker and 20.6 g/g for 0.5 wt.% crosslinker. Capillary micromechanical experiments showed that a reduction in crosslinking density decreased the PAM elastic modulus from 1100 ± 310 kPa to 110 ± 15 kPa. Yet all PAM maintained mechanical integrity when mixed with cement and cured under a compressive load. SEM analysis showed hydrated product formation within cement voids left behind by dehydrated PAM. PAM with reduced crosslinking densities resulted in a 26 ± 7.3% increase in hydrated product formation within these voids. Cement paste density and compressive strength was not compromised by the addition of PAM.

Keywords: Superabsorbent polymer · Cement microstructure · Internal curing · Chemical admixtures

1 Introduction

The chemistry of superabsorbent polymer (SAP) particles, including monomer chemistry and crosslinking density, can alter the material's swelling capacity (Eichenbaum et al. 1999; Quintero et al. 2010; Zhu et al. 2015; Krafcik and Erk 2016; Krafcik et al. 2017; Krafcik et al. 2018). The chemical composition of SAP particles is typically comprised of acrylamide (AM) monomer, acrylic acid (AA) monomer, or a mixture of the two. Particles containing AA display increased swelling capacity in pure water compared with AM-based particles due to the greater concentration of anionic carboxylate (COO^-) groups in the polymer network from the AA monomer. However in pore solution, these carboxylate groups cause rapid deswelling of the particle due to the formation of ionic crosslinks with cations naturally present in pore solution (Erk and Bose 2018). Rapid deswelling may not be desirable in high performance concrete applications because if the release of water is premature (*e.g.*, before concrete

W. P. Boshoff et al. (Eds.): SAP 2019, RILEM Bookseries 24, pp. 20–28, 2020.
https://doi.org/10.1007/978-3-030-33342-3_3

placement), mitigation of self-desiccation and autogenous shrinkage will be limited. Due to the lack of carboxylate groups in the polymer network, the swelling response of AM-based particles is much less sensitive to the presence of cations in pore solution – rapid deswelling is typically not observed. Thus, AM-based particles with sufficient swelling capacities may be better suited for most internal curing applications compared with AA-based particles and AA-AM copolymer particles.

One strategy to increase the swelling capacity of SAP particles is to decrease the density of chemical crosslinks in the polymer network (Erk and Bose 2018), achieved by reducing the concentration of crosslinker in the synthesis reaction. However, the increased swelling capacity of lightly crosslinked particles is at the expense of a reduction in particle toughness, as there is more water and correspondingly less polymer to withstand external forces. Thus, SAP particles with sufficient swelling behavior in pore fluid may be prone to physical deformation and damage when incorporated into concrete; *e.g.*, large SAP particles may break down into smaller particles, which could reduce mixture workability, change desorption kinetics, and ultimately decrease their performance as internal curing agents.

The hypothesis investigated here is that by using only AM monomer and altering the crosslinking density, a SAP particle that is chemically resistant to cations and has an increased swelling capacity while maintaining sufficient toughness can be synthesized and further utilized as an effective internal curing agent for cement. To investigate this, polyacrylamide microspheres were synthesized with varying amounts of crosslinker to directly investigate the impact of crosslinking density on particle swelling capacity and mechanical properties. Particles were added to cement paste to determine their impact on the microstructure, density, and strength of internally cured cement paste.

2 Materials and Methods

Polyacrylamide microspheres (herein referred to as "PAM") were synthesized following previous work (Davis et al. 2018) and formulations can be found in Table 1. The density of chemical crosslinks in PAM was varied by changing the concentration of crosslinker (N,N'-methylbisacrylamide, MBAM) in the reaction to be 0.5, 1, and 2% by weight of monomer. The resulting three PAM formulations are subsequently referred to herein as 0.5X, 1X and 2X, denoting crosslinker ("X") concentration. A constant mixing speed of 500 rpm was used. PAM swelling capacity, size, and morphology were characterized using optical microscopy. Gravimetric and volumetric swelling capacities were calculated following previous work (Davis et al. 2018). Swelling capacities were determined in deionized (DI) water and cement pore solution. Cement pore solution was prepared by mixing 10 g of ordinary Portland cement with 100 mL of DI water. This solution was stirred for 30 s and allowed to sit until sedimentation of the cement particles occurred (approximately 1 min). Capillary micromechanics was used to characterize PAM compressive, shear, and elastic moduli (Wyss et al. 2010). In this technique, back-pressure was applied to force the PAM microspheres into the tapered walls of a glass capillary tube. Optical microscopy was then used to measure deformation as a function of applied pressure.

Cement pastes were prepared with and without PAM. A water-to-cement (w/c) ratio of 0.35 was used for all samples. Dry PAM were added to Portland cement at 0.2% by weight of cement (bwoc), followed by the addition of tap water and Glenium 3030 NS full-range water reducing agent (BASF, Germany) at 0.7 mL per 100 g of cement. Cement paste batches were mixed for 1 min, set aside for 30 s, and then mixed again for 1 min. This mixing procedure was used to mix samples by hand and using a vacuum mixer (to differentiate between the effect of entrapped air voids). To simulate the curing of cement under compressive load, custom-made piston and cylinder cement molds were designed and fabricated from PVC. Exact dimensions of these molds can be found in Fig. 1. To apply a pressure during curing, weights were stacked on top of the piston until a desired pressure was reached. For each sample cured under pressure, a control from the same cement batch was also cast and cured without load for comparison. All samples were cured for 72 h at a temperature of 23 ± 1 °C.

Table 1. Formulations for the Aqueous Phase of PAM.

PAM formulation	AM (g)	DI Water (ml)	MBAM (g)	MBAM (wt.%)	NaPS (ml)	TMED (ml)
0.5X	2.4	12	0.0125	0.5	1	1.3
1X	2.4	12	0.025	1	1	1.3
2X	2.4	12	0.05	2	1	1.3

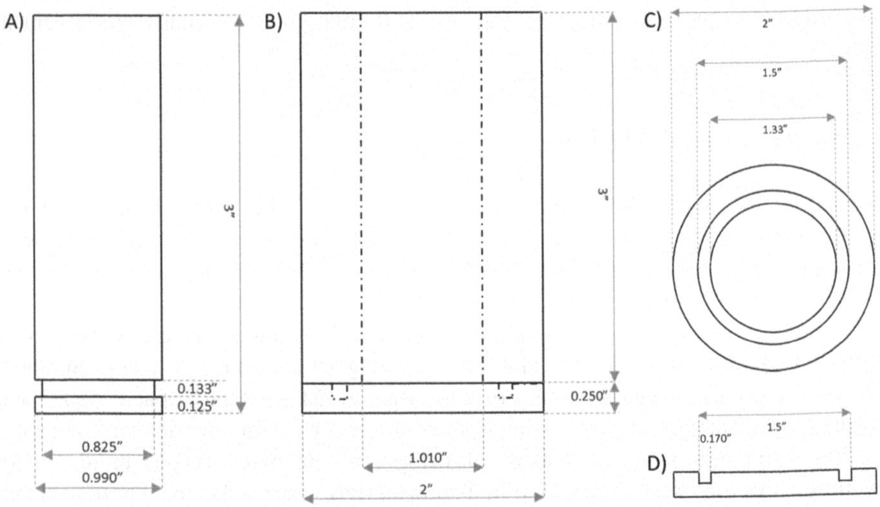

Fig. 1. Schematic of cement mold for curing under a specified load. This diagram depicts (A) side view of the mold piston, (B) side view of the cylinder with removable bottom plate, (C) top view of the removable bottom plate, and (D) side view of removable bottom plate.

Once cured, samples were demolded, cut, and soaked in methanol for 24 h to stop further hydration of the cement. The samples were vacuum dried for 48 h, mounted in epoxy, and polished. A NanoScience Instruments Phenom Desktop SEM (Phoenix, AZ) was used to collect backscattered electron micrographs of the cement microstructure and to perform energy dispersive X-ray spectroscopy for elemental composition analysis. Apparent density measurements were made using the ASTM C127 standard. Compression tests were performed in triplicate after 72 h of curing using an Insight 820.300-SL (MTS Systems Corporation) at a constant strain of 1 mm/min. A maximum compressive strength was found for each sample and was reported as an average with an error of one standard deviation.

3 Results and Discussion

3.1 PAM Characterization

PAM size, equilibrium swelling capacity, and morphology can be seen in Fig. 2. The size of dry PAM did not have a significant dependence on crosslinker concentration (Fig. 2A). The size of hydrated PAM was reduced as the crosslinker concentration increased (from 0.5X to 1X to 2X) as crosslinks reduce expansion of the solvated polymer network and inhibit flexibility. Thus, a lower density of chemical crosslinks in the polymer network results in a larger hydrated size (Eichenbaum et al. 1999). equilibrium swelling occurred immediately for 2X and 1X formulations due to their reduced swelling capacities while the larger swelling capacity of the 0.5X formulation required approximately 18 h to reach equilibrium. As expected for an AM-based hydrogel particle, similar swelling capacities and rates were found in DI water and pore solution for all PAM. For 0.5X, the equilibrium swelling capacity was 21 g/g, more than double the swelling capacity of 1X (8.4 g/g) and four times the swelling capacity of 2X (5.1 g/g) – see Fig. 2B. Equilibrium swelling capacities using the volumetric method resulted in lower values compared to the gravimetric method, consistent with previous work (Davis, Kelly, and Erk 2018).

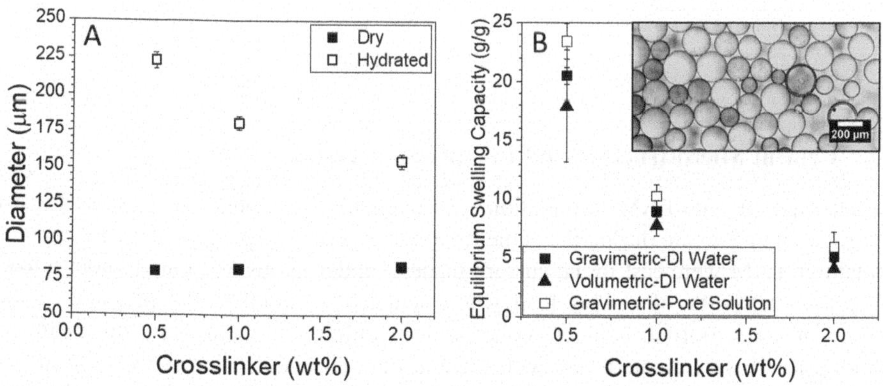

Fig. 2. (A) Average particle size for dry and hydrated PAM (DI water) using optical microscopy. (B) Average equilibrium swelling capacities in DI water using the gravimetric and volumetric methods and in pore solution. Micrograph of hydrated 2X. Error bars represent 95% confidence intervals.

Capillary micromechanics experiments were conducted for 2X and 0.5X formulations. Micrographs from these experiments can be found in Fig. 3. For 2X, more pressure was required to reach the same deformation as 0.5X (*e.g.*, 40 and 120 kPa compared to 4 and 10 kPa, respectively). Table 2 shows the average compressive, shear, and elastic moduli along with the Poisson's ratio. Compared with the 2X formulation, the elastic modulus of 0.5X was reduced by a factor of ten. As crosslinker concentration is reduced, there are fewer chemical crosslinks to restrict expansion of the polymer network, allowing 0.5X to absorb more water at the expense of moduli. The calculated Poisson's ratio decreased for 2X, suggesting a reduction in ability to deform elastically. An observed manifestation of this was the greater likelihood of 2X to visibly fracture during the micromechanical tests compared with 0.5X.

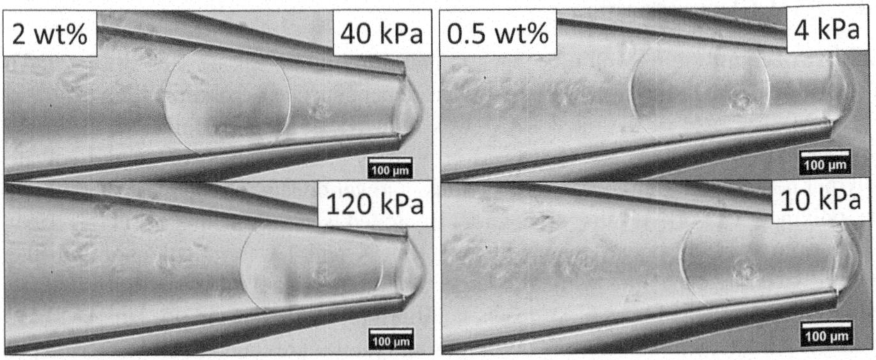

Fig. 3. Micrographs of capillary micromechanics tests for (left) 2X and (right) 0.5X formulations, with applied pressures noted on each image.

Table 2. Average compressive (K), shear (G), elastic (E) moduli, and Poisson's ratio (ν), reported with an error of one standard deviation from the mean.

MBAM (wt.%)	K (kPa)	G (kPa)	E (kPa)	ν
0.5 (0.5X)	160 ± 33	41 ± 6.4	110 ± 15	0.38 ± 0.036
2 (2X)	860 ± 190	430 ± 130	1100 ± 310	0.29 ± 0.033

3.2 Cement Microstructure and Compression Testing

To determine if both PAM formulations can maintain their spherical shape and resist deformation or fracture during the cement mixing and casting process, cement samples containing PAM were cast in the custom-made cylinder molds and cured under a load of approximately 86 kPa. This load was calculated to simulate the pressure at the bottom of a ten-foot-tall concrete column. Two different cement paste mixes incorporating PAM were created, one with 2X and the other containing 0.5X, both at dry dosages of 0.2% bwoc (note that both PAM formulations have similar dry size and density). Figure 4 shows microstructures for vacuum mixed 0.5X-cement that was cured without a load (Fig. 4A) and with a load of 86 kPa (Fig. 4B). Even though 0.5X deformed significantly at pressures of 10 kPa (see Fig. 3), there were no visible signs of deformation within the cured cement paste. To confirm this, an aspect ratio of horizontal to vertical pore diameters was calculated for over 50 voids per sample, and an average value of 1.0 ± 0.06 was calculated. Evidence of deformation or fracture was not observed for either PAM formulation. These testes were also carried out for hand mixed cement paste and did not show signs of PAM particle deformation. Thus, even though PAM became significantly softer as crosslinker concentration was reduced, the mechanical integrity of PAM within the paste was not impacted.

Microstructural analysis of the 0.5X-cement samples in Fig. 4 shows a matrix of calcium silicate hydrate (CSH) with unhydrated and partially hydrated cement grains. In the areas immediately surrounding the voids remaining from dehydrated PAM, an increase in hydrated product was observed. Inside these voids, hydrated product platelets were observed to grow inward from the apparent edges of the void. This hydrated product is most likely calcium hydroxide (CH) due to the crystal morphology and minimal (~2 wt.%) silicon found during elemental analysis, although it may also contain some CSH. Calcium silicates (C_2S and C_3S) in the cement react with water to form CSH and CH (Chourasia 2015). Due to the low w/c ratio in these mixtures, the lack of water reduces the amount of hydrated product formed in early stages. The swollen PAM serves as a reservoir of water which enhances formation of CH in PAM-related voids. When comparing the microstructures of cement containing 2X and 0.5X, an increase in CH formation was observed inside the 0.5X voids – see Figs. 5B and C. As 0.5X has a larger swelling capacity than 2X, the 0.5X formulation is expected to absorb more water and retain it for longer periods of time, supplying ample water to facilitate CH formation. To directly quantify the amount of CH formation within PAM-related voids, the area fraction of CH growth inside over 150 voids was measured and is reported in Fig. 5A. The average CH growth area fraction for 2X and 0.5X was 21 ± 2.9% and 47 ± 4.4%, respectively, a statistically significant difference (p-value < 0.00001). By reducing the crosslinker concentration, there was a 26 ± 7.3% increase in CH formation.

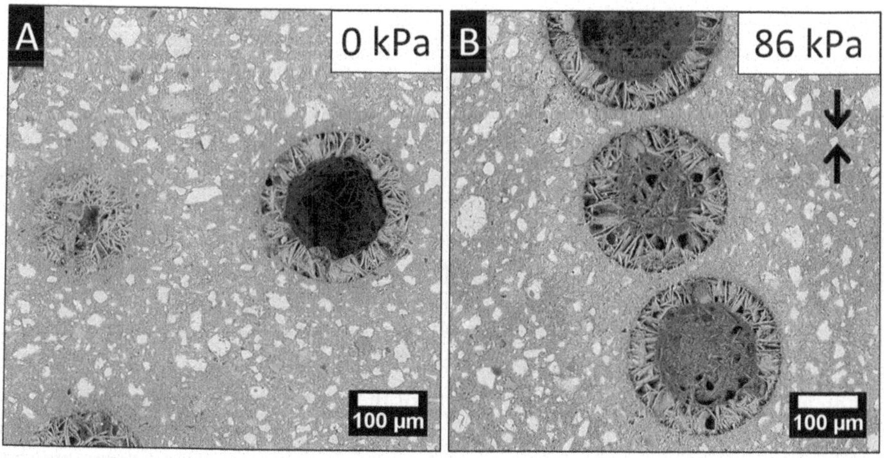

Fig. 4. Micrographs of 0.5X-related voids in cement paste cured for 72 h (A) unloaded and (B) with a load of 86 kPa. Arrows indicate loading direction.

Density measurements were made for vacuum mixed and hand mixed cement paste samples without PAM (control) and containing 2X and 0.5X. Vacuum mixing resulted in a higher density (2.047 ± 0.010 g/cm^3) than hand mixing (2.018 ± 0.014 g/cm^3) regardless of PAM content. While the method used for mixing the paste samples had an impact on density, the addition of PAM did not alter the apparent density by more than one standard deviation from the control. Thus, even though there are voids left behind by PAM, some of this porosity was replaced with CH (as seen in Fig. 5) and the presence of PAM ultimately created more hydrated product which has a higher density than unhydrated cement grains (Neville 1995).

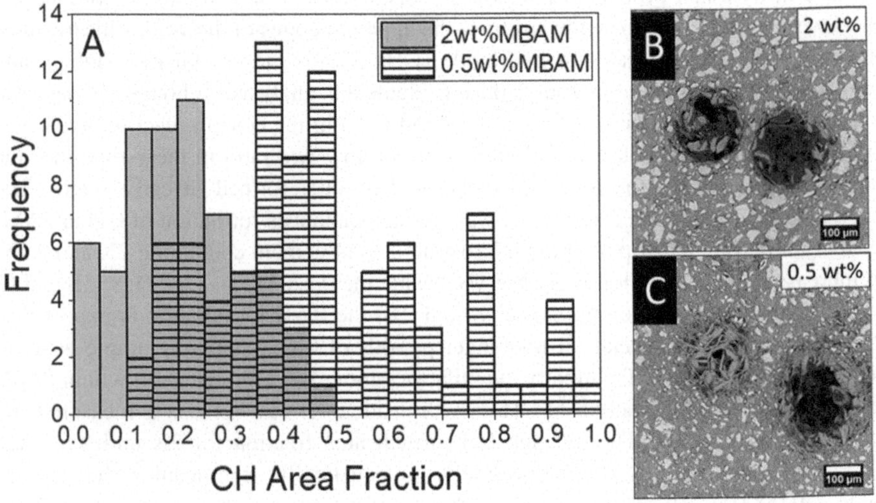

Fig. 5. (A) CH area fraction inside PAM-related voids in cement paste containing 0.5X and 2X. Micrographs of hand mixed cement containing (B) 2X and (C) 0.5X after 72 h of curing.

Average compressive strength measurements after 72 h of curing for vacuum mixed and hand mixed cement samples containing no PAM (control), 2X, and 0.5X are reported in Fig. 6A. Vacuum mixed samples displayed the greatest compressive strengths due to a reduction in porosity as confirmed by density measurements. While sensitive to the presence of porosity, the compressive strength values were apparently not sensitive to the addition of PAM at the selected dosage (0.2% bwoc). All samples were found to fail in similar manner, undergoing splitting failure mode, with several vertical fractures extending from top to bottom as seen in Fig. 6B (Neville 1995).

It can be concluded that the addition of relatively soft PAM did not have a negative impact on the compressive strength of cement paste. Thus, PAM-based internal curing agents can be designed to have improved swelling characteristics, promote CH formation, and maintain the mechanical strength of the cement paste. Future work should investigate the use of PAM microspheres with a smaller particle size. Reducing the particle size can be achieved by increasing the mixing speed during synthesis and has

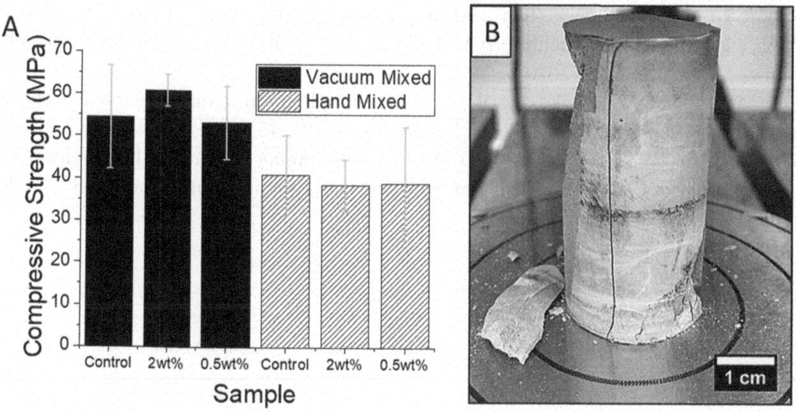

Fig. 6. (A) 72-h strengths of cement paste with and without PAM. Error bars represent one standard deviation from the mean. (B) Image of fractured cement paste specimen after compressive failure for a vacuum mixed control sample.

been demonstrated to reach hydrated particle sizes of less than 50 μm (Davis et al. 2018) compared to the 150 μm particles used in this study. By decreasing the particle size, improvements in cementitious composite strength may be achieved from more complete "refilling" of PAM-related voids with hydrated product.

References

Chourasia, A.: Protection of bio-deteriorated reinforced concrete using concrete sealers. Int. J. Mater. Chem. Phys. **1**(1), 11–19 (2015)

Davis, C.R., Kelly, S.L., Erk, K.A.: Comparing laser diffraction and optical microscopy for characterizing superabsorbent polymer particle morphology, size, and swelling capacity. J. Appl. Polym. Sci. **135**(14), 1–10 (2018)

Eichenbaum, G.M., et al.: Investigation of the swelling response and loading of ionic microgels with drugs and proteins: the dependence on cross-link density. Macromolecules **32**, 4867–4878 (1999)

Erk, K.A., Bose, B.: Using polymer science to improve concrete: superabsorbent polymer hydrogels in highly alkaline environments. In: Gels and Other Soft Amourphous Solids, American Chemical Society, pp. 333–356 (2018)

Krafcik, M.J., Bose, B., Erk, K.A.: Synthesis and characterization of polymer-silica composite hydrogel particles and influence of hydrogel composition on cement paste microstructure. Adv. Civ. Eng. Mater. **7**(4), 590–612 (2018)

Krafcik, M.J., Erk, K.A.: Characterization of superabsorbent poly(sodium-acrylate acrylamide) hydrogels and influence of chemical structure on internally cured mortar. Mater. Struct. **49**(11), 4765–4778 (2016)

Krafcik, M.J., Macke, N.D., Erk, K.A.: Improved concrete materials with hydrogel-based internal curing agents. Gels **3**(46), 1–18 (2017)

Neville, A.: Properties of Concrete, 4th edn. Prentice Hall, Harlow (1995)

Quintero, S.M.M., et al.: Swelling and morphological properties of poly (vinyl alcohol) (PVA) and poly (acrylic acid) (PAA) hydrogels in solution with high salt concentration. Polymer **51**(4), 953–958 (2010)

Wyss, H.M., Franke, T., Mele, E., Weitz, D.A.: Capillary micromechanics: measuring the elasticity of microscopic soft objects. Soft Matter **6**(18), 4550–4555 (2010)

Zhu, Q., Barney, C.W., Erk, K.A.: Effect of ionic crosslinking on the swelling and mechanical response of model superabsorbent polymer hydrogels for internally cured concrete. Mater. Struct. **48**(7), 2261–2276 (2015)

X-Ray Micro Tomography of Water Absorption by Superabsorbent Polymers in Mortar

Claudia Romero Rodriguez[1](✉), Maxim Deprez[2], Fernando F. de Mendonca Filho[1], Stefanie van Offenwert[2], Veerle Cnudde[2,3], Erik Schlangen[1], and Branko Šavija[1]

[1] Microlab, Department of 3MD, Faculty of Civil Engineering and Geosciences, Delft University of Technology, Stevinweg 1, 2628 CN Delft, The Netherlands
{c.romerorodriguez,F.Filho,Erik.Schlangen, B.Savija}@tudelft.nl

[2] PProGRess/UGCT, Geology Department, Faculty of Sciences, Ghent University, Krijgslaan 281 S8, 9000 Ghent, Belgium
{Maxim.Deprez,Stefanie.VanOffenwert, Veerle.Cnudde}@UGent.be

[3] Department of Earth Sciences, Faculty of Geosciences, Utrecht University, Princetonlaan 8A, 3584CD Utrecht, The Netherlands

Abstract. Superabsorbent Polymers (SAP) have been recently subject of investigation as smart admixtures for cement-based materials. The properties of these polymers enable their use for internal curing, increasing freeze/thaw resistance, boosting autogenous self-healing and providing a crack self-sealing effect in cementitious composites. Except for the earliest application, the functioning of these beneficial effects invloves the absorption by the polymers of ingress water in the hardened cementitious matrix and later release, as well as their capacity to complete multiple absorption/desorption cycles. In this work, the absorption of water in mortar with superabsorbent polymers is monitored during the first 60 min of absorption through micro-CT. The experimental series included the presence of cracks. The registration and differentiation of sub-minute (18 s) scans enabled the individuation of bulk water content distribution in the mortar with a resolution of 55 μm. The swollen volume of SAP could also be quantified and studied in time. The results point out that although embedded SAP absorb water from the matrix, this absorption is slow and reduced with respect to water absorption during mixing for the used SAP. Same effect is observed for SAP in the cracks.

Keywords: SAP · Mortar · X-ray micro computed tomography · Concrete durability

1 Introduction

Most of durability problems in cementitious composites involve the ingress of water into the matrix, as well as of harmful species in solution. Carbonation of cement, frost damage, chloride ingress, etc. are vivid examples of such problems. When cracks are

W. P. Boshoff et al. (Eds.): SAP 2019, RILEM Bookseries 24, pp. 29–37, 2020.
https://doi.org/10.1007/978-3-030-33342-3_4

present in the material, further acceleration of the degradation mechanisms listed above happens due to the additional surfaces from which deleterious substances can penetrate. In recent years, some studies have pointed out the use of superabsorbent polymers (SAP) to improve the durability of cement-based materials (Jensen and Hansens 2002). These admixtures are polyelectrolyte gels which absorb water many times their own weight. When SAP are added into fresh cementitious mixtures, they absorb water and swell, when the material sets and dries, the stored water is released and the SAP shrink leaving behind a macropore.

Internal curing of concrete with SAP has been proven to reduce shrinkage cracking significantly (Geiker et al. 2004) with respective positive implications for the durability of the studied materials. Same type of admixture was shown to be beneficial for increasing the frost resistance of concrete due to the creation of a uniformly distributed macropores system (Mechterine et al. 2017). Also crack self-sealing and self-healing effects have been associated to the presence of these polymeric particles (Lee et al. 2010; Snoeck et al. 2016). Moreover, there exist some studies that point out beneficial side effects of embedded SAP in the resistance of the material against carbonation and chloride ingress (Beushausen et al. 2014; Dang et al. 2017). Whereas some other researches show slightly worsened performances against carbonation (Reinhardt and Assman 2009).

In this work the authors studied the absorption behavior of embedded and in-crack SAP during capillary water absorption of cement-based materials through micro-CT (Cnudde and Bonne 2013). Such information could be used to understand and unveil the potential of these particles to improve the durability of cement-based materials.

2 Methods

2.1 Materials and Sample Preparation

Two mortars with water-to-cement ratio of 0.45 were made with and without superabsorbent polymers. Mix designs of the mortars are reported in Table 1. CEM I 52.5 R, from ENCI Netherlands, tap water, superplasticizer Master Glenium 51 from BASF Netherlands and dry quartz aggregates 0.125/2 mm from Dekker Grondstoffen were used for the preparation of the mortars. A commercially available SAP Floset 27 cc, in this paper denominated as SAP F, was used in the SAP mortar. This consisted of cross-linked copolymer of acrylamide and acrylate, supplied from SNF SAS (Andrezieux, France). Particle size distribution of the dry SAP is shown in Fig. 1 after measurement in ethanol via laser diffraction. For the design of the mixtures, attention was paid to compensate for the water absorbed by the SAP during mixing by adding extra water in the mortar with SAP. This quantity was determined by adding tentative water amounts until matching the same flow table test results of 185–190 mm. This way it was determined that SAP F absorption capacity during mixing, Abs_{mix}, was 20 g_{water}/g_{SAP}. Same amount of superplasticizer was added to both mixtures to prevent air bubble formation. The sand-to-cement ratio was kept the same for both reference and SAP mortar at 3.27. The mixtures differ on the total amount of mortar to compensate for the volume occupied by swollen SAP during mixing.

Table 1. Mix design of mortars (quantities in [kg/m^3].

Component	4REF	4F0.5
CEM I 52.5 R	527	499
Water	237	224
Additional water	–	50
SAP	–	2.5
Superplasticizer	1.2	1.2
Aggregates	1720	1628
1–2 mm	528	500
0.5–1 mm	444	420
0.25–0.5 mm	374	354
0.125–0.25 mm	374	354

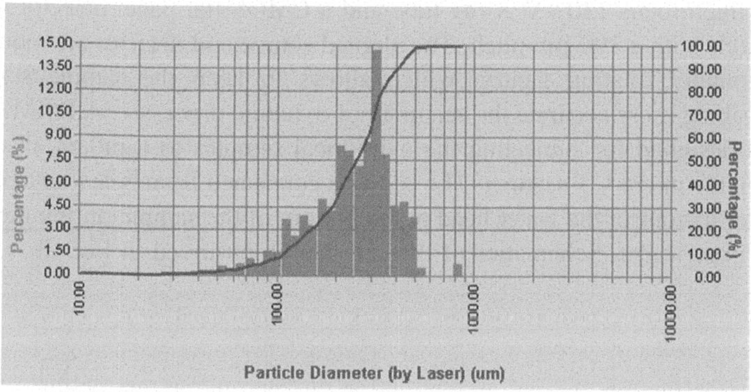

Fig. 1. Particle size distribution of SAP F by laser diffraction.

A Hobart planetary mixer with a capacity of 5 l was used to prepare the fresh mortar. All dry components, cement, sand and SAP (when applicable), were mixed for 1 min at speed 1. Previously mixed tap water and superplasticizer were added during the next 30 s while the dry components were still mixing and successively the whole mix was left for other extra 30 s at speed 1. The mixer was stopped for 1 min, time during which the walls and bottom of the bowl were scraped and mixed by hand in the fluid mortar. Next, the fresh mortar was mixed for 1 min and 30 s at speed 1 and 2, respectively. The mortar was left to rest for 10 min while covered by plastic foil to prevent water evaporation. This waiting time was necessary for the achievement of absorption equilibrium by SAP F.

The mortars were cast into the moulds in two layers and put 15 s in the vibrating table for each layer. Cylindrical moulds with diameter of 16 mm and height of 32 mm were employed. Two diametrically opposed groves with 2 mm side ran along the height of the mould. The samples were covered with plastic foil and left to set and harden for 24 h in laboratory conditions. After 24 h the samples were demoulded and stored in a fog room at 20 ± 1 °C and 95% Relative Humidity for 28 days. At 21 days they were sawn in smaller cylinders of 10 mm height and returned to the fog room.

At 28 days of age, the cylinders were put in an oven at 40 °C until constant weight attainment for approximately one week. All the surfaces were then sealed with duct tape until testing. To create the cracked samples, prior to the test, some wrapped cylinders were split in Brazilian Tensile Test configuration. The two parts were put back together by inserting a prismatic rod within the grooved space with width equal to 2 mm plus the desired crack width, 300 μm, and then by bridging the surfaces with bi-component glue Pleximon. The scheme is shown in Fig. 2(a). In this way the crack width was controlled to a certain extent.

2.2 Micro-CT Differential Dynamic Scanning

The Environmental micro-CT scanner (EMCT) from the Centre of X-ray tomography of Ghent University (UGCT) (Dierick et al. 2014) was employed to monitor the absorption of demineralized water in the mortar. The scanner consists of a standard directional microfocus 130 kV X-ray tube and a CMOS flat panel detector with 1316 by 1312 pixels with a 100 μm pitch. The aligned source and detector are mounted on a rigid horizontally rotating gantry, which allows to keep the sample stage static, therefore making more accurate the scanning of dynamic processes. A PMMA cell was specifically designed for subjecting the cylindrical samples to capillary absorption of water within the micro-CT scanner. The cell was connected from below to a pump via a hose in order to control the water head at the bottom of the sample during the capillary absorption experiment. Schematics of the cell can be observed in Fig. 2b.

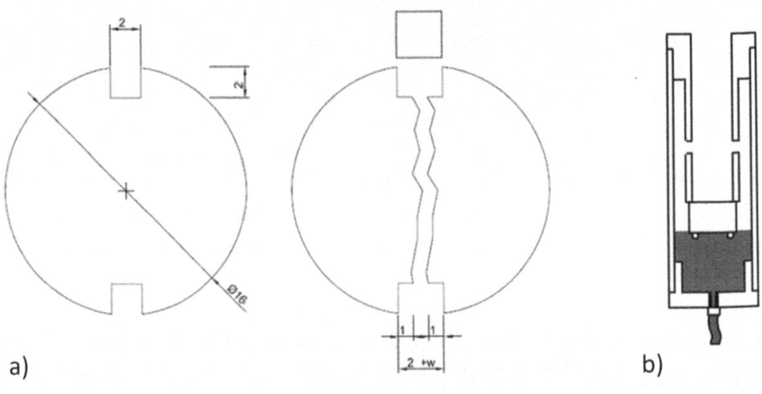

Fig. 2. Schematics of the sample and crack generation (Dimensions in mm).

Due to the poor attenuation contrast between water, air, SAP and cementitious materials and to the spatial resolution not being enough to resolve the pore space of cement-based composites, herein we employed a differential X-ray scanning procedure (Boone et al. 2014). The samples were scanned at the initial "dry" state and during saturation. The arithmetic difference between grey-value (GV) images at different states and normalization with respect to air and water GV resulted in the qualitative and quantitative monitoring of water absorption in the samples (Cui et al. 2018). The dry state tomography was acquired with accelerating voltage of 120 kV and current of 80 μA, 2200 projections and exposure time of 80 ms for a final spatial resolution of 14 μm. The dynamic wet tomography was acquired with accelerating voltage of 120 kV and 133 μA current, 600 projections and exposure time of 30 ms for a final spatial resolution of 28 μm in binning mode 2x2. The latter were acquired continuously during the first 10 min of water absorption and every 10 min until 60 min of absorption.

The acquired projections were reconstructed in a 3D volume employing Tescan XRE reconstruction software Octopus Reconstruction® (Vlassenbroeck et al. 2006) and corrected for ring artifacts, spots and beam hardening. Different scans of the same sample were registered through DataViewer, available open source from Bruker. All image analysis was performed through the open source freeware ImageJ. Median filter was applied prior to subtraction of the stacks to avoid noise propagation.

Two segmentation procedures were used to separate (1) embedded SAP at the swollen state from mortar matrix and (2) SAP in the crack from water. Segmentation (1) was implemented on the wet stack through simple thresholding operation since there is enough contrast between air and water. Segmentation (2) required the use of Trainable Weka Segmentation plugin in ImageJ (Arganda-Carreras et al. 2017) where the characteristics of swollen SAP were trained from the swollen gel in the macropores. A post segmentation algorithm was implemented to filter segmented particles smaller than the minimum size of SAP at the dry state in order to exclude obvious segmentation errors.

3 Results

From the mix design and estimated amount of absorbed water from the rheological measurements, the expected volume fraction of SAP macropores was 5%. We could measure the real value via treatment of the dry state scan data in which SAP macropores and air voids were segmented through segmentation (1) described in the previous section. The air voids were filtered out by imposing that the sphericity of the segmented particles was to be smaller than 0.90. The counting of remaining objects in the stacks yielded an average SAP macropore total volume of 5.36% which was in agreement to the estimated volume. From the particle analysis performed in the segmented SAP macropores it emerged that the swollen particles had an average sieve diameter of 368 μm vs. an estimated sieve size, d_{mix}, of 480 μm. The latter was calculated through Eq. 1 by assuming a spherical particle shape and that the density of the swollen SAP, ρ_{mix}, is that of water:

$$d_{mix} = d_{dry}\sqrt{\frac{\rho_{dry}}{\rho_{mix}}Abs_{mix}} \tag{1}$$

Where Abs_mix is the absorption capacity of the SAP during mixing of mortar [g_wat/g_SAP] and d_dry is the diameter of SAP at the dry state. Figure 3a shows the 3D renders of absorbed water in mortar matrix and Fig. 3b in SAP after 60 min of capillary absorption of water. It can be observed that at the arrival of the waterfront at a certain height in the sample, SAP don't swell immediately as can be seen from the difference in height of the wetting front and upper swollen SAP particles. Reasons for this occurrence could be the low level of saturation of the surrounding matrix at the wetting front position which results in a slower percolation towards the macropore.

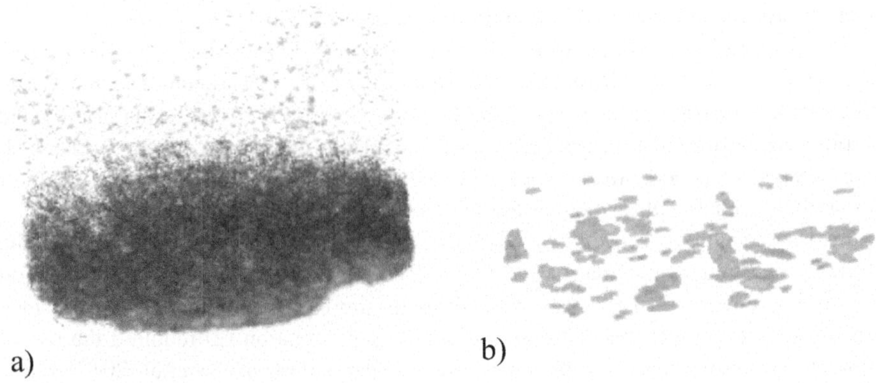

Fig. 3. 3D renderization of water absorption of (a) mortar matrix in sound sample and (b) swollen SAP in sound mortar sample.

Also a rough quantification of the capillary absorbed water was done via treatment of the differential micro-CT data. Noise in the stacks due to i.e. scattering was limited by imposing a cutoff to the calculated water content in the voxels minor or equal to the bulk porosity of the sample, measured a posteriori through gravimetry. In Fig. 4, a graph is shown of the water absorption for 4REF and 4F0.5 mor (quantified in the sole mortar phase from tomography of 4F0.5) and for 4F0.5 mor + SAP (quantified in both mortar and SAP phases from tomography of 4F0.5). From the graph it results evident that the mortar with SAP takes up more water than its reference, due to the absorption of water by the SAP, since the matrix absorption of 4F0.5 was very similar to the reference mortar absorption. This was previously proposed elsewhere (Rodriguez et al. 2018) from numerical simulations.

In Fig. 5 swollen SAP in a portion of a crack are shown. These were segmented using Segmentation (2) described before. Average sieve size measured from the segmentation resulted 644 μm. This means that during the capillary absorption experiment the SAP in the crack swelled to a sieve diameter 3.2 times the diameter of the SAP during mixing, much less than the estimated through free absorption capacity in demineralized water (7 times) (Pelto et al. 2017). This disagreement between the two values has been found in other studies regarding self-sealing before (Rodriguez et al. 2019). In a study from (Lee et al. 2018), the authors explain the changes in absorption capacity of SAP by the absorption of Ca2+ ions into the polymers during mixing of fresh cementitious mixtures.

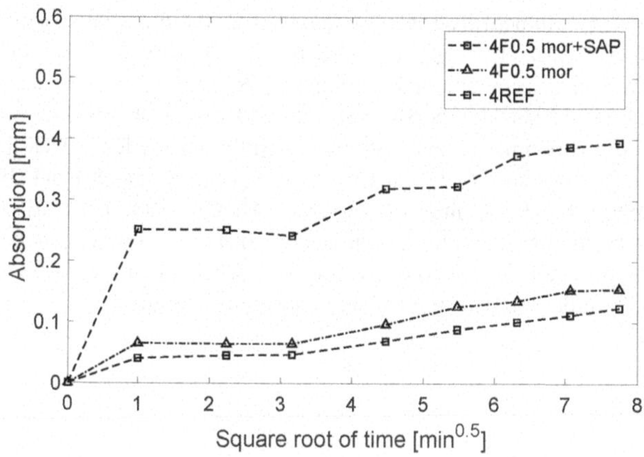

Fig. 4. Water absorption during sorptivity experiments in plain mortar 4REF and SAP-containing mortar 4F0.5 quantified through X-ray micro tomography.

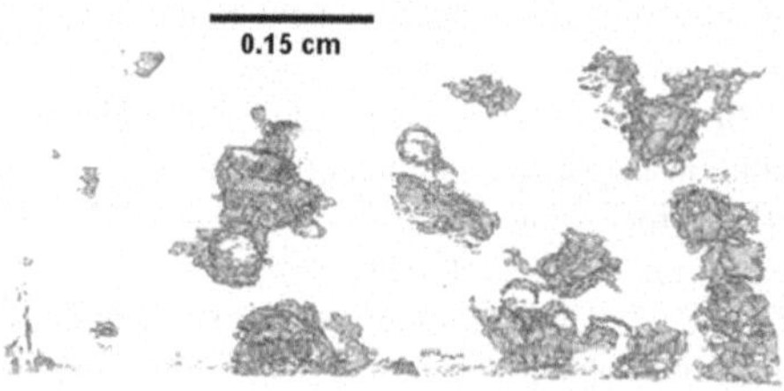

Fig. 5. Swollen SAP particles in the crack during capillary water absorption in mortar 4F0.5 segmented from X-ray micro tomography stack.

4 Conclusion

In this study, dynamic differential dynamic micro CT was used to monitor capillary water absorption in mortar with and without SAP. From the experimental results, the following conclusions are drawn;

- Differential dynamic micro CT is a powerful tool to monitor water absorption in cement-based materials.
- Some morphological data from the resulting composite can be studied: water absorption by the particles during mixing and during capillary water absorption.
- Water absorption due to SAP absorption can be separated from the total water absorption of the composite. We show direct evidence of the additional water absorbed by the embedded SAP. There is potential to study internal curing through this technique.
- Quantification of water retained by SAP in the crack can help in giving indications of potential of self-sealing and improved self-healing.

Acknowledgments. C. Romero Rodriguez acknowledges the financial support from the Construction Technology Research Program funded by the Ministry of Land, Infrastructure and Transport of the Korean Government under the grant 17SCIP-B103706-03 and also the financial aid of the Cost Action 15202 under the Short Scientific Mission with reference number 39002, in the realization of the experiments. The authors gratefully acknowledge UGCT for the use of micro-CT scanner facilities and the Ghent University Special Research Fund (BOF-UGent) for the financial support to the Centre of Expertise UGCT (BOF.EXP.2017.007). FWO is acknowledged for funding Maxim Deprez (project 3.G.0041.15). The authors wish to thank Dr. Didier Snoeck for providing a generous quantity of SAP F for the experiments and to Iván Josipovic and Dr. Tim de Kock for their highly appreciated feedback.

References

Arganda-Carreras, I., Kaynig, V., Rueden, C., Eliceiri, K.W., Schindelin, J., Cardona, A., Sebastian Seung, H.: Trainable Weka Segmentation: a machine learning tool for microscopy pixel classification. Bioinformatics **33**(15), 2424–2426 (2017)

Beushausen, H., Gillmer, M., Alexander, M.: The influence of superabsorbent polymers on strength and durability properties of blended cement mortars. Cement Concr. Compos. **52**, 73–80 (2014)

Boone, M., De Kock, T., Masschaele, B., De Schryver, T., Van Hoorebeke, L., Cnudde, V.: 4D mapping of fluid distribution in porous sedimentary rocks using X-ray micro-CT differential imaging. In: IMA 2014: Proceedings of the 21st General Meeting of the International Mineralogical Association, pp. 292–292 (2014)

Cui, D., Sun, W., Wang, Q., Gu, C.: Use of tomography to estimate the representative elementary volume in mortars stained with potassium iodide. Mater. Des. **147**, 80–91 (2018). https://doi.org/10.1016/j.matdes.2018.03.029

Cnudde, V., Boone, M.: High-resolution X-ray computed tomography in geosciences: a review of the current technology and applications. Earth Sci. Rev. **123**, 1–17(2013)

Dierick, M., Van Loo, D., Masschaele, B., Van den Bulcke, J., Van Acker, J., Cnudde, V., Van Hoorebeke, L.: Recent micro-CT scanner developments at UGCT. Nucl. Instrum. Methods Phys. Res. Sect. B **324**, 35–40 (2014)

Geiker, M.R., Bentz, D.P., Jensen, O.M.: Mitigating autogenous shrinkage by internal curing, pp. 143–154. ACI Special Publications (2004)

Jensen, O.M., Hansen, P.F.: Water-entrained cement-based materials II. Experimental observations. Cem. Concr. Res. **32**(6), 973–978 (2002)

Lee, H., Wong, H.S., Buenfeld, N.R.: Potential of superabsorbent polymer for self-sealing cracks in concrete. Adv. Appl. Ceram. **109**(5) (2010). https://doi.org/10.1179/174367609X459559

Lee, H., Wong, H., Buenfeld, N.: Effect of alkalinity and calcium concentration of pore solution on the swelling and ionic exchange of superabsorbent polymers in cement paste. Cement Concr. Compos. **88**, 150–164 (2018). https://doi.org/10.1016/j.cemconcomp.2018.02.005

Mechtcherine, V., Schröfl, C., Wyrzykowski, M., Gorges, M., Lura, P., Cusson, D., Margeson, J., De Belie, N., Snoeck, D., Ichimiya, K., Igarashi, S.I., Weiss, J.: Effect of superabsorbent polymers (SAP) on the freeze–thaw resistance of concrete: results of a RILEM interlaboratory study. Mater. Struc. **50**(1) (2017). https://doi.org/10.1617/s11527-016-0868-7

Pelto, J., Leivo, M., Gruyaert, E., Debbaut, B., Snoeck, D., De Belie, N.: Application of encapsulated superabsorbent polymers in cementitious materials for stimulated autogenous healing. Smart Mater. Struct. **26**(10), 105043 (2017). https://doi.org/10.1088/1361-665X/aa8497

Reinhardt, H.W., Assmann, A.: Enhanced durability of concrete by superabsorbent polymers. In: Brittle Matrix Composites, vol. 9, pp. 291–300. Elsevier (2009). https://doi.org/10.1533/9781845697754.291

Rodríguez, C.R., Figueiredo, S.C., Schlangen, E., Snoeck, D.: Modeling water absorption in cement-based composites with SAP additions. In: Computational Modelling of Concrete Structures: Proceedings of the Conference on Computational Modelling of Concrete and Concrete Structures (EURO-C 2018), Bad Hofgastein, Austria, February 26-March 1 2018, p. 295. CRC Press (2018)

Rodríguez, C.R., Figueiredo, S.C., Snoeck, D., Deprez, M., Schlangen, E., Šavija, B.: Numerical investigation of crack self-sealing in cement-based composites with superabsorbent polymers (2019, Manuscript submitted for publication)

Snoeck, D., Dewanckele, J., Cnudde, V., De Belie, N.: X-ray computed microtomography to study autogenous healing of cementitious materials promoted by superabsorbent polymers. Cement Concr. Compos. **65**, 83–93 (2016). https://doi.org/10.1016/j.cemconcomp.2015.10.016

Vlassenbroeck, J., Masschaele, B., Cnudde, V., Dierick, M., Pieters, K., Van Hoorebeke, L., Jacobs, P.: Advances in X-ray Tomography for Geomaterials. Presented at the 2nd International Workshop on the Application of X-ray CT for Geomaterials, pp. 167–173 (2006)

Characterization of Neutral Versus Anionic Superabsorbent Polymers (SAPs) in Ion-Rich Solutions for Their Use as Internal Curing Agents

Parawee Tangkokiat[1], Thanapat Thanapornpavornkul[1], Surapit Muangkaew[1], Wanwipa Siriwatwechakul[1(✉)], Jirawan Siramanont[2], and Chalermwut Snguanyat[2]

[1] School of Bio-Chemical Engineering and Technology, Sirindhorn International Institute of Technology, Thammasat University, Pathum Thani 10210, Thailand
wanwipa@siit.tu.ac.th

[2] SCG Cement Co., Ltd., Kaengkhoi, Saraburi 18260, Thailand
chalerms@scg.com

Abstract. Predominant SAPs used for internal curing are anionic gels, such as polyacrylamide, polyacrylic acid, and polyacrylate. They contain or can be hydrolyzed to form carboxylate functional groups. This allows them to retain water up to a few hundred times their dry weight, especially in a high pH environment. SAPs' swelling behavior is desirable, but other characteristics of SAPs are important for their application as internal curing agents. These anionic SAPs are not only sensitive to the pH environment but also the types of ions. SAPs show less swelling in a calcium-rich solution than in a sodium-rich solution at the same pH. This has direct implications for internal curing because the cement pore solution is rich in various kinds of cations.

In this study, two types of SAPs are developed. One is a neutral SAP in which its swelling ability is independent of its ionic environment, and the other is a semi-anionic SAP, which contains an anionic moiety (polyacrylamide). As expected, the SAPs with anionic moieties show a higher degree of swelling in all environments. The cement pastes dosed with the semi-anionic SAP show a lower calcium concentration in the pore solution. These cement pastes show a lower initial rate of hydration, observed from isothermal calorimetry. The results suggest that the lower calcium content in the pore solution due to calcium absorption into the semi-anionic SAPs may interfere with the initial cement hydration.

Keywords: Neutral SAPs · Anionic SAPs · Hydrogels · Cement pore solution · Cement hydration

1 Introduction

Internal curing is considered an effective method in assuring full hydration of cementitious materials, especially in high-performance concrete (HPC) with a low water-to-cement ratio (w/c < 0.4) (de Sensale and Goncalves 2014). HPC is described

W. P. Boshoff et al. (Eds.): SAP 2019, RILEM Bookseries 24, pp. 38–45, 2020.
https://doi.org/10.1007/978-3-030-33342-3_5

as concrete that has superior mechanical and durability properties in the hardened state. However, the use of a low water-to-cement ratio can easily cause self-desiccation and autogenous shrinkage at the early stage of cement hydration. To prevent the shrinkage in HPC, supplying additional water materials such as superabsorbent polymers (SAPs) is used to compensate for the water within the cement paste during cement hydration (Jensen and Hansen 2002; Siramanont et al. 2010).

SAPs are networks of cross-linked polymers, which have the ability to absorb many times their own weight of liquid from the environment (Jensen and Hansen 2001). SAPs act as an internal water reservoir for the cement paste matrix. This water is consumed during cement hydration, mitigating self-desiccation (Mignon et al. 2017). Thus the potential use of SAPs is often dominated by their swelling behavior. Previous studies showed that the SAP swelling behavior depends on many factors, including the degree of crosslinking, chemical structure of the polymer, and environmental effects such as temperature, pH, and ionic concentration (Siriwatwechakul et al. 2010).

SAPs can be classified into two groups which are ionic and non-ionic based on the nature of the side chains. The predominant SAPs used for internal curing are anionic, such as polyacrylamide, polyacrylic acid and polyacrylate (Siramanont et al. 2010; Schröfl et al. 2012; Siriwatwechakul et al. 2012; Lee et al. 2018). These SAPs contain amide groups which can interact with water to form carboxylate groups with hydrogen that can be deprotonated, leaving anionic charges along the polymer chains. These charges provide the repulsive force on the polymer chains, rendering SAPs highly swollen in an aqueous environment, especially at high pH (Peppas et al. 2000; Siriwatwechakul et al. 2010).

The ability to swell in aqueous solution is not the only important characteristic for internal curing agents. Previous studies reported that the swelling ratio of polyacrylate SAPs was sensitive to the pH of their environment, and the type of ions in the surrounding solution (Siriwatwechakul et al. 2012; Lee et al. 2018). Siriwatwechakul et al. found that polyacrylate showed less swelling in a calcium-rich solution than in a sodium-rich solution at the same pH (Siriwatwechakul et al. 2012). Lee et al. showed that ion absorption was a dynamic process, in which Ca^{2+} can replace other monovalent cations, and the bound Ca^{2+} can also be released when the aqueous environment changed (Lee et al. 2018). These results help in understanding hydration chemistry because the pore solution is rich in various kinds of cations, ranging from monovalent to trivalent cations.

In this study, we synthesized two types of SAPs. One was a neutral SAP made from polyethylene glycol (PEG). It is referred to as a polyethylene glycol dimethacrylate (PEGDMA) hydrogel. The other was a semi-anionic SAP made from PEG copolymerized with polyacrylamide, PEGDMA-co-AM. Their swelling behavior was examined in an ion-rich environment using the teabag method. The ion composition of the solution surrounding SAPs was analyzed by inductively coupled plasma mass (ICP-MS) spectroscopy. Subsequently, two types of SAPs were incorporated into the cement paste with the dosage 0.1% and 0.5% (w/w cement). The pore solutions were collected every 30 min for two hours, to understand the ionic environment in the pore solution at different times. The hydration of cement in the presence of SAPs was also monitored by using an isothermal calorimeter. By understanding the effects of SAPs, the outcome of this work will allow the engineering of SAPs with properties that are suitable for internal curing.

2 Experimental

2.1 Materials

Polyethylene glycol MW4000 (PEG4000) was supplied by Chemipan (Bangkok, Thailand). Methacrylic anhydride (MA), Triethylamine (TEA) and *N,N'*-methylene-bisacrylamide (NMBA) were supplied by Merck (USA). *N,N,N'N'*-Tetra-methylenediamine (TEMED) was purchased from Sigma-Aldrich (St. Luis, MO, USA). Ammonium persulfate (APS), Potassium carbonate (K_2CO_3), Sodium hydroxide (NaOH), Calcium hydroxide $Ca(OH)_2$ and Dichloromethane (DCM) were obtained from Ajax (NSW, Australia).

2.2 SAP Preparation

Preparation of PEGDMA

Polyethylene glycol dimethacrylate (PEGDMA) was prepared by the esterification of methacrylic anhydride. The reaction was performed by dissolving 50 g of PEG4000 in 150 mL of DCM in a three-neck round-bottom flask equipped with a magnetic stirrer. The solution was purged with N_2 gas to remove oxygen which can inhibit the reaction and stirred until all PEG was dissolved. MA (4.1 mL, 2.2 molar equivalent to PEG) and 2 mL of TEA, with a ratio of 1 mL TEA to 75 mL of DCM, were added to the flask. The reaction was performed for 72 h at room temperature under a nitrogen atmosphere.

After 72 h, the reaction mixture was transferred into a separatory funnel. Then, K_2CO_3 (1.5 M) was added to neutralize the methacrylic acid by-product. The added volume of K_2CO_3 solution was half of the total volume of the reaction mixture. The mixture was allowed to phase separate by gravity, overnight. The bottom phase contained PEGDMA and was collected for further purification. DCM was evaporated to concentrate the PEGDMA solution, and PEGDMA was precipitated using 300 mL of diethyl ether to obtain solid PEGDMA powder. The solid was dried overnight in a vacuum oven at room temperature. PEGDMA powder was kept for further characterization by FT-IR and NMR.

Preparation of PEGDMA Hydrogels

Two types of SAPs were prepared, based on PEGDMA (see Table 1). The hydrogels were prepared by using *N,N'*-methylenebisacrylamide (NMBA) as a crosslinker by free radical polymerization with ammonium persulfate (APS) as an initiator. All the monomer concentrations were kept at 8% (w/w), and the crosslinker to PEG monomer ratio was kept at 1:1000 (by weight). The monomers were dissolved in deionized (DI) water and were purged with N_2 gas for 30 min. APS and TEMED, used as an accelerator, were added to the solution to initiate polymerization at room temperature. The mixture was allowed to form hydrogels overnight. After 24 h, the hydrogels were collected and washed with DI water to remove the unreacted monomers.

Table 1. Recipe for preparing PEGDMA Hydrogels in 30 mL solution

Sample No.	Sample name	Monomer				Crosslinking agent (NMBA) (mg)
		PEGDMA (μmol)	AM (μmol)	PEGDMA (g)	AM (mg)	
1	PEGDMA Hydrogels	1395	0	2.40	0	24
2	PEGDMA-co-AM Hydrogels	1255	1255	2.16	90	22

2.3 SAP Characterizations

Fourier-Transform Infrared Spectroscopy (FTIR) and Nuclear Magnetic Resonance Spectroscopy (NMR)

NMR and FTIR were used to ensure that PEGDMA contains the functional groups C=O and C=C from attaching methacrylic anhydride to the PEG chain. The C=O and C=C stretching bands can be observed in the range of 1650–1780 cm^{-1} and 1600–1680 cm^{-1}, respectively.

Swelling Ratio of PEGDMA and PEGDMA-co-AM Hydrogels in Aqueous Solution

The swelling ability of PEGDMA and PEGDMA-co-AM were measured using the teabag method. Dry hydrogels (1 g) were placed in 50 × 50 mm teabags and soaked into 50 mL of four different aqueous solutions: DI water (pH 7), 0.01 M NaOH (pH 12), 0.01 M $Ca(OH)_2$ (pH 12), and saturated cement solution (pH 13). The saturated cement solution was prepared by stirring 10 g of OPC (type I) in 100 mL DI water for one hour, and subsequently removing the suspended particles by filtration. All of the solutions were kept at room temperature for 24 h, after which each teabag was collected and placed on a tissue paper to absorb the excess water before measuring the weight of the hydrogels. Each measurement was done in 5 replicates, and the swelling ratio (s) was calculated by Eq. (1).

$$s = \frac{(M_3 - M_2 - M_1)}{M_1} \tag{1}$$

where M_3 is the total weight of hydrogel with the wet teabag, M_2 is the weight of wet teabag and M_1 is the weight of the dry hydrogels. In addition, the ion contents in the aqueous solutions before and after swelling ratio measurement were analyzed by using inductively coupled plasma (ICP).

Isothermal Calorimetry

The hydration of cement in the presence of PEGDMA and PEGDMA-co-AM hydrogels was evaluated by using an isothermal calorimeter. In this study, an 8-channel

TAM Air (TA Instruments) was used for the isothermal calorimetry experiments. The samples were prepared according to the design mix in Table 2. In summary, cement pastes were prepared by mixing Ordinary Portland Cement (OPC) with PEGDMA and PEGDMA-co-AM hydrogels (0.1 wt.% and 0.5 wt.%) at a water-to-cement ratio of 0.485 by using a high-speed mixer at 300 rpm for one minute. Mixing was continued at speed of 800 rpm for another two minutes. The samples were then placed in glass ampoules and inserted into the TAM Air to capture the early hydration reaction at 25 ± 0.01 °C for 48 h, using air as a reference.

Table 2. Cement paste mixture composition for isothermal calorimetry.

Sample	Contents	SAP type	SAP dosage (wt.% cement)
I	OPC Type I		0
II	OPC Type I + 0.1 wt.% PEGDMA	PEGDMA	0.1%
III	OPC Type I + 0.5 wt.% PEGDMA	PEGDMA	0.5%
IV	OPC Type I + 0.1 wt.% PEGDMA-co-AM	PEGDMA-co-AM	0.1%
V	OPC Type I + 0.5 wt.% PEGDMA-co-AM	PEGDMA-co-AM	0.5%

Ion Concentration in Pore Solution

The samples were prepared by placing 10 g of OPC with hydrogels (dosage according to Table 2) in a mixer. Then, 100 mL of deionized water was added to the mixer, and the paste was stirred at 500 rpm for 2 min before collecting the first sample. Five mL of each sample was collected at 30 min, 1 h, 1.5 h and 2 h. The cement paste was filtrated by using a syringe filter with a pore diameter of 0.2 μm. The ion contents (sodium, calcium, silicon, aluminum, and iron) in the pore solution were measured using ICP.

3 Result and Discussion

3.1 SAP Characterizations

The synthesis of PEGDMA and PEGDMA-co-AM hydrogels was confirmed by NMR and FTIR. The swelling ratio of the hydrogels in four different solutions are shown in Fig. 1. The swelling ratio of semi-anionic PEGDMA-co-AM hydrogels is higher than the neutral hydrogels in all aqueous environments. In an ion rich environment, the swelling ratio is lower than that in the aqueous solution at a similar pH.

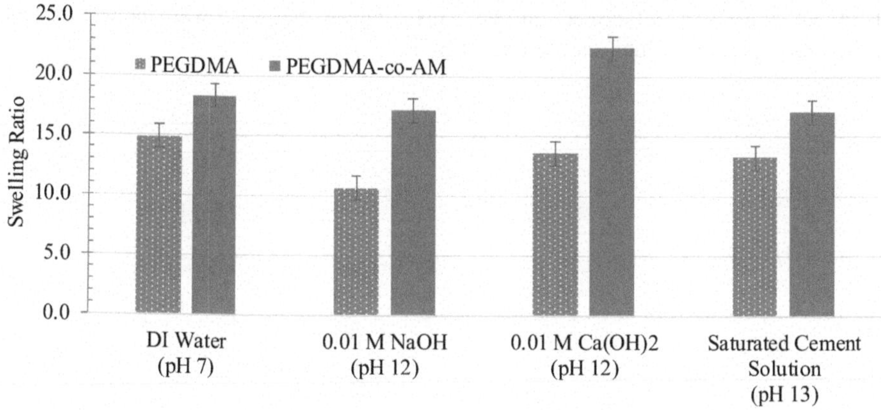

Fig. 1. Swelling ratio of neutral PEGDMA vs. semi-anionic PEGDMA-co-AM hydrogels in four different solutions.

The time evolution of the calcium concentration in the pore solution is plotted in Fig. 2. The cement paste dosed with PEGDMA-co-AM SAP (Sample IV and Sample V) show a lower calcium concentration in the pore solution. This result is consistent with previous studies that polyacrylate/polyacrylamide SAPs can effectively bind to calcium ions (Siriwatwechakul et al. 2010); thus, less calcium is dissolved in the pore solution.

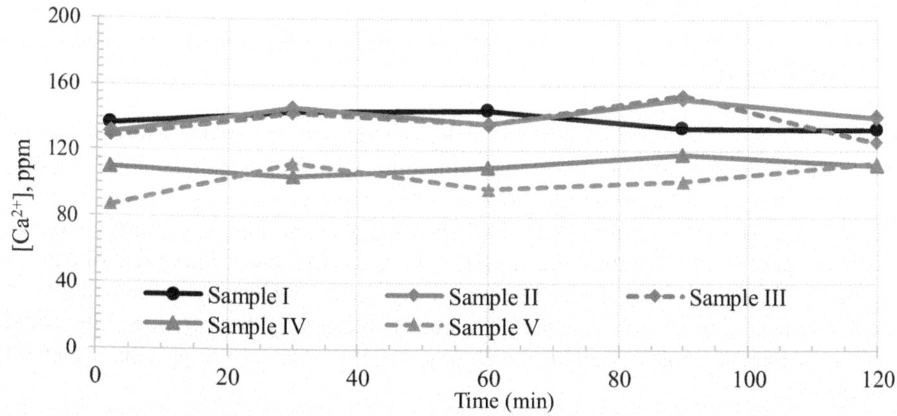

Fig. 2. Calcium ion concentration in pore solutions. The compositions of the cement paste are detailed in Table 2.

The heat flow curves in Fig. 3 show the impact of the SAP on the hydration of the OPC type I cement. The ascending slope of the first peak is related to the acceleration period, and associated with the precipitation of hydration products mainly C-S-H and CH. The cement paste dosed with PEGDMA-co-AM SAP (Sample IV and Sample V) visibly show a lower initial reaction rate. This suggests that the lower calcium content in the pore solution influences the rate of hydration.

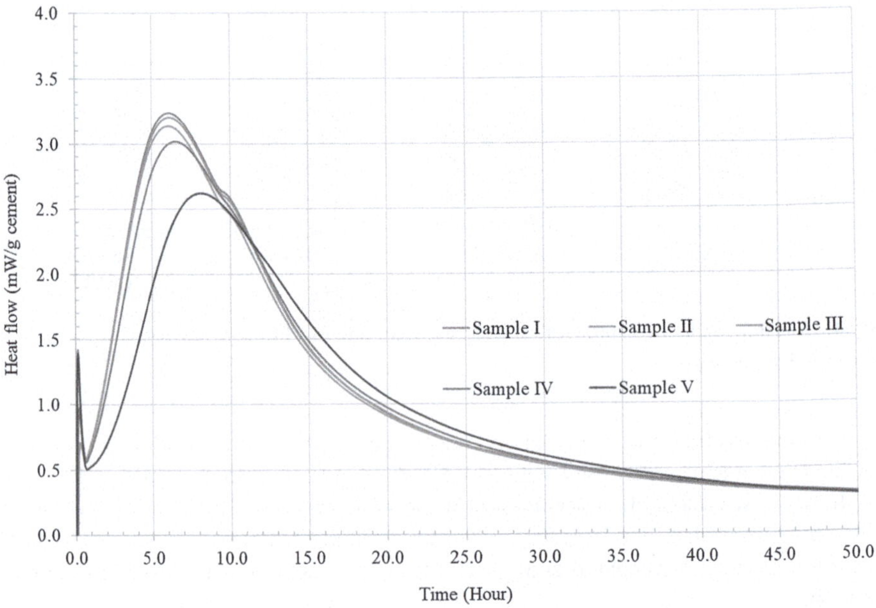

Fig. 3. Rate of heat evolved from cement pastes at a W/C ratio of 0.485. The compositions of the cement paste are detailed in Table 2.

4 Conclusion

In this study, we presented two types of SAPs: neutral SAP referred to as polyethylene glycol dimethacrylate (PEGDMA) and semi-anionic SAP made from PEG copolymerized with polyacrylamide (PEGDMA-co-AM). There is a consistent trend for these two sets of SAPs. The semi-anionic SAP shows higher swelling with fewer calcium ions in the pore solution of the cement paste. This leads to a lower initial rate of cement hydration. The results show that as internal curing agents, SAPs may not only act as an internal source of water supply, and also directly interfere with hydration. This could result in a more complex strength development rather than just supplying water for hydration.

References

de Sensale, G.R., Goncalves, A.F.: Effects of fine LWA and SAP as internal water curing agents. Int. J. Concr. Struct. Mater. **8**(3), 229–238 (2014)

Jensen, O.M., Hansen, P.F.: Water-entrained cement-based materials: I. Principles and theoretical background. Cem. Concr. Res. **31**(4), 647–654 (2001)

Jensen, O.M., Hansen, P.F.: Water-entrained cement-based materials II. Experimental observations. Cem. Concr. Res. **32**(6), 973–978 (2002)

Lee, H.X.D., et al.: Effect of alkalinity and calcium concentration of pore solution on the swelling and ionic exchange of superabsorbent polymers in cement paste. Cement Concr. Compos. **88**, 150–164 (2018)

Mignon, A., et al.: Development of amine-based pH-responsive superabsorbent polymers for mortar applications. Constr. Build. Mater. **132**, 556–564 (2017)

Peppas, N.A., et al.: Hydrogels in pharmaceutical formulations. Eur. J. Pharm. Biopharm. **50**(1), 27–46 (2000)

Schröfl, C., et al.: Relation between the molecular structure and the efficiency of superabsorbent polymers (SAP) as concrete admixture to mitigate autogenous shrinkage. Cem. Concr. Res. **42**(6), 865–873 (2012)

Siramanont, J., et al.: The impact of SAP structure on the effectiveness of internal curing. In: International RILEM Conference on Use of Superabsorbent Polymers and Other New Additives in Concrete, Lynby, Denmark, RILEM Publications SARL (2010)

Siriwatwechakul, W., et al.: Superabsorbent polymer structures. In: International RILEM Conference on Use of Superabsorbent Polymers and Other New Additives in Concrete, RILEM Publications SARL (2010)

Siriwatwechakul, W., et al.: Behavior of superabsorbent polymers in calcium- and sodium-rich solutions. J. Mater. Civ. Eng. **24**(8), 976–980 (2012)

Evaluation of Absorption Capacity and Spacing of Superabsorbent Polymer Particles in Cement Paste

Soushi Yamashita and Shin-ichi Igarashi(✉)

Department of Civil and Environmental Engineering, Kanazawa University, Kakuma-machi, Kanazawa 920-1192, Japan
soushi1205@stu.kanazawa-u.ac.jp,
igarashi@se.kanazawa-u.ac.jp

Abstract. The absorption capacity of superabsorbent polymers (SAP) in a real cement environment is a need-to-know property as a new admixture for concrete. In this study, fundamental stereology techniques were applied to cross sections of cement pastes in which solution-polymerized SAP particles were embedded. Its absorption capacity was estimated from the area fraction and particle size distribution of the SAP in the 2D planes. Further, the representative spacing between the SAP particles was also evaluated using point process statistics. The absorption capacity estimated was greater than that obtained by the tea-bag method using a cement filtrate. The SAP seems to absorb mixing water quickly in the initial short time during mixing. The number density of the SAP particles in the cross sections was also greater than the estimation calculated from the absorption capacity and the particle size distribution of initial dry SAP. This fact suggests that the SAP particles broke away during mixing. The centroids of SAP particles were distributed as to form a regular pattern. A procedure to evaluate the median distance from a given location to the surface of the nearest SAP particle was proposed by combining the mean diameter of swollen SAP profiles and the point process G- and F-functions. The distances between SAP particles were found at most a few mm for the mass fraction of 0.58% against cement.

Keywords: Absorption · Particle size distribution · Image analysis · Point process statistics

1 Introduction

It's been a long time since a pioneering work on superabsorbent polymer (SAP) as an internal curing admixture was published by Jensen and Hansen (2001). For the last two decades, it has been well recognized that SAP can be used as not only an internal curing ingredient for mitigating autogenous shrinkage but also an additive for providing concrete with additional features such as frost resistance (Mechtcherine et al. 2017; Wyrzykowski et al. 2018). Nowadays, SAP is further expected to be used as a reducer of plastic shrinkage and a modifier of rheological properties at fresh concrete (Mechtcherine and Reinhardt 2012).

W. P. Boshoff et al. (Eds.): SAP 2019, RILEM Bookseries 24, pp. 46–54, 2020.
https://doi.org/10.1007/978-3-030-33342-3_6

Those promising functions of SAP are, needless to say, based on its quite high absorbing capacity (Schröfl et al. 2017). However, in spite of its importance, there haven't been any standard test methods to evaluate the absorption capacity in a real cement environment. The absorption capacity of SAP is greatly reduced when it is soaked in an alkaline solution. In particular, Ca^{2+} ion in a pore solution adversely affects the absorption behavior in concrete. A test method that simulates a high alkaline environment in concrete has been recently proposed by RILEM TC 260 RSC (Snoeck et al. 2018). In the proposed method, a prescribed amount of SAP absorbs a cement filtrate, following the general procedure of a tea-bag method or a newly proposed filtration method.

The proposed method is useful for comparing the sorption kinetics among different SAP products and judging suitability of a SAP for a cement mixture. Mix proportions of concretes with SAPs can be designed based on the test result of each SAP. However, it should be noted that an absorption capacity estimated from the sorption test may not be necessarily the same as a real capacity in concrete. In the test, plenty of alkaline solution is present, and a SAP can freely absorb it without any obstacles. However, in concrete, there is a limited amount of mixing water around SAP particles. Solid particles of cement and aggregate may also disturb the free absorption and subsequent expansion of SAP particles. Further, the composition and concentration of the mixing water around SAP change continuously. These differences in the surrounding chemical and physical conditions from the absorption test could affect a real absorbing behavior of SAP in concrete.

In this study, an absorption capacity of a SAP in a real cement paste environment was evaluated by image analysis. The number density of SAP particles in cross sections of cement paste specimens is discussed from the perspective of applicability of a fundamental stereology formula for the numbers of particles in 2D and 3D unit volumes. Further, a characteristic spacing between expanded SAP particles is evaluated using their particle size distribution and the point process statistics of second-order stereology.

2 Experimental

2.1 Materials and Production of Specimens Containing SAP

Ordinary Portland cement was used. The SAP used (dry density 1.4 g/cm^3) was produced by a solution polymerized method. Its absorption behavior determined by the test method proposed by RILEM TC 260 RSC is shown in Fig. 1. The prescribed amount of SAP was incorporated into a cement paste matrix with w/c of 0.50. However, its exact absorption capacity in the cement paste is not known at first. Thus, the SAP was assumed to have a half capacity of the capacity determined by the tea-bag method. The amount of SAP was determined as to have the volume fraction of 10% to the whole volume when the SAP fully soaked up mixing water in fresh cement paste. The mix proportion is shown in Table 1. The cement paste containing the SAP was prepared in accordance with Japanese Industrial Standard. Small prismatic specimens of $2 \times 2 \times 3$ cm^3 were produced.

2.2 Determination of Particle Size Distribution of SAP

(1) Dry SAP particles

Particle size distribution of the SAP was determined by a microscopy method. An appropriate amount of SAP was spread over a black sheet of paper. They were observed with a common stereomicroscope. The SAP particles were directly segmented from the microscope images. Some particles that touched each other were manually separated. Following the ordinary procedure for particle analysis, the equivalent circle diameters of the SAP particles were obtained.

(2) Swollen SAP particles in cement paste

The cement paste specimens containing the SAP were demolded at 24 h after casting. They were immediately cut to obtain about ten slices for observing the cross sections. Their surfaces were then dried by a solvent exchange procedure. The dried surfaces were coated with a low viscosity epoxy resin with red dye. After the resin hardened at a room temperature, the surfaces were polished to expose cross sections of SAP particles (Fig. 2(a)). Red particles in an observation field were converted to points of their centroids (Fig. 2(b)). This procedure was also applied to plain cement paste specimens without SAP. Red spots in the plain cement paste were regarded as large capillary pores or some other defects. A threshold diameter to discriminate those red spots was determined by their particle size distribution. Applying the threshold diameter to the cross sections including SAP profiles, red regions that were not considered as SAP particles were deleted. The total number of SAP particles segmented in the cross sections was seven hundred. It was assumed that their SAP profiles were representative of all the possible sections cut through the specimens. Based on this assumption and common practice, the 2D particle size distribution on the SAP profiles was regarded as 3D particle size distribution in the cement paste. The particles of swollen SAP were assumed as spheres of water. The particle size distribution of SAP in the cement paste matrix was calculated as mass fractions in each class of diameters.

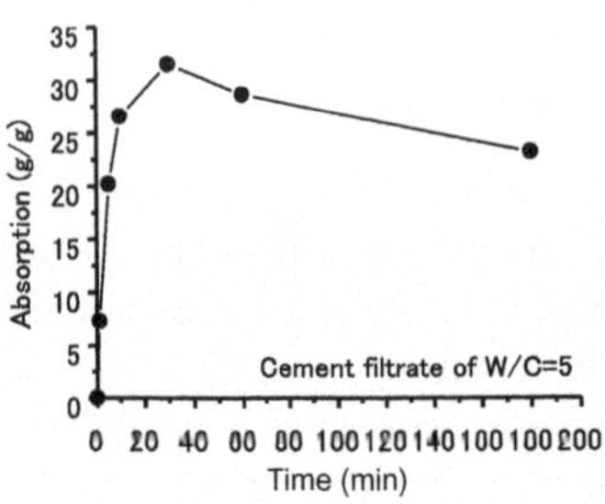

Fig. 1. Change in absorption capacity of SAP with time

Table 1. Mix proportion of cement paste with SAP

Cement: Water: SAP	1.0: 0.59: 5.8×10^{-3}

(by mass ratio)

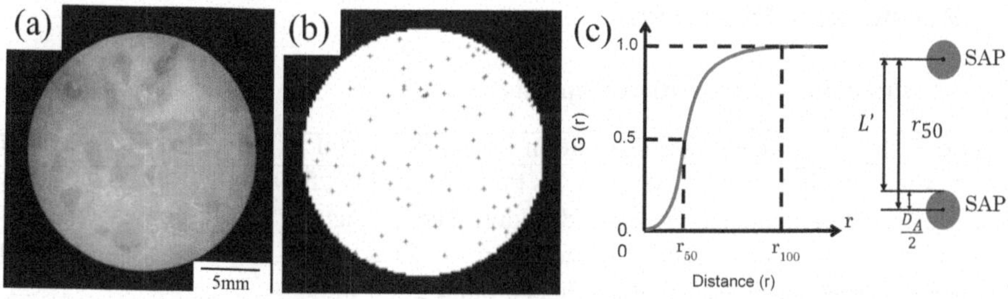

Fig. 2. SAP particles as elements of a spatial point process: (a) dyed SAP particles (b) a point pattern converted from an image (c) definition of characteristic distance r_{50} from the nearest neighbor distance distribution function

2.3 Nearest Neighbor Distance Distribution Between SAP Particles and a Distance from a Given Location to SAP Particles

Coordinates $\boldsymbol{x}_i$ (i = 1, …, n) of centroids of each SAP particle were determined by image analysis. This set of points was regarded as a spatial point process $X = \{\boldsymbol{x}_i : i = 1, \ldots, n\}$. To determine a characteristic distance between SAP particles, the nearest neighbor distance distribution function was calculated from the point process. For a point $\boldsymbol{x}_i \in X$, its nearest neighbor distance $d_i = min_{i \neq j} \|\boldsymbol{x}_i - \boldsymbol{x}_j\|$ is written as $d_i = d(\boldsymbol{x}_i, X \backslash \boldsymbol{x}_i)$. Then, for any $r \geq 0$, the nearest neighbor distance distribution function G(r) is defined by the following probability equation:

$$\mathrm{G(r)} = \Pr\{d_i = d(\boldsymbol{x}_i, X \backslash \boldsymbol{x}_i) \leq r | \boldsymbol{x}_i \in X\} \tag{1}$$

The nearest neighbor distance distribution function is a cumulative probability function of the distance r. It converges to one at long distances (Fig. 2(c)). The distance which corresponds to the probability of 0.5 (the second quartile) is defined as a median distance r_{50} (Fig. 2(c)).

The spherical contact distribution function F(r) is also obtained for the point process X. Its definition is given by Eq. (2).

$$\mathrm{F(r)} = \Pr\{d(u, X) \leq r | u \notin X\} \tag{2}$$

Where u is an arbitrary reference point which does not belong to the point process X. The values of F(r) are probabilities finding a point of the point process X within a distance r from a given location. If a point process is completely random, G(r) is the same as F(r). Using the functions of G(r) and F(r), J(r) function of SAP point patterns is also calculated. The J(r) function is defined as Eq. (3). It enables to discriminate ingeniously between clustering and regularity.

$$\mathrm{J(r)} = (1 - G(r))/(1 - F(r)) \tag{3}$$

3 Results and Discussion

3.1 Estimation of Absorption Capacity

Figure 3 shows particle size distribution of dry SAP and expanded SAP in cement paste. The diameters of dry SAP particles range from about 500 to 1800 μm (Fig. 3(a)). Its mass median diameter is about 860 μm. The diameter range becomes wider in cement paste. It approximately ranges from 1000 to 4200 μm. The mass median diameter after absorbing mixing water is around 2600 μm.

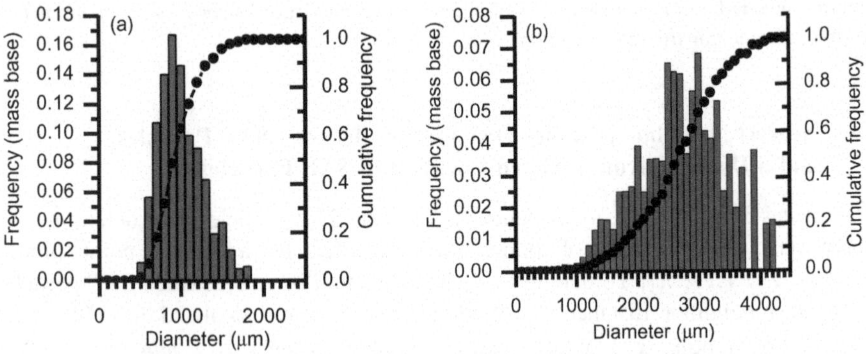

Fig. 3. Particle size distribution of SAP: (a) Dry SAP before mixing (b) SAP profiles in cement paste

Table 2. Volume fractions of SAP particles (%)

Before absorbing	After absorbing
0.46	31.6

Table 2 shows the volume fractions of SAP. The volume fraction after absorbing water was evaluated by the area fraction of SAP particles. The fraction of SAP in cement paste is different from the fraction of 10% that was initially aimed at. If all the swollen SAP particles are assumed as spheres of water, the absorption capacity of SAP can be calculated. Using the mix proportion in Table 1, the absorption capacity is estimated at 47.5 g/g. This is, as expected, much greater than the capacity (15 g/g) assumed in advance. Furthermore, this is also greater than the capacity evaluated by the tea-bag method (Fig. 1). As mentioned before, an absorption capacity of SAP in real cement mixtures is generally supposed to decrease compared to the absorption capacity measured by the sorption test. The amount of water available for absorbing is limited in the mixture and expansion of SAP may be restricted due to the presence of solid particles in the vicinity of SAP particles. However, contrary to the expectation, no reduction in absorption capacity was seen at all. It suggests that SAP could quickly absorb much amount of mixing water of which alkalinity was not so high at the beginning of mixing. Furthermore, there were no aggregate particles in the cement paste specimens. As a result, the SAP could absorb water freely as to exhibit the high absorption of 47.5 g/g. Figure 4 shows SAP particles in the cement paste immediately after mixing. They progressively expanded with time as if there were no obstacles around them. It should be noted that the cross sections (Fig. 2) were observed at 24 h. Thus the SAP particles could have already released its internal water. Even so, the large red trace of particles suggests that the SAP has previously expanded to that degree at the initial period of time.

3.2 Number Intensity of SAP Estimated by Fundamental Stereology Formulae

Using the particle size distribution of dry SAP and a mass of SAP used for a batch of cement paste mixture, the number of SAP particles N_V per unit volume of the cement paste can be calculated. If SAP particles are assumed to disperse randomly in 3D space, the number of SAP per unit area in a plane section, N_A can be related to N_V by the following Eq. (4).

$$N_A = E[H]N_V \tag{4}$$

Where E[H] is the mean height of the SAP particles. The height of a SAP is defined as its projected length onto a line normal to the cross section plane.

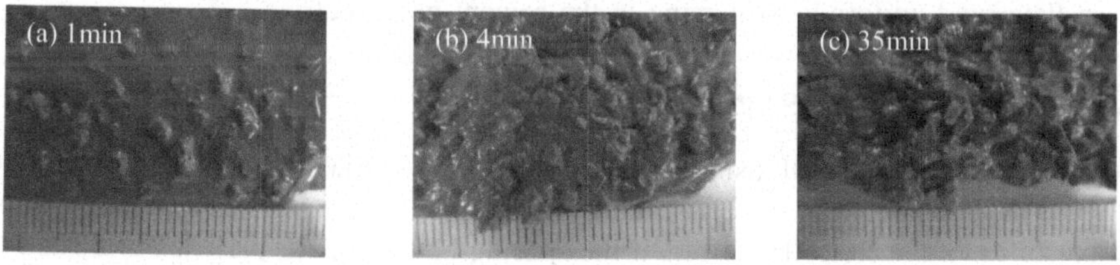

Fig. 4. Expansion of SAP particles with time (one scale = 1 mm)

Using the mass of SAP for a specimen and its particle size distribution at a dry state, the mass fractions of SAP in each class of diameters d_i $(i = 1, \cdots, n)$ are easily calculated. If all the SAP particles are assumed spherical, the number of SAP particles n_i belonging to a class i can be calculated. Further, regardless of particle sizes, all the SAP particles are assumed to have the same absorption capacity that is determined by the area fraction of SAP particles in cement paste. Then a difference in the volume of a SAP particle before and after absorbing water is expressed by Eqs. (5a) and (5b).

$$\frac{4}{3}\pi\left(\frac{d_i}{2}\right)^3 \rho_S a = \frac{4}{3}\pi\left(\frac{D_i}{2}\right)^3 \rho_W \tag{5a}$$

$$D_i = d_i^3\sqrt{\frac{a\rho_S}{\rho_W}} \tag{5b}$$

Where a is the absorption capacity, ρ_S and ρ_W are densities of SAP and water, respectively. D_i is a diameter after expansion. Substituting the absorption capacity of 47.5 g/g into Eqs. (5a) and (5b), a dry SAP particle expands about 4.1 times the length of its initial diameter. As mentioned before, the particle size distribution determined from many 2D SAP profiles in the cross sections is regarded as 3D particle size distribution. The mass median diameters before and after absorbing are 860 μm and 2600 μm, respectively. The ratio of diameters is about 3.0, which is not greatly different from the estimated ratio of 4.1.

Furthermore, when all the particles belonging to each class of diameters are assumed to have the same absorption of 47.5 g/g, then it is possible to calculate the diameters of each particle after absorbing water. The calculation gives about 2600 μm as the mean diameter E[H] of expanded SAP particles. Substituting this diameter and the number density N_V into Eq. (4), the number of SAP particles in unit area, N_A is estimated at about 5.3/cm^2. The number actually counted was 14.8/cm^2. The number of SAP in the observation area is tallied without any edge corrections. Therefore, the number of 14.8 may be an overestimation of the real number. Nevertheless, it seems that the difference is relatively large between the estimation and the observation. This fact suggests that SAP particles were broken off into several pieces during mixing. Actually, extremely great particles of SAP were not observed, as shown in Figs. 2 and 3. The maximum diameter was about 4000 μm at most. This size may result from the original diameter of 1000 μm, which is far smaller than the maximum. It should be noted that the break of SAP particles could be advantageous in terms of not leaving great flaws and uniformly dispersing in the cement paste matrix.

3.3 Distance Between SAP Particles

Whatever the purposes of using SAP, its optimum spatial arrangement should be attained as far as possible since an excess amount of SAP adversely affects strength or permeability of concrete. Therefore, if the spacing between SAP particles or distances from the furthest location to the nearest surface of SAP is known, it enables to determine the amount of SAP that depends on a purpose of SAP usage. Figure 5 shows the nearest neighbor distance distribution function G(r) of SAP particles. It converges to the cumulative probability of 1.0 at about 3100 μm. The function is regarded as a regularly arranged or repulsive pattern since it is plotted lower than the Poisson point process. The SAP particles have their own sizes so that points of the centroids of the particles cannot enter within ranges of their diameters. Thus the function has an x-intercept, which is called a hard-core distance. The most frequent distance is found at around 2000 μm. If the distance corresponding to the cumulative probability of 0.50 is regarded as a characteristic distance r_{50}, it is about 1800 μm.

Figure 6 shows the spherical contact distribution function F(r). It follows the pattern of completely random up to about 1000 μm. Then the function is plotted higher than the Poisson process in the range from 1000 to 3700 μm. This suggests that the pattern is regular or repulsive arrangement of points in the range. This range is almost the same as the range where the G(r) increases rapidly. Figure 7 is J(r) function which is obtained by combining the G(r) and F(r) functions. It is trivial that SAP particles form a regular pattern at long distance. This also results from the sizes of SAP particles themselves.

When SAP is used as an admixture for mitigating autogenous shrinkage or providing frost resistance, the distance between SAP particles and the range where a SAP particle can cover its surrounding matrix as a protected region is of interest. They depend on the absorption capacity and spatial distribution of SAP particles. In this study, the amount of SAP is about 0.6% by mass of cement, and the estimated absorption is about 47.5 g/g. The median distance of F(r) function is about 1200 μm. The maximum distance from an arbitrary location to the center of the nearest SAP particle is about 4000 μm. Using these estimations and the mean diameter of about 1300 μm, a SAP particle should cover a cement paste region within about 3400 μm from its surface at the maximum as a protected region. This length may be attainable by ordinary SAP mixtures, especially in a high w/c if water released from SAP can travel a few mm (Wyrzykowski et al. 2012). Of course, the amount of SAP depends on its purpose of usage and physical properties. However, this procedure to estimate a characteristic length of the particle systems may be convenient for using a SAP as a functional admixture in concrete.

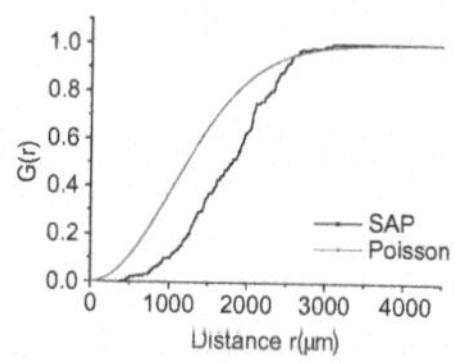

Fig. 5. Nearest neighbor distance distribution function of SAP

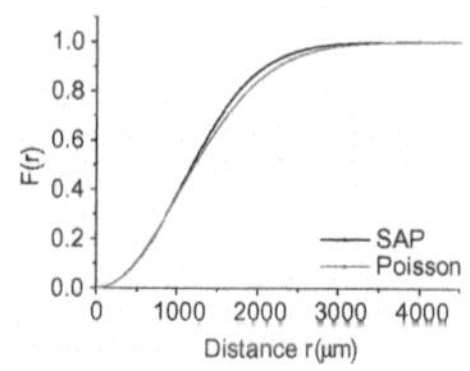

Fig. 6. Spherical contact distribution function of SAP

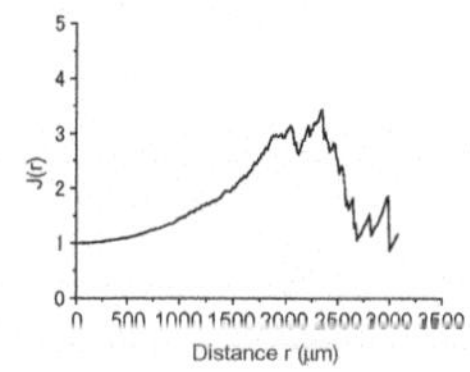

Fig. 7. J-function of SAP

4 Conclusions

The major results obtained in this study are as follows;

(1) The absorption capacity of SAP used in this study was greater than that estimated by the tea-bag method using a cement filtrate.
(2) The number density of SAP particles in 2D planes was greater than that estimated by combination of the absorption capacity and the fundamental formula of stereology. This suggests SAP particles broke away during mixing.
(3) A method to estimate characteristic distances between SAP particles was proposed using the point process statistics and the particle size distribution of expanded SAP.
(4) For the mix proportion in this study, SAP particles formed a dispersion system that was expected to cover the surrounding matrix of a few mm from their surfaces as a protected region.

References

Jensen, O.M., Hansen, P.F.: Water-entrained cement-based materials: I. Principles and theoretical background. Cem. Concr. Res. **31**(4), 647–654 (2001)

Mechtcherine, V., Reinhardt, H.W.: Application of Superabsorbent Polymers (SAP) in Concrete Construction, State-of-the-Art Report Prepared by Technical Committee 225-SAP. Springer, Dordrecht (2012)

Mechtcherine, V., et al.: Effect of superabsorbent polymers (SAP) on the freeze - thaw resistance of concrete: results of a RILEM interlaboratory study. Mater. Struct. **50**, 14 (2017)

Schröfl, C., Snoeck, D., Mechtcherine, V.: A review of characterization methods for superabsorbent polymer (SAP) samples to be used in cement-based construction materials: report of the RILEM TC 260-RSC. Mater. Struct. **50**, 197 (2017)

Snoeck, D., Schröfl, C., Mechtcherine, V.: Recommendation of RILEM TC 260-RSC: testing sorption by superabsorbent polymers (SAP) prior to implementation in cement-based materials. Mater. Struct. **51**, 116 (2018)

Wyrzykowski, M., Lura, P., Pesavento, F., Garwin, D.: Modeling of water migration during internal curing with superabsorbent polymers. J. Mater. Civ. Eng. **24**, 1006 (2012)

Wyrzykowski, M., Igarashi, S., Lura, P., Mechtcherine, V.: Recommendation of RILEM TC 260-RSC: using superabsorbent polymers (SAP) to mitigate autogenous shrinkage. Mater. Struct. **51**, 135 (2018)

Mechanical Properties and Durability

Effect of SAP on the Freeze-Thaw Resistance of Concrete: Tests According to Russian Standards

Vyacheslav R. Falikman(✉)

Structural Concrete Association, 22-2 Ryazansky Prospect, 109428 Moscow, Russia
vfalikman@yandex.ru

Abstract. Russian standardized method for an estimation of concrete frost resistance is characterized by number of cycles of freezing and thawing of specimens under standard test conditions without essential strength decrease. The frost resistance tests were carried out in accordance with GOST 10060-2012 and GOST 26134-2016. Two SAP-types were selected for study. One basic and two accelerated freezing-and-thawing test methods were used. Cubic concrete specimens which were subjected to freezing-and-thawing tests had dimensions of 100 × 100 × 100 mm. Frost-resistance factor K_{fr} of concretes, i.e. the ratio of strength of key samples after a given number of alternate freezing and thawing cycles to the strength of control samples, was calculated for each sample after test cycles. Control value was estimated as 0.95. For tests in salt water, K_{fr} was even over 1 sometimes that indicates to the available reserve of durability. In contrast, the frost resistance of some samples does not exceed 200 cycles depending on the type of SAP. A different picture was observed when testing concrete in fresh water. In this case, the frost resistance of concrete with SAP was always higher than that of the control samples. It seems that SAP reduces the strength of concrete at the age of 28 days compared to the check sample without SAP by 10 … 13%; in the process of testing, concrete with SAP is gaining strength, resulting in an K_{fr} increase. Thus, the procedure prescribed by the Russian standard should be clarified in respect to testing of concretes with SAP.

Keywords: Superabsorbent polymers (SAP) · Concrete · Freezing-and-thawing tests · Standards

1 Introduction

Superabsorbent polymers (SAP) are cross-linked polyelectrolytes, which swell upon contact with water or aqueous solutions resulting in formation of hydrogel (Horie et al. 2004; Kabiri 2003). They were proposed for the first time by Jensen (Jensen and Hansen 2001) as novel chemical admixture for controlling autogenous shrinkage in HPC as internal curing agent. Later, Jensen and Hansen suggested that SAP could be used to produce concrete with predefined size and spacing of air inclusions (Jensen and

W. P. Boshoff et al. (Eds.): SAP 2019, RILEM Bookseries 24, pp. 57–64, 2020.
https://doi.org/10.1007/978-3-030-33342-3_7

Hansen 2002). Normally, this is a way to protect concrete against freeze-thaw action (Laustsen et al. 2015).

The increasing interest in the use of SAP as a concrete admixture and the need for more intense scientific researches had led to the creation of RILEM Technical Committee 225-SAP "Application of Superabsorbent Polymers in Concrete Construction" in 2007. A key issue its activity was to compile a State-of-the-Art Report finalized in 2012 (RILEM STAR 2012). According to StAR, the main applications of SAP include modifying the rheology of fresh concrete, mitigation of plastic and autogenous shrinkage as well as triggering self-sealing and self-healing of cement based construction materials. Next RILEM Technical Committee 260-RSC "Recommendations for Use of Superabsorbent Polymers in Concrete Construction" was created aiming to draw together the facts for a successful of proper application SAP material in concrete mix designs and proof of its performance and to develop practical recommendation for constructors. One of the most important topics of TC activity was to propose Recommendations for using superabsorbent polymers (SAP) for improving freeze-thaw resistance of cement-based materials for construction sector. These Recommendations base on an interlaboratory experimental study of international research groups tested SAP materials in terms of their influences on freeze-thaw and deicing agent resistance of ordinary concrete (Mechtcherine et al. 2018). Russian freeze - thaw resistance tests were performed according to local standards, both with deicing salts (NaCl) and pure water, what expands the range of previous studies within the interlaboratory experimental study. Russian standards differ markedly from other national and international standards.

2 Russian Standard on Frost Resistance

The CIS (Russian) standard GOST 10060 (2012) establishes several methods for an estimation (determining) the concrete frost resistance during repeated freezing and thawing. These standardized methods are characterized by number of cycles of freezing and thawing of specimens without essential strength decrease under standard test conditions. As usual, three methods are used: one basic and two accelerated ones. If the results of concrete frost resistance estimation determined by the basic and accelerated methods differ, the results obtained by the basic method are taken as final.

According to standard, after 28th day of curing, during 3 days before the frost resistant tests have been started, concrete specimens are saturated in fresh water (basic 1st method) or in 5% salt-water at temperature 18 ± 2 °C (2nd and 3rd test methods). Compressive strength and mass measurements are performed after 2–4 h, when concrete specimens were saturated. Specimens freezing are carried out in the climatic chamber at temperature minus (18 ± 2) °C (basic and so called 2nd test method) and at temperature minus (50 ± 5) °C (3rd method). When 1st and 2nd test method are used, the time required for the specimen gets frozen in the freezer is 2,5 h and thawing of the specimen in salt-water (18 ± 2 °C) requires a minimum of 2 h. Compressive strength and mass measurements are performed in 2–4 h after thawing. When 3^{d} test method is used, the time required for the specimen gets frozen in the freezer was (2,5 ± 0,5) h to reach the temperature minus (50 ± 5) °C. After (2,5 ± 0,5) h temperature-matched curing, the temperature in chamber has to be raised up to −10 °C during (1,5 ± 0,5) h, samples have to be withdrawn, and then thawing of the specimen in salt-water (18 ± 2 °C) requires a minimum of (2,5 ± 0,5) h. Compressive strength and mass measurements were performed in 2–4 h after thawing. After each standardized number of freezing/thawing cycles, fresh water and NaCl solutions are renewed.

After a certain number of cycles, compressive strength and mass losses of the specimens are measured. If the changes in compressive strength do not exceed 5% and the weight changes do not exceed 3%, the sample passes the number of cycles and the frost resistance tests shall continue. Concrete grade on frost resistance (from 25 to 1000) is fixed according to GOST requirements.

3 Materials and Test Methods

Materials. Raw materials available on market were used. Portland cement of CEM I 32.5 grade under GOST 10178 (1985) was produced by Stary Oskol cement plant (Table 1).

Table 1. Cement mineral composition by Bogue.

Cement	C_3S	C_2S	C_3A	C_4AF	R_2O
Stary Oskol CEM I 32,5 R	65,2	13,9	8,5	12,7	0,69

The high-range water-reducing admixture (HRWRA) - superplastisizer was a commercial product based on β-naphthalene sulfonate (PNS type). Its dosage by mass of cement (bwoc) could be adjusted on demand.

Synthetic air-entraining admixture (AEA) entrained very finely distributed air micro-pores (<0.3 mm) into the concrete was used for comparison. AEA was added during mixing or to the concrete mixture together with the added water.

Two SAP types (named as SAP-3 and SAP-4, correspondingly) were selected for the test. They were shipped by the TC. Both of them were "retentive" with respect to their sorptivity in time in extracted cement pore solution, as determined by the so called "tea-bag method" (Schrofl et al. 2012). According to received information, the polymers have irregular particle shapes as they were produced via the bulk polymerization technique, followed by crushing into single particles.

As aggregates, locally typical aggregates for ordinary concrete such as quartz sand corresponding to requirements of GOST 8736 (2014) and crushed granite aggregates with a diameter of up to 20 mm corresponding to requirements of GOST 26663 (2015) were used. Aggregate fractions had to be composed in such a way that the resulting grading curve would be according to DIN 1045-2 (2008).

Experimental Methods. The frost resistance tests were carried out in accordance with GOST 10060 (2012). Slump and bleeding of concrete mix were tested according to Russian standard GOST 10181 (2014). Some properties of concrete mix were determined in accordance with EN 12350 (2010). Concrete strength at the age of 28 days was tested according to GOST 10180 (2012) and EN 12390 (2009) with cube specimens 10 × 10 × 10 cm.

Proportioning of Concrete and Mixing Procedure. The mixture proportions of concrete used in experiments were chosen according to TC recommendation to provide ordinary concrete mix for road construction (Table 2).

Table 2. Mixture proportions of concrete used.

	Mix design, kg per 1 m^3 of concrete mixture
PC CEM I 32.5 R	365
Water	164
Quartz sand 0.06/0.2	137
Quartz sand 0/2	432
Quartz sand 2/5	275
Granite 5/20	1050
Superplasticizer	0,6% by cement mass

Dry SAP was mixed with cement to promote uniform distribution in the concrete. All dry constituents were placed in a pan mixer (batch size 30 l). Water was added, and the mix was mixed for 2 min. After that superplasticizer was added, and the mix was mixed again for 2 min more. This mixing time provided sufficient time for SAP water absorption during mixing, even for the coarsest SAP size fraction.

4 Results and Discussion

The dosages of all admixtures had been adjusted to meet the requirements of workability as well as the air content in the AEA-mixtures. In Table 2, the results of air content measurements in fresh mixtures are presented. As it could be seen, the addition of SAP did not lead to additional air entrainment. When SAPs are added together with additional water, the effect of reduction of mechanical properties is roughly similar to that obtained by simply increasing W/C. On the contrary, air entrainment leads to a clear reduction of mechanical properties for a given W/C. This negative effect is considerably higher than increasing W/C together with adding SAP (Table 3).

Basic (in fresh water) and two accelerated (with deicing salt) freezing-and-thawing test methods were used according to procedure described above, and compressive strength losses of the specimens were measured. Concrete specimens which were subjected to freezing-and-thawing tests had cubic shape with internal dimensions of 100 × 100 × 100 mm. Six control and twelve tested specimens per each mixture were prepared.

After a certain number of cycles, compressive strength and mass changes of the specimens were fixed. If the compressive strength changes didn't exceed 5% and weight changes didn't exceed 3%, it is considered that the specimen has passed number of cycles and frost resistance tests to be continued. Concrete grade on frost resistance is fixed according to GOST requirements.

Table 3. Test results.

№	Admixture and dosage	W/C	Slump, cm	Concrete mix density, kg/m^3	Air content, %	Compressive strength at age of 28 days, MPa
1	0,6% FK 48	0,45	5	2416	1,6	55,3
2	0,6% FK 48 0,15% SAP-3	0,5	4	2384	1,8	53,4
3	0,6% FK 48 0,15% SAP-4	0,5	3	2378	1,7	51,4
4	0,6% FK 48	0,5	15	2396	1,7	47,6
5	0,6% FK 48 0,4% Air 202[a]	0,45	7	2318	5,1	42,9

[a]10% solution of AEA.

Frost-resistance factor K_{fr} of concretes, i.e. the ratio of strength of key samples after a given number of alternate freezing and thawing cycles to the strength of control samples, was calculated for each sample after test cycles. Control value was estimated as 0.95. Sometimes K_{fr} was even over 1 that indicates to the available reserve of durability. In contrast, the frost resistance of some samples does not exceed 200–250 cycles depending on the type of SAP (Fig. 1).

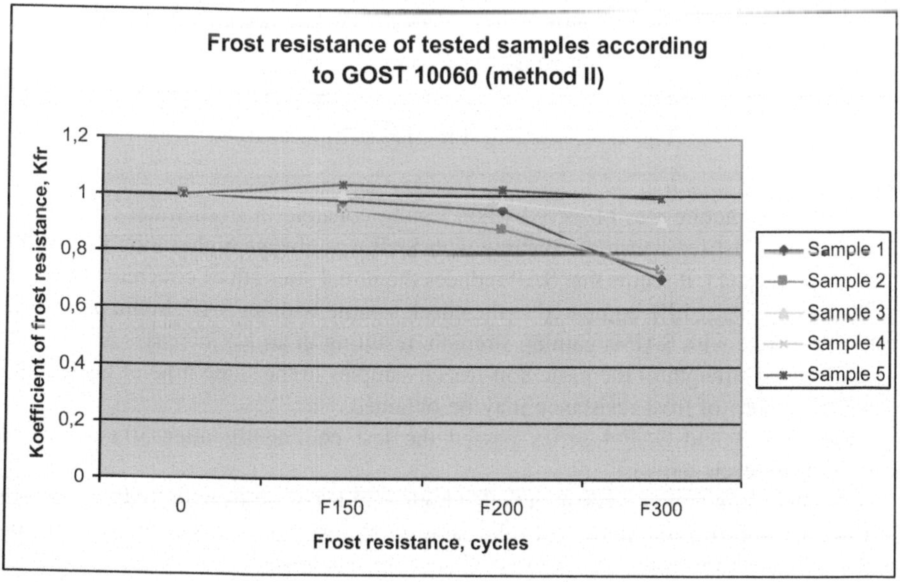

Fig. 1. Concrete frost resistance (II method).

In accelerated tests at temperature minus (50 ± 5) °C with deicing salt only air-entraining admixtures increase the frost resistance of concrete. Frost resistance of concrete with SAP does not exceed 200 cycles, and for samples SAP-3 it is even lower – less than 200 cycles, while the grade of concrete on frost resistance with AEA exceeds 300 (Fig. 2). It is possible that such harsh testing conditions do not allow the development of the pore structure of concrete with SAP.

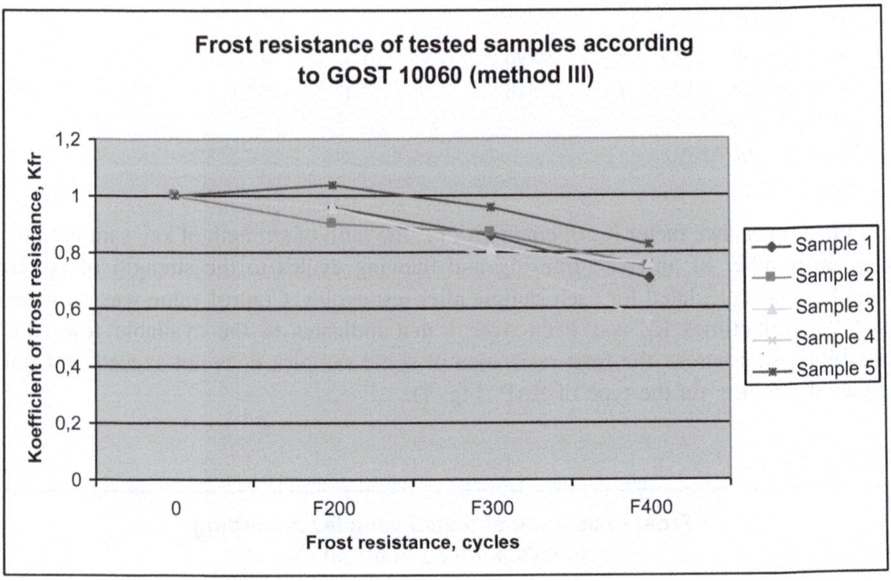

Fig. 2. Concrete frost resistance (III method).

A different picture was observed when testing concrete in fresh water (Fig. 3). In this case, the frost resistance of concrete with SAP was always higher than that of the control samples (1). It seems that SAP reduces the initial strength of concrete at the age of 28 days by 10 … 13% compared to the check sample without SAP; in the process of testing, concrete with SAP is gaining strength, resulting in an K_{fr} increase. As a result, an understated strength of the basic non-frozen samples in the same time of testing and very high grades of frost resistance may be obtained.

Both SAP-3 and SAP-4 series passed the test confidently after 309 cycles (4 samples from each series).

It is interesting to note nevertheless that both samples of series with SAP-3 and with SAP-4 after 400 freezing-thawing cycles showed almost the same results as after 300 cycles. At the same time, the greatest change in the lower limit of the confidence interval according to GOST 10060 for the samples of the 2 and 3 series was occurred after 300 cycles, when a drop of 9.5 and 7.9 MPa was recorded respectively. Apparently, the real frost resistance of these series should be F200, if the control samples were been tested simultaneously with the main samples after 200 cycles.

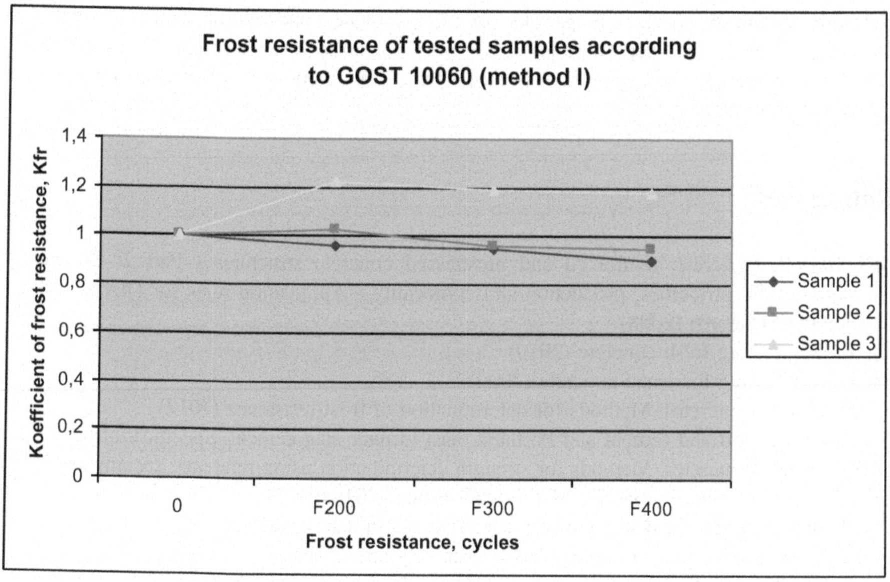

Fig. 3. Concrete frost resistance (I method).

On this way, concrete with SAP-3 has not passed 300 cycles, and with SAP-4 barely kept within the required values.

Looking at the results, the procedure prescribed by the Russian standard should be clarified in respect to testing of concretes with SAP. Probably, it is necessary to develop a special separate test method for the concrete frost resistance with SAP taking into account the strength development in concrete with SAP in the process of testing in fresh water or modify it in this case specifying the duration and conditions of an additional curing.

5 Conclusions

The results of the accomplished research have demonstrated that the application of SAP provides certain control of the properties of concrete. It seems that SAP voids can improve frost resistance but this effect is less pronounced than the effect of air-entraining admixtures. It is possible that a greater advantage could be obtained in special technologies, such as shotcrete. Today it is almost universally recognized that during hardening and/or desiccation process SAP particles release absorbed water and leave behind cavities with sizes of the swollen SAP particles; these cavities act like air voids.

In future research involving addition of SAP aiming to increase concrete durability, it should be necessary to update the existing standards taking into account the structure of SAP particles and behavior of SAP in the cementitious environment.

Acknowledgements. Author is grateful to all colleagues and among them especially to Tatyana L. Zimina and Oleg O. Kalashnikov from Scientific Research Center "Construction", and also to Vadim G. Solovyev from Moscow State University of Civil Engineering for their valuable contributions.

References

DIN 1045-2: Concrete, reinforced and prestressed concrete structures - Part 2: Concrete - Specification, properties, production and conformity - Application rules for DIN EN 206-1 (Foreign Standard) (2008)

EN 12350: Testing fresh concrete (2010)

EN 12390: Testing hardened concrete (2009)

GOST 10060: Concretes. Methods for determination of frost-resistance (2012)

GOST 10178: Portland cement and Portland blast furnace slag cement. Specifications (1985)

GOST 10180: Concretes. Methods for strength determination using reference specimens (2012)

GOST 10181: Concrete mixtures. Methods of testing (2014)

GOST 26663: Normal and fine grained concretes. Specifications (2015)

GOST 8736: Sand for construction works. Specifications (2014)

Horie, K., Báron, M., Fox, R.B., He, J., Hess, M., Kahovec, J., Kitayama, T., Kubisa, P., Maréchal, E., Mormann, W., Stepto, R.F.T., Tabak, D., Vohlídal, J., Wilks, E.S., Work, W.J.: Definitions of terms relating to reactions of polymers and to functional polymeric materials (IUPAC Recommendations 2003). Pure Appl. Chem. **76**(4), 889–906 (2004)

Jensen, O.M., Hansen, P.F.: Water-entrained cement-based materials: II. Experimental observations. Cem. Concr. Res. **32**, 973–978 (2002)

Jensen, O.M., Hansen, P.F.: Water-entrained cement-based materials: I. Principles and theoretical background. Cem. Concr. Res. **31**, 647–654 (2001)

Kabiri, K.: Synthesis of fast-swelling superabsorbent hydrogels: effect of crosslinker type and concentration on porosity and absorption rate. Eur. Polym. J. **39**(7), 1341–1348 (2003)

Laustsen, S., Hasholt, M.T., Jensen, O.M.: Void structure of concrete with superabsorbent polymers and its relation to frost resistance of concrete. Mater. Struct. **48**, 357–368 (2015)

Mechtcherine, V., Snoeck, D., Schrofl, C., De Belie, N., Klemm, A.J., Ichimiya, K., Moon, J., Wyrzykowski, M., Lura, P., Toropovs, N., Assmann, A., Igarashi, S., De La Varga, I., Almeida, F.C.R., Erk, K.A., Ribeiro, A.B., Custodio. J., Reinhardt, H.W., Falikman, V.: Testing superabsorbent polymer (SAP) sorption properties prior to implementation in concrete: results of a RILEM Round-Robin Test. Mater. Structur. **50**, 14 (2018). https://doi.org/10.1617/s11527-018-1149-4

RILEM State-of-the-Art Report Prepared by Technical Committee 225-SAP: In: Mechtcherine, V., Reinhardt, H.W. (eds.) Application of Superabsorbent Polymers (SAP) in Concrete Construction, 166p. Springer, Heidelberg (2012)

Schrofl, C., Mechtcherine, V., Gorges, M.: Relation between the molecular structure and the efficiency of superabsorbent polymers (SAP) as concrete admixture to mitigate autogenous shrinkage. Cem. Concr. Res. **42**(6), 865–873 (2012)

Influence of Superabsorbent Polymers on Properties of High-Performance Concrete with Active Supplementary Cementitious Materials of Nigeria

David O. Nduka[1(✉)], Babatunde J. Olawuyi[2], Timothy O. Mosaku[1], and Opeyemi Joshua[1]

[1] Department of Building Technology, College of Science and Technology, Covenant University, Ota, Ogun State, Nigeria
{david.nduka, timothy.mosaku, ope.joshua}@covenantuniversity.edu.ng

[2] Department of Building, School of Environmental Technology, Federal University of Technology, P.M.B 65, Minna, Nigeria
babatunde@futminna.edu.ng

Abstract. Concrete of strengths classes ≥ C55/67 referred to as high strength or high-performance concrete (HSC/HPC) are noted to be generally of low water/binder (W/B), made from binary or ternary cements with silica fume (SF) being a necessary constituent, and often requiring internal curing. Non-availability and high cost of SF in most sub-Saharan Africa like Nigeria however makes HSC/HPC production in this region very difficult and hence the continued search for alternative supplementary cementitious materials (SCM) with good performance properties as constituents of ternary/binary cements in HPC. This study thereby examines the strength properties of metastable calcined clay (MCC) based HPC cured internally with superabsorbent polymer (SAP) 0.2–0.3% (by weight of binder (b_{wob})). HPC mixtures of varied MCC and Rice husk ash (RHA) contents containing two SAP grain sizes labelled (SP_1 < 300 μm and SP_2 < 600 μm) were cast in 100 mm cubes and cured for varying ages (7, 14, 28 and 56 days) before testing. The hardened specimens were subjected to compressive strength and water absorption tests at the varied curing ages for the performance assessment of the binder types and SAP grain sizes in HPC with age. This study revealed the possibility of achieving Class 1 HPC (50–75 N/mm^2) utilizing industry manufactured calcined clay and locally produced RHA in Nigeria. The compressive strength of HPCs increased as the curing age increases for both SCM type, SAP contents and grain sizes. RHA based HPCs however showed better strength performance at the early ages than the MCC based. SAP addition in MCC based HPCs led to slight decrease in compressive strength as the SAP contents increased while the RHA based HPCs on the other hand, revealed slight increase in compressive strength with increase in SAP contents.

Keywords: Superabsorbent polymers · Metastable calcined clay · Rice husk ash · High-performance concrete · Strength properties

W. P. Boshoff et al. (Eds.): SAP 2019, RILEM Bookseries 24, pp. 65–74, 2020.
https://doi.org/10.1007/978-3-030-33342-3_8

1 Introduction

Within the past four decades, there is a growing acceptance of high-performance concrete (HPC) in construction facilities across the globe. The acceptance of HPC is due its better performance in mechanical strength, durability, flow-ability and economic cost amongst others when compared with normal strength concrete (Han et al. 2018; Nduka et al. 2018). ACI (1999) defined HPC "as concrete meeting special combinations of performance and uniformity requirements that cannot always be achieved routinely using traditional constituents and normal mixing, placing, and curing practice". The applications of this type of concrete have majorly been witnessed in the construction of tunnels, precast pylons, bridges, shotcrete repairs, tall buildings, parking garages and more (Aītcin 2004). Thus, the applicability of HPC in developing country like Nigeria would improve the future performances of infrastructure projects.

HPC is essentially a concrete with low water-binder ratio (W/B) ranging from 0.2–0.38. The substantial amount of cement and supplementary cementitious materials (SCMs) inherent in the mix results in increased temperature upon water addition and densification within the concrete area. Nduka et al. (2018) observed issues of concern like difficulty in curing vertical members, inaccessible locations in buildings and poor workmanship when external curing methods are used in HPC structures. Consequently, to mitigate these challenges in concrete production, internal curing (IC) with SAP has gained tremendous attention in literature and practice in producing HPC. SAP is largely polymers that has capacity to store and release large amounts of water. SAP's utilisation as an IC-agent in HPC is due to its ability for quick absorption of water and slow release of water during the hydration process of cement paste (Yang et al. 2019). It has been adjudged to be efficient in reducing autogenous shrinkage and penetration of deleterious ions in HPC mortars and concrete.

Among the various constituents of HPC, SCMs of pozzolanic nature play significant roles in meeting HPC requirements. SCMs are majorly siliceous/aluminous finely divided solid minerals that are used to substitute part of PC in concrete and mortar production and in the presence of water will react chemically with $Ca(OH)_2$ to form cementitious product (Zhou et al. 2017). Significances of SCMs include the refinement and improvement of pore size distribution, capillary pores and interfacial transition zone of hardened concrete, reduction of large pores, and improved density of cementitious products, improved workability in fresh state and lowering of W/B in HPC mix (Han et al. 2018; Busari et al. 2019). Silica fume (SF) is a prominent SCM used in HPC but its availability, economics and service considerations are pertinent issues of concern in HPC applications (Aītcin 2004). This thereby draws attention to abundant clay and some possible agricultural waste ashes like rice husk ash (RHA) available in this region towards filling the gap.

A number of studies have investigated the blend of industrial based pozzolans like SF, fly ash and slags (ground granulated blast furnace slag and corex slag) in HPC production. Studies on substitution of PC with metastable calcined clay (MCC) in achieving Class 1 (50–75 MPa) HPC is however scarce in literature. The present study thereby examines the possibility of incorporating commercially produced MCC as an alternative to SF in HPC production within Nigeria. It further compares the influence of SAP in MCC and RHA based HPCs.

2 Experimental Procedure

2.1 Materials

The materials used for the study are SAP, natural sand, crushed granite stone, PC (CEM II 42.5 N), MCC, RHA, water and superplasticizer. Two grain sizes of SAP tagged "FLOSET 27CS $\leq$ 300 μm (SP_1) and FLOSET 27CC $\leq$ 600 μm (SP_2)" as described in earlier publication (Olawuyi and Boshoff 2017) at varied SAP contents (0%, 0.2% and 0.3%) by weight of binder (b_{wob}) were used for the study. The absorption capacity determined by tea-bag test as reported in Olawuyi and Boshoff (2013) is 250 g/g in distilled water and 25 g/g in cement pore solution (CPS) for both the SP_1 and SP_2.

CEM II/B-LL, 42.5 N conforming to BS EN 197-1 (2011) and NIS 444-1 (2003) was used as the main binder (PC). MCC purchased from the Pozzolana Cement Plant in Ota, Ogun State and RHA obtained from Rice husk acquired from rice mill in Minna, Niger State, Nigeria served as the SCMs for this study. The rice husk was calcined to ash at 700 °C in a controlled incinerator and pulverised using a grinding mill at Building Laboratory, Federal University of Technology, Minna. Blends of the binders (as presented in Table 1 shows the two reference HPC mixtures while SAP at both 0.2% and 0.3% contents were incorporated with additional water of 12.5 g/g provided for SAP absorption for required specimen production.

The natural sand used at air dry condition had minimum particle size of 300 μm (i.e. all the particles smaller than 300 μm removed using the sieving method) in compliance with requirement for fine aggregate specification for HPC production (Neville 2012). The sand has the following physical characteristics: Fineness Modulus – FM = 2.87, Coefficient of uniformity – C_u = 1.23, Coefficient of curvature – C_c = 3.28 and dust content = 0.1%. This conforms to medium sand classification according to Shetty (2004). Crushed granite stone that passed through 13.50 mm sieve size and retained on 9.50 mm sieve size was used as course aggregate in compliance with typical HPC mixes found in literature (Aītcin 2004; Neville 2012; Olawuyi and Boshoff 2018). The crushed granite was used as saturated surface dry conditions after been washed to eliminate fine content that will likely increase water demand. Result of the physical properties tests on the constituent materials are found in Sect. 3.1 (Table 3).

A polycarboxylic ether (PCE) polymer-based superplasticizer supplied by BASF Limited was used as the superplasticizer and administered at constant concentration of 1.5% b_{wob}. Potable water as specified by BS EN 1008 (2002) available within the concrete laboratory of Department of Building Technology, Covenant University, Ota was used for mixing. Additional water for SAP based on 12.5 g/g of SAP (Olawuyi and Boshoff 2013) was also introduced.

2.2 Methods

2.2.1 Properties of Constituent Materials

The oxide composition of the cementitious materials (PC, MCC and RHA in powdered form) was examined at the Laboratory of the Ewekoro Factory of Lafarge Plc. About 100 g of the materials were packaged in a sealed polythene sheet and sent to the Laboratory for XRF analysis. The result is as presented in Sect. 3.3. The particle size distribution of the aggregate's samples (i.e. the sieved sand and granite stone) was determined by wet sieving while the specific gravity of the aggregate and binders were also determined in the Building laboratory of Covenant University. Section 3.3 further present and discuss the result of the physical properties of the constituent materials.

2.2.2 HPC Specimen Preparation

HPC target strength of C55/67 at 28 day with 0.3 W/B mix design utilizing British method was adopted as reference mix with the other mixes having different SAP contents (0.2% and 0.3%) for the two SAP grain sizes of SP_1 and SP_2. The total mass of the binders in the mixes include 485 kg/m^3 (90% b_{wob}) of PC, 55 kg/m^3 (10% b_{wob}) of MCC or RHA respectively. The details of the mix constituents for the reference HPC mixtures is presented in Table 1.

Table 1. Reference mixes composition

Constituents	Reference mixes (kg/m^3)	
	MCCC	RHAC
Water	156	156
PC (CEM II 42.5 N)	485	485
CA (13 mm maximum)	1050	1050
FA (retained on 300 μm sieve)	700	700
MCC	55	0
RHA	0	55
Superplasticizer (Masterglenium Sky 504 - 1.5% b_{wob})	8.1	8.1
Water/binder (W/B)[a]	0.3	0.3

Legend
MCCC = 10% MCC based HPC
RHAC = 10% RHA based HPC
CA = Coarse Aggregate
FA = Fine Aggregate
[a]W/B = (water + liquid content of superplasticizer)/(PC + MCC + RHA)

The varied SAP contents were incorporated for the two SAP grain sizes (SP_1 and SP_2) studied with additional water at 12.5 g/g provided for SAP absorption and the effect on the respective mixture's workability measured via slump flow table test in accordance with EN 12350:5, 2000. The result is presented in Sect. 3.3. The cast 100 mm cubes HPCs were de-moulded after 24 h and cured by full immersion in water with additional IC provided through the SAP incorporated for the respective ages before testing.

2.2.3 Water Absorption

The water absorption test was carried out on triplicate samples of 100 mm cube and the mean reported. The specimens were fully submerged in water for 56 days, then weighed and afterwards oven dried to a permanent mass at the temperature of 115 °C. The weight differences were then used in calculating the values of water absorption using Eq. 1. The terms w_1 and w_2 refers to dry and wet weight respectively.

$$\frac{W2 - W1}{W1} * 100 \tag{1}$$

2.2.4 Compressive Strength

HPC samples were made and cured in compliance with BS EN standards (BS EN 12350-1 & 5 2000; 12390-1 & 2 2000; 12390-3 2002) for compressive strength. The compressive strength tests were performed on 120 samples using 2000 kN loading capacity Eccles Compressive Testing Machine at 0.5 N/mm^2 loading rate. The concrete specimens were also weighted and dimension measured immediately after de-moulding and at the respective curing ages (7, 14, 28 & 56 days) after removal from curing tank for determination of the de-moulded density and final density of the hardened HPCs.

3 Results and Discussion

3.1 Physical and Chemical Properties of Constituent Materials

Result of the XRF analysis conducted on the PC, MCC and RHA powders are as presented in Table 2. The result reveal both MCC and RHA to be of good silica (SiO_2) content (55% and 94% respectively). The RHA is majorly SiO_2 (94%) with a silica-sesquioxide (S-S) ratio (SR) of 49.3 and aluminium-sesquioxide ratio (AR) of 1.1 is observed to be a very strong and reactive Class F Pozzolan in conformance to ASTM C618.

The SiO_2 content of the MCC is also far above the 35% minimum specified by the standard for a Class F Pozzolan. The sum of silica, alumina and ferric oxides (SiO_2+Al_2O_3+Fe_2O_3) for both the RHA (96%) and MCC (82%) is above the 70% specified for the Class of Pozzolan in ASTM C618. PC on the other hand is majorly calcium oxide (CaO - 64%). This is in agreement with oxides composition for CEM II Portland cement found in literature (Neville 2012).

Table 3 reveals the fine aggregate is observed to have a coefficient of uniformity (C_u) of 2.39, coefficient of curvature (C_c) of 0.94 and fineness modulus (FM) of 2.87 implying a medium sand of Shetty (2004) classification while the coarse aggregate used for the study is a uniformly graded stone. The details of physical properties as shown in Table 3 affirms that the fine and coarse aggregates are suitable for HPC production.

Table 2. Oxide composition of binder constituents

Oxides (%)	MCC	RHA	PC
SiO_2	55.04	93.60	21.50
Al_2O_3	24.49	1.00	5.20
Fe_2O_3	2.14	0.90	1.20
CaO	0.52	1.30	64.00
MgO	0.26	1.20	2.90
SO_3	0.02	0.10	4.50
Na_2O	0.05	1.70	0.60
K_2O	0.17	0.20	0.10
Minor Oxides	2.32	0.00	0.00
LOI	14.99	0.00	0.00
$SiO_2+Al_2O_3+Fe_2O_3$	81.67	95.50	27.90
SR	2.07	49.26	3.36
AR	11.44	1.11	4.33
Total	100.00	100.00	100.00

Table 3. Physical properties of materials

Item	Sand	Granite	RHA	MCC	PC
D_{10}	360	10000			
D_{30}	540	11000			
D_{60}	860	13000			
C_u	2.39	1.30			
C_c	0.94	0.93			
FM	2.87	–			
SG	2.65	2.7	2.3	2.5	3.4

3.2 Fresh Properties of HPC

Results of the flow table test conducted on the HPCs are shown in Table 4. The results revealed that all the mixtures are within the target limits of 600–700 mm slump flow. The variations as observed are very minor (665 to 675 mm) for all mixtures. This implies irrespective of the SAP grain sizes or contents, the mixtures have similar consistency and workability.

Table 4. Fresh properties of HPC mixtures

Specimen label	MCCC					RHAC				
	1	2	3	4	5	1	2	3	4	5
Constituents (kg/m3)	Ref	SP_1		SP_2		Ref	SP_1		SP_2	
SAP contents (%bwob)	0	0.2	0.3	0.2	0.3	0	0.2	0.3	0.2	0.3
Water	156	156	156	156	156	156	156	156	156	156
PC (CEM 42.5 N)	485	485	485	485	485	485	485	485	485	485
MC (10% bwob)	55	55	55	55	55	0	0	0	0	0
RHA (10% bwob)	0	0	0	0	0	55	55	55	55	55
CA (13 mm maximum)	1050	1050	1050	1050	1050	1050	1050	1050	1050	1050
FA (retained on 300 μm)	700	700	700	700	700	700	700	700	700	700
SAP	0	1.08	1.62	1.08	1.62	0	1.08	1.62	1.08	1.62
S/plasticizer	8.1	8.1	8.1	8.1	8.1	8.1	8.1	8.1	8.1	8.1
Additional Water (12.5 g/g of SAP)	0	13.5	20.3	13.5	20.3	0	13.5	20.3	13.5	20.3
Total W/B	0.3	0.3	0.3	0.3	0.3	0.3	0.3	0.3	0.3	0.3
Slump Flow	670	665	665	670	665	675	675	670	675	675
Designed Density	2454	2469	2476	2469	2476	2454	2469	2476	2469	2476
De-moulded Density	2494	2473	2452	2457	2462	2436	2458	2487	2442	2443

Legend
SP_1 = SAP grain size type 1 ($\leq$600 μm)
SP_2 = SAP grain size type II ($\leq$300 μm)

3.3 Hardened Properties of HPCs

3.3.1 Compressive Strength of HPCs

Table 5 shows the compressive strength of various HPCs containing equal amount of CEM II 42.5 N, 10% MCC/RHA and SAP at varying contents and two grain sizes. All the HPCs revealed strength increases with increase in curing age with the RHACs showing better strength performance at the early ages. At 7-day curing, the reference MCC based HPC ($MCCC_1$) slightly outperformed reference RHA based HPC ($RHAC_1$) by 8.6% in compressive strength while SAP addition in MCCCs ($MCCC_2$ to $MCCC_5$) led to slight decrease in compressive strength as the SAP contents increased. On the other hand, SAP addition in RHACs ($RHAC_2$ to $RHAC_5$) resulted in slight increase in compressive strength with increase in SAP contents.

Thus, RHACs with SAP performed slightly better in strength than MCCCs at the early age. Similar trend was observed at the 14th day of curing. The 14th day strength of $RHAC_1$ is 4.37% above that of $MCCC_1$ while MCCCs containing SAP experienced slight strength decrease with SP_2 inclusion revealing better performance than SP_1. RHACs also had reduced compressive strengths at all SAP contents except $RHAC_3$ with 0.3% SP_1 content.

Both $MCCC_1$ and $RHAC_1$ at 28 days curing was noted to be at similar value of compressive strength (47.09 N/mm^2 and 47.10 N/mm^2 respectively). $MCCC_2$ to $MCCC_5$ mixes indicated increased in compressive strength over the control $MCCC_1$ by 3.8% while $RHAC_2$ to $RHAC_5$ also increased by 4.6% in compressive strengths over $RHAC_1$ control. The CS_{28} Factor (in Table 5) further revealed that SAP inclusion in all the HPC specimens resulted in slight compressive strength increases at the 28th day curing. Consequently, $RHAC_2$ to $RHAC_5$ showed better strength development with SAP inclusion than the $MCCC_2$ to $MCCC_5$. Thus, $RHAC_2$ to $RHAC_5$ slightly outperformed $MCCC_2$ to $MCCC_5$ by 0.86% with almost equal compressive strength for both SAP grain sizes at varying percentages.

Table 5. Compressive strength and CS_{28} factor

Specimen/age (days)	Compressive strength (N/mm^2)				CS_{28} factor			
	7	14	28	56	7	14	28	56
$MCCC_1$	42.97	43.15	47.09	51.51	0.91	0.92	1.00	1.09
$MCCC_2$	41.04	42.87	47.18	60.28	0.87	0.91	1.00	1.28
$MCCC_3$	40.46	41.32	47.61	60.53	0.86	0.88	1.01	1.29
$MCCC_4$	41.37	43.67	52.66	60.49	0.88	0.93	1.12	1.28
$MCCC_5$	40.67	43.81	48.28	62.72	0.86	0.93	1.03	1.33
$RHAC_1$	39.28	45.12	47.10	60.30	0.83	0.96	1.00	1.28
$RHAC_2$	42.49	44.44	49.88	61.60	0.90	0.94	1.06	1.31
$RHAC_3$	42.37	47.19	49.29	61.91	0.90	1.00	1.05	1.31
$RHAC_4$	40.47	42.12	49.09	61.63	0.86	0.89	1.04	1.31
$RHAC_5$	42.77	42.54	49.16	61.63	0.91	0.90	1.04	1.31

Legend

CS_{28} = the compressive strength value at respective curing ages as compared to the 28th day strength.

Similarly, 56 days curing period showed that $RHAC_1$ compressive strength outperformed $MCCC_1$ by 17.3%. There is no much significant difference in compressive strength value of the MCCCs at a round up value of 61 N/mm^2 (for $MCCC_2$ to $MCCC_5$) that can be adduced to SAP contents nor grain size. $MCCC_4$ however exhibits the highest compressive strength value of 62 N/mm^2. SAP addition was hence observed to about 20% strength value above the reference $MCCC_1$ strength value. In the same vein, the RHACs containing SAP ($RHAC_2$ to $RHAC_5$) gave a round up compressive strength value of 62 N/mm^2 (1.63% above the reference $RHAC_1$) at all SAP content and grain size, implying no significance influence in compressive strength value attributable to SAP size nor contents.

It can therefore be inferred that the two SCMs (MCC and RHA) is effective in HPC production with similar compressive strength values achieved while SAP addition in the HPCs resulted in slight compressive strength increase at no significant influence by SAP grain size nor contents within the limits examined.

3.3.2 Water Absorption of HPCs

The water absorption of the reference HPC mixes ($MCCC_1$ and $RHAC_1$ respectively) as determined on the 56th day is low (Table 6). The absorbed water is 0.61 (MCCC1) and 0.48 (RHAC1) respectively. The RHACs containing SAP had same water absorption value (0.61%) except $RHAC_4$ with a slightly higher value (0.73). The water absorption values or MCCCs containing SAP ($MCCC_2$ to $MCCC_5$), was observed to be slightly higher with no particular SAP grain size nor contents. Both MCCCs and RHACs mixes are however observed to be generally of water absorption values below 5% irrespective of the SAP contents and grain sizes added, an indication of dense parking of the HPCs (i.e. low permeability). This implies the two SCMs when used to produce HPCs will give good resistance to aggressive chemical attacks in concrete and hence ensure good durability and long service life. SAP addition in the HPCs has little or no effect in their ability to resist aggressive chemicals.

Table 6. Water absorption of HPCs

Specimen		$MCCC_1$	$MCCC_2$	$MCCC_3$	$MCCC_4$	$MCCC_5$	$RHAC_1$	$RHAC_2$	$RHAC_3$	$RHAC_4$	$RHAC_5$
Mass (kg)	Wet	2.494	2.473	2.452	2.457	2.462	2.436	2.458	2.487	2.442	2.443
	Dry	2.479	2.436	2.436	2.442	2.439	2.424	2.443	2.472	2.424	2.428
Water absorption (%)		0.61	1.49	0.65	0.61	0.93	0.48	0.61	0.61	0.73	0.61

4 Conclusion

This study revealed the possibility of achieving Class 1 HPC (50–75 N/mm^2) utilizing industry manufactured calcined clay and locally produced rice husk ash in Nigeria. The HPCs made from binary blends of CEM II/B-LL 43.5 N and MCC/ RHA had Masterglenium Sky 504 (a PCE) as superplasticizer with SAP of two grain sizes (SP_1 and SP_2) and contents (0.2% and 0.3%) incorporated as IC-agent. Inferences drawn from the study can thereby be highlighted as follows:

1. Both MCC and RHA sample used are good Class F Pozzolan having requisite physical and chemical properties as specified in the standards (ASTM C618)
2. All the HPCs (MCCCs and RHACs) studied are of similar range of workability (slump flow values of 665 to 675 mm) irrespective of the SCMs, SAP grain sizes and contents added.
3. The compressive strength of HPCs increased as the curing age increases for both SCM type, SAP contents and grain sizes. RHACs however showed better strength performance at the early ages than the MCCCs.
4. SAP addition in MCCCs led to slight decrease in compressive strength as the SAP contents increased while the RHACs on the other hand, revealed slight increase in compressive strength with increase in SAP contents.
5. SAP's influence on compressive strength of HPC at the later age is noted to be slightly positive. The increases in strength between 28th and 56th curing days depicts that the test for more than 28th days age is better seen as the actual properties of the tested concrete. The long term strength gain can be attributed to the latent reactions of pozzolanic materials (MCC and RHA) which is enhanced by internal curing provided by SAP.
6. The water absorption of the reference HPC mixes ($MCCC_1$ and $RHAC_1$ respectively) as determined on the 56th day is low (0.61 and 0.48). SAP addition resulted in higher water absorption with some inconsistencies. The values however being lower than 5% irrespective of the SAP contents and grain sizes added. This is an indication of dense parking of the HPCs (i.e. low permeability).

Acknowledgement. Authors acknowledge the followings: Covenant University Centre for Research, Innovation and Discovery (CUCRID); Mr. Guillaume Jeanson (Construction Product Manager) SNF Floerger - ZAC de Milieux, 42163 ANDREZIEUX Cedex – FRANCE; the management of Armorsil Manufacturing Incorporation, Nigeria Building and Roads Research Institute (NBRRI) and Ewekoro Factory of Lafarge Plc., Nigeria for the assistance received in conference support fees, materials procurement, use of facilities, softwares and time input in the analysis.

References

ACI THPC/TAC: ACI defines high performance concrete, (the Technical Activities Committee Report (Chairman - H.G. Russell)). American Concrete Institute, USA (1999)

Aĩtcin, P.C.: High Performance Concrete. Taylor & Francis e-Library, New York (2004)

British Standard Institution – BSI: Cement – composition, specifications and conformity criteria for common cements, BS EN 197: Part 1, London, BSI (2000)

BSI: Testing of fresh concrete, BS EN 12350: Part 1, Sampling, London (2000)

BSI: Testing of fresh concrete, BS EN 12350: Part 5, Flow Table Test, London(2000)

BSI: Testing of hardened concrete, BS EN 12390: Part 1, shape, dimension and other requirement for specimens and mould, London (2000)

BSI: Testing of hardened concrete, BS EN 12390: Part 2, making and curing specimen for strength tests, London (2000)

BSI: Testing of hardened concrete, BS EN 12390: Part 3, compressive strength test specimens, London (2002)

Busari, A.A., Akinmusuru, J., Dahunsi, B.: Response surface analysis of the compressive strength of metakaolin self-compacting concrete. Adv. Sci. Technol. Res. J. **13**(2), 7–13 (2019)

Han, C., Shen, W., Ji, X., Wang, Z., Ding, Q., Xu, G., Tang, X.: Behavior of high-performance concrete pastes with different mineral admixtures in simulated seawater environment. Constr. Build. Mater. **187**, 426–438 (2018)

Nduka, D., Ameh, J., Joshua, O., Ojelabi, R.: Awareness and benefits of self-curing concrete in construction projects: builders and civil engineers perceptions. Buildings **8**(8), 109 (2018)

Neville, A.M.: Properties of Concrete, 5th edn. Pearson Educational Limited, Harlow (2012)

Olawuyi, B.J., Boshoff, W.P.: Compressive strength of high-performance concrete with absorption capacity of Super-Absorbing-Polymers (SAP). In: Proceedings of the Research and Application in Structural Engineering, Mechanics and Computation, Cape Town, South Africa, pp. 2–4 (2013)

Olawuyi, B.J., Boshoff, W.P.: Influence of SAP content and curing age on air void distribution of high-performance concrete using 3D volume analysis. Constr. Build. Mater. **135**, 580–589 (2017)

Olawuyi, B.J., Boshoff, W.P.: Influence of superabsorbent polymer on the splitting tensile strength and fracture energy of high-performance concrete. In: MATEC Web of Conferences, vol. 199, p. 11004. EDP Sciences (2018)

Shetty, M.S.: Concrete Technology - Theory and Practice. S. Chand and Company Limited, New Delhi (2004)

Yang, J., Liu, L., Liao, Q., Wu, J., Li, J., Zhang, L.: Effect of superabsorbent polymers on the drying and autogenous shrinkage properties of self-levelling mortar. Constr. Build. Mater. **201**, 401–407 (2019)

Zhou, D., Wang, R., Tyrer, M., Wong, H., Cheeseman, C.: Sustainable infrastructure development through use of calcined excavated waste clay as a supplementary cementitious material. J. Clean. Prod. **168**, 1180–1192 (2017)

Comparative Study of Superabsorbent Polymers and Pre-soaked Pumice as Internal Curing Agents in Rice Husk Ash Based High-Performance Concrete

B. J. Olawuyi[1,2(✉)], R. O. Saka[1], D. O. Nduka[2], and A. J. Babafemi[3]

[1] Department of Building, School of Environmental Technology, Federal University of Technology, P.M.B 65, Minna, Nigeria
{babatunde, saka.rasaq}@futminna.edu.ng

[2] Department of Building Technology, College of Science and Technology, Covenant University, Ota, Ogun State, Nigeria
david.nduka@covenantuniversity.edu.ng

[3] Department of Building, Faculty of Environmental Design and Management, Obafemi Awolowo University, Ile-Ife, Nigeria
ajbabafemi@oauife.edu.ng

Abstract. Utilisation of superabsorbent polymers (SAP) and pre-soaked lightweight aggregates (LWA) as internal curing (IC) agents for the mitigation of autogenous shrinkage and micro-cracking of high strength/high-performance concrete (HSC/HPC) have been well researched and documented in literature. Rice husk ash (RHA) on the other hand has been adjudged to be of good pozzolanic activity and a possible alternative to silica fume (SF) in low water/binder (W/B) concrete production. An experimental comparative study was conducted in the current work to assess the effectiveness of the two known IC-agents on rice husk ash (RHA) based HPC. HPC mixtures of $f_{c,cube28} = 60$ MPa minimum target strength produced and internally cured with 0.3% content of SAP by weight of binder (b_{wob}) and varied content of pre-soaked pumice (5 to 10% in steps of 2.5%) by weight of coarse aggregate (b_{wocg}) were cast using 100 mm cubes samples. Thereafter, the samples were cured for 7, 14, 28 and 56 days by water immersion before subjecting them to compressive strength test. The results showed 0.2% b_{wob} SAP HPC ($SHPC_1$) to be the best performed internally cured HPC at the early ages with similar long-term strength values as 5 and 7.5% b_{wocg} saturated pumiced HPC ($PHPC_{1\&2}$). The study thereby recommends SAP content of 0.2% b_{wob} and saturated pumice content up to 7.5% b_{wocg} for use as IC-agent in HPC.

Keywords: Superabsorbent polymers (SAP) · Pre-soaked lightweight aggregates (LWA) · Rice husk ash (RHA) · Compressive strength · High-performance concrete (HPC)

W. P. Boshoff et al. (Eds.): SAP 2019, RILEM Bookseries 24, pp. 75–84, 2020.
https://doi.org/10.1007/978-3-030-33342-3_9

1 Introduction

High-performance concrete (HPC) utilisation in building construction is getting popular in the recent years due to the unique properties of low water/binder (W/B), high strength, high modulus of elasticity and possibility of having relatively thin sections for construction of tunnels, precast pylons, bridges, shotcrete repairs, tall buildings, parking garages and more (Aïtcin 2004; Orosz 2017). HPC is defined by American Concrete Institute, ACI (1999) "as concrete meeting special combinations of performance and uniformity requirements that cannot always be achieved routinely using traditional constituents and normal mixing, placing, and curing practice". Production of this low W/B concrete is known to often require incorporating other cementitious material especially silica fume (SF) – a scarce and expensive commodity in sub-Saharan Africa especially Nigeria. The non-availability of SF therefore necessitates consideration for rice husk ash (RHA) which has been adjudged to be of good pozzolanic activity and possible alternative to silica fume.

RHA is an agro-industrial waste obtained through calcination of rice husk in a controlled environment. De Sensale and Viacava (2018) estimated RHA generation around the globe to about 20 million tonnes where the husk accounts for about 20% by weight of rice paddy and after burning the husk, ash obtained is in the range of 20–25% of the total weight of the husk (Kaur et al. 2018). RHA is known to be of low density and very high silica content when calcined in controlled setting. Its specific gravity ranges from 2.05–2.53 which is lower than the 3.5 value of PC (Fapohunda et al. 2017). The work of Fapohunda et al. (2017) reported the cumulative useful oxide (SiO_2+Al_2O_3+Fe_2O_3) of RHA calcined in a controlled incinerator to be above 70% conforming to Class F fly ash of ASTM C618 (2015). Limitation of RHA in Nigeria at reaching the optimum pozzolanic reactivity in concrete products was also noted and this was attributed to poor calcination methods in an uncontrolled incineration facility. Largely, the burning temperature in a suitable incinerator is an important factor that affect the amorphous state of RHA in concrete and mortar applications.

HPC however comes with attendant autogenous shrinkage and micro-cracking mitigation for which superabsorbent polymers (SAP) and pre-soaked lightweight aggregates (LWA) used as internal curing is found as possible solution, well researched and documented. Previous studies (Persson 1997; Kovler and Jensen 2005; Bentz and Weiss 2011; Di Bella et al. 2012, 2016) have empirically revealed that HPC is essentially a concrete with low W/B ranging from 0.2–0.38. The substantial amount of cement and supplementary cementitious materials (SCMs) inherent in the mix results in increased temperature upon water addition and densification within the concrete area. Savva et al. (2018) inferred a direct relationship between ambient temperature and pozzolanic activity of cementitious materials. The authors posit that cementitious grains are usually influenced by ambient temperature; hence a fast reaction which hinders uniform distribution of hydration products leading to an increase in porosity of hydrated gel. In the same vein, inclusion of SCM furthers the propagation of autogenous shrinkage, chemical shrinkage and self-desiccation due to combined effects of hydration and pozzolanic reaction necessitating higher moisture demand in the concrete (Wu et al. 2017). Nduka et al. (2018) also observed issues of concern such as like difficulty in curing vertical members, inaccessible locations in buildings and poor workmanship when external curing methods are used in HPC structures. Consequently, to mitigate these challenges in concrete production, an innovative curing technique termed "internal curing (IC)" has gained tremendous attention in literature and practice in producing HPC.

Traditionally, two types of curing systems (internal and external) exists in maintaining the internal temperature and relative humidity of all sections of concrete types. Internal curing (IC) is achieved by incorporating materials that have the capacity to absorb and desorb water during the plastic and hardened state of designed concrete respectively (Tu et al. 2019). The dispersed materials provide additional water internally to the cementitious materials thereby furthering hydration process and subsequently reduced self-desiccation otherwise known as autogenous shrinkage. Two notable materials (Light weight aggregates and SAP) have been used in past studies and practice in advancing internal curing techniques. Studies examining the effectiveness of the two IC-agents on RHA based HPC mix subjected to same curing regimes is scarce in literature. The current study is thereby an attempt to fill this gap on performance assessment of both SAP and pre-soaked pumice as IC-agent in RHA based HPC.

2 Experimental Procedure

2.1 Materials

The materials used for the study are SAP, pumice stone, crushed granite stone, natural sand, cement (CEM II 42.5 N), rice husk ash (RHA), water and superplasticizer. The SAP is a thermoset polymer specifically the covalently cross-linked polymers of acrylic acid and acrylamide, neutralised by alkali hydroxide produced by SNF Floerger in France. The SAP "FLOSET 27CC" of grain size $\geq$ 600 μm as described in Olawuyi and Boshoff (2017) was incorporated at 0.2% and 0.3% contents (by weight of binder – b_{wob}) as IC-agent as a comparative study to saturated pre-soaked pumice (13 mm maximum grain size) at varied contents (5%, 7.5% and 10%) of coarse aggregate. SAP absorption capacity of 12.5 g/g as recommended in Olawuyi (2016) was adopted for additional water provision in this study.

CEM II/A-LL, 42.5 N (3X) from Dangote Cement, Obajana Plant conforming to BS EN 197-1 (2011) and NIS 444-1 (2003) was used as the main binder (PC) and RHA obtained from Rice husk acquired from rice mill in Minna, Niger State, Nigeria served as the SCM for this study. The rice husk was calcined to ash at 700 °C in a controlled incinerator and pulverised using a grinding mill at Building Laboratory, Federal University of Technology (FUT), Minna. The RHA powder was white in colour - an indication of complete burning of all carbon and impurities within the husk. The powdered PC and RHA were then packaged and sent to Ewekoro factory of Lafarge Cement for X-ray florescence (XRF) analysis to determine their oxide composition. Blends of the binders (as presented in Table 1 shows the two reference HPC mixtures).

The natural sand used at air dry condition had minimum particle size of 300 μm (i.e. all the particles smaller than 300 μm removed using the sieving method) in compliance with requirement for fine aggregate specification for HPC production (Aïtcin 1998; Beushausen and Dehn 2009; Neville 2012). The sand has the following physical characteristics: Fineness Modulus – FM = 2.87, Coefficient of uniformity – C_u = 1.23, Coefficient of curvature – C_c = 3.28 and dust content = 0.1%. This conforms to medium sand classification according to Shetty (2004). Crushed granite stone and the pumice that passed through 13 mm sieve size and retained on 9.50 mm sieve size was used as course aggregate in compliance with typical HPC mixes found in literature (Aïtcin 2004; Beushausen and Dehn 2009; Neville 2012; Olawuyi and Boshoff 2018). The crushed granite was used at saturated surface dry conditions after been washed to eliminate fine content that will likely increase water demand. The pumice was soaked in water for 24 h, removed and allowed to drain to saturated surface dry condition before use. Result of the physical properties tests on the constituent materials are found in Sect. 3.1 (Fig. 1 and Table 3).

Hydroplast-300 – a polycarboxylic ether (PCE) polymer based superplasticizer supplied by Armorsil Manufacturing Incorporation was used as the superplasticizer and administered at constant concentration of 1.5% b_{wob}.

Clean potable water studied by Ogunbayo et al. (2018) and as specified by BS EN 1008 (2002) available within the concrete laboratory of Department of Building Technology, Covenant University, Ota was used for mixing. Additional water for SAP absorption was based on 12.5 g/g of SAP (Olawuyi and Boshoff 2013) was also introduced.

2.2 Methods

2.2.1 Properties of Constituent Materials

The oxide composition of the cementitious materials (PC and RHA in powdered form) was examined at the Laboratory of the Ewekoro Factory of Lafarge Plc. About 100 g of the materials were packaged in a sealed polythene sheet and sent to the Laboratory for XRF analysis. The result is as presented in Sect. 3. The particle size distribution of the aggregate's samples (i.e. the sieved sand and granite stone) was determined by wet sieving while the specific gravity of the aggregate and binders were also determined in the Building laboratory of FUT, Minna. Section 3 further present and discuss the result of the physical properties of the constituent materials.

2.2.2 HPC Specimen Production

HPC target strength of C55/67 at 28 day with 0.3 W/B mix design utilizing British method was adopted as reference mix with the other mixes having different SAP contents (0.2% and 0.3%). The total mass of the binders in the mixes include 485 kg/m^3 (90% b_{wob}) of PC and 55 kg/m^3 (10% b_{wob}) of RHA. The details of the mix constituents for the HPC mixtures is presented in Table 1.

The different SAP contents (0.2 & 0.3b_{wob}) were included for the SAP internally cured HPCs ($SHPC_1$ and $SHPC_2$) with additional water at 12.5 g/g provided for SAP absorption while the pre-soaked saturated surface dry pumice was measured and added at required contents (5, 7.5 and 10% by weight of coarse aggregate) for the Pumice internally cured HPCs ($PHPC_1$, $PHPC_2$ and $PHPC_3$ respectively). The cast 100 mm cubes HPCs were de-moulded after 24 h and cured by full immersion for the respective ages before testing.

Table 1. Mix composition

Materials	Mixture types (kg/m^3)					
	Control	HPC with SAP		HPC with pre-soaked Pumice		
	CHPC	$SHPC_1$	$SHPC_2$	$PHPC_1$	$PHPC_2$	$PHPC_3$
CEM II 42.5 N	485	485	485	485	485	485
Rice Husk Ash-RHA (10%)	55	55	55	55	55	55
Superplasticizer (1.5% b_{wob})	8.1	8.1	8.1	8.1	8.1	8.1
Fine sand	700	700	700	700	700	700
Coarse aggregate	1050	1050	1050	997.5	971.25	945
Pre-soaked Pumice				52.5	78.75	105
SAP (0.2% & 0.3% b_{wob})		1.08	1.62			
Water	156	156	156	156	156	156
Additional water for SAP		12.5 g/g of SAP				
W/B	0.3	0.3	0.3	0.3	0.3	0.3

Legend
RHA = Rice Husk Ash
CHPC = Control HPC
SHPC = SAP internally cured HPC
PHPC = Pumice cured HPC

2.2.3 Compressive Strength and Density

HPC samples were made and cured in compliance with BS EN standards (BS EN 12350-1 & 5 2000; 12390-1 & 2 2000; 12390-3 2002) for compressive strength. The compressive strength tests were performed on 108 samples using 2000kN loading capacity ELE Compressive Strength Testing Machine at 0.5 N/mm^2 loading rate. The concrete specimens were also weighted and dimension measured immediately after de-moulding and at the respective curing ages (7, 14, 28, 56 & 90 days) after removal from curing tank for determination of the de-moulded density and final density of the hardened HPCs.

3 Results and Discussion

3.1 Physical and Chemical Properties

Result of the XRF analysis conducted on the PC and RHA powders are as presented in Table 2. The result reveal RHA is majorly SiO_2 (94%) with a silica-sesquioxide (S-S) ratio (SR) of 49.3 and aluminium sesquioxide ratio (AR) of 1.1 implying a very strong and reactive Class F Pozzolan in conformance to ASTM C618. The sum of silica, alumina and ferric oxides (SiO_2+Al_2O_3+Fe_2O_3) for the RHA (96%) is above the 70% specified for the Class of Pozzolan in ASTM C618. PC on the other hand is majorly calcium oxide (CaO - 64%). This is in agreement with oxides composition for CEM II Portland cement found in literature (Neville 2012; Mehta and Monteiro 2014).

Table 3 reveals the fine aggregate sample has coefficient of uniformity (Cu) of 2.39, coefficient of curvature (Cc) of 0.94 and fineness modulus (FM) of 2.87 implying a medium sand of Shetty (2004) classification while the granite stone and the pumice (i.e. coarse aggregates) used for the study are both a uniformly graded stones. The details of physical properties as shown in Table 3 affirms that the fine and coarse aggregates are suitable for HPC production.

Table 2. XRF result for RHA and PC

Oxides (%)	RHA	PC
SiO_2	93.6	21.5
Al_2O_3	1	5.2
Fe_2O_3	0.9	1.2
CaO	1.3	64
MgO	1.2	2.9
SO_3	0.1	4.5
Na_2O	1.7	0.6
K_2O	0.2	0.1
Minor Oxides	0	0
LOI	0	0
$SiO_2+Al_2O_3+Fe_2O_3$	95.5	27.9
SR	49.26	3.36
AR	1.11	4.33
Total	100	100

Table 3. Physical properties of materials

Item	Sand	Granite	Pumice	RHA	PC
D10	360	10000	10000		
D30	540	11000	11000		
D60	860	13000	13000		
Cu	2.39	1.30	1.30		
Cc	0.94	0.93	0.93		
FM	2.87				
SG	2.65	2.70	1.90	2.30	3.40

3.2 Compressive Strength of HPCs

Table 4 and Fig. 1 show the compressive strength of various HPCs containing equivalent quantity of CEM II 42.5 N and 10% RHA. The control HPC mixture without an IC-agent is referred to as CHPC, the mixtures with SAP incorporated are SHPCs while the HPC mixture containing varying contents of pre-soaked pumice is the PHPCs. All the HPCs show strength increases as curing age increases; the CHPC had the best early-age strength performance followed by $SHPC_1$ while the $PHPC_3$ had the lowest 7-day strength (38% below the CHPC). Comparable strength development was noticed at the 14-day of curing. $PHPC_{1\&2}$ had same strength value (48 N/mm^2) while the strength of $PHPC_3$ is 38 N/mm^2. The $SHPC_1$ strength of 55 N/mm^2 is the closest to that of the control (CHPC, 57 N/mm^2) and is 20% above that of $SHPC_2$ (47 N/mm^2) at the 14th day.

Table 4. Compressive strength of HPCs and CS_{28} factor

Specimen/age (Days)	Compressive strength (N/mm^2)						$f_{cu(28)}$ factor					
	7	14	28	56	90	120	7	14	28	56	90	120
CHPC	54.00	57.00	59.83	62.67	65.33	71.00	0.90	0.95	1.00	1.05	1.09	1.19
$PHPC_1$	46.20	48.22	56.80	62.33	65.63	68.20	0.77	0.81	0.95	1.04	1.10	1.14
$PHPC_2$	45.10	48.03	56.47	62.70	68.57	71.50	0.75	0.80	0.94	1.05	1.15	1.19
$PHPC_3$	33.37	36.30	46.20	51.80	54.73	57.47	0.56	0.61	0.77	0.87	0.91	0.96
$SHPC_1$	51.33	55.00	57.92	60.00	65.42	69.58	0.86	0.92	0.97	1.00	1.09	1.16
$SHPC_2$	42.50	42.92	46.67	49.17	51.67	57.08	0.71	0.72	0.78	0.82	0.86	0.95

Legend:
CHPC – Control HPC mixture
$PHPC_1$ – 5 % Saturated Pumice HPC
$PHPC_2$ – 7.5% b_{woca} Saturated Pumice HPC
$PHPC_3$ – 10% b_{woca} Saturated Pumice HPC
$SHPC_1$ – 0.2%n b_{wob} SAP cured HPC
$SHPC_2$ – 0.3% b_{wob} SAP cured HPC
$f_{cu(28)}$ = the compressive strength value at respective curing ages as compared to the 28th day strength.

The 28-day strength for both $PHPC_1$ and $PHPC_2$ maintains similar values (56.8 N/mm^2 and 56.5 N/mm^2 respectively) while $PHPC_3$ still had the lowest strength (46.2 N/mm^2) amongst the PHPCs. $SHPC_1$ maintained a close strength value (58 N/mm^2) to the CHPC (60 N/mm^2) while $SHPC_2$ had strength of 47 N/mm^2 (22% below that of CHPC). The trend of up 28-day strength revealed $PHPC_1$&$_2$ and $SHPC_1$ had similar level of performance in strength development 95% to 97% of the control (CHPC) while $PHPC_3$ and $SHPC_2$ remains at 77 and 78% value of CHPC ($f_{cu(28)}$ Factor, Table 4).

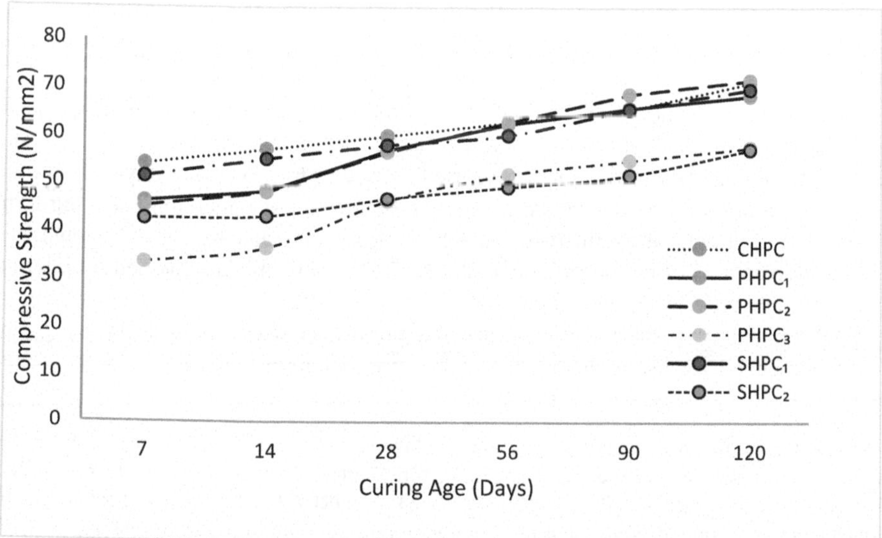

Fig. 1. Compressive strength of internally cured HPCs

Similar trend was maintained for long-term strength (56th to 120th day strength) with all the HPCs except $SHPC_2$ and $PHPC_3$ having values greater than the 28-day strength of CHPC at 120 days age. $PHPC_2$ (71.5 N/mm^2) met up with the control (CHPC, 71 N/mm^2) on the 120th day. $SHPC_1$ is the best performed internally cured HPC at the early ages and maintained good strength values at the later ages too. The $f_{cu\ (28)}$ factor (Table 3) further showed that pumice inclusion up to 7.5% b_{wocg} has no negative influence on the strength of the RHA based HPC while the optimum SAP content for the concrete without any adverse effect is 0.2% b_{wob}.

4 Conclusion and Recommendation

The current study investigated the feasibility of actualising 60 N/mm^2 RHA based HPC using locally available light weight aggregate (pumice) and SAP as IC- agent. The HPCs were made from binary binder made of CEM II/A-LL 42.5N and 10% RHA b_{wob}. The workabilities of the HPCs was improved using Hydroplast-300 (a PCE) as superplasticizer. SAP and Pre-soaked Pumice at varying percentage replacements were added as IC- agents.

Inferences drawn from the study can be summarised as follows:

1. The RHA sample used was a good Class F Pozzolan having requisite physical and chemical properties as specified in ASTM C618 and subsequently attained the target compressive strength of 60 N/mm^2.
2. The compressive strength of HPCs increased as the curing age increases for both IC-agent types and SAP contents. PHPCs and SHPCs, however, showed lower strength performance at the early ages.
3. The optimum pumice content for no adverse effect on compressive strength of the RHA based HPC is 7.5% b_{wocg} while the 0.2% b_{wob} was observed as optimum SAP content. SHPC1 (at 0.2% b_{wob} SAP content) proved as the best internally cured HPC for early age strength and showed similar long-term strength values as $PHPC_{1\&2}$.
4. Both pre-soaked pumice (up to 7.5% b_{wocg}) and SAP (0.2% b_{wob}) as IC-agent gave strength of similar values as the RHA based HPCs without the IC-agents. The comparable compressive strength recorded between 56–120 curing days' shows the long-term strength as the actual properties of the tested concrete. The long-term strength gain can be ascribed to the latter release of water by the IC agents furthering reactions of the pozzolanic material (RHA), which is augmented by internal curing delivered by pumice and SAP.
5. SAP content of 0.2% b_{wob} and saturated pumice content up to 7.5% b_{wocg} are recommended for use as IC-agent in HPC at no strength loss.

Acknowledgement. We acknowledge The followings: Covenant University Centre for Research, Innovation and Discovery (CUCRID); Mr. Guillaume Jeanson (Construction Product Manager) SNF Floerger - ZAC de Milieux, 42163 ANDREZIEUX Cedex – FRANCE; the management of Armorsil Manufacturing Incorporation, and Ewekoro Factory of Lafarge Plc Nigeria for the assistance received in materials procurement, use of facilities, softwares and time input in the analysis.

References

ACI THPC/TAC: ACI defines high performance concrete, (the Technical Activities Committee Report (Chairman - H.G. Russell)). U.S.A: American Concrete Institute (1999)

Aïtcin, P.C.: High Performance Concrete. CRC Press, Boca Raton (1998)

Aïtcin, P.C.: High Performance Concrete. Taylor & Francis e-Library, New York (2004)

Bentz, D.P., Weiss, W.J.: Internal Curing: A 2010 State-of-the-Art Review. US Department of Commerce, National Institute of Standards and Technology, Gaithersburg (2011)

Beushausen, H., Dehn, F.: High-performance concrete. In: Fulton's Concrete Technology, 9th edn., pp. 297-304. Cement and Concrete Institute, Midrand (2009)

British Standard Institution – BSI: Cement – composition, specifications and conformity criteria for common cements, BS EN 197: Part 1, London (2000)

BSI: Testing of fresh concrete, BS EN 12350: Part 1, Sampling, London (2000)

BSI: Testing of fresh concrete, BS EN 12350: Part 5, Flow Table Test, London (2000)

BSI: Testing of hardened concrete, BS EN 12390: Part 1, shape, dimension and other requirement for specimens and mould, London (2000)

BSI: Testing of hardened concrete, BS EN 12390: Part 2, making and curing specimen for strength tests, London (2000)

BSI: Testing of hardened concrete, BS EN 12390: Part 3, compressive strength test specimens, London (2002)

BSI: Eurocode 2 – Design of concrete structures – Part 1-1: General rules and rules for buildings, London (2004)

de Sensale, G.R., Viacava, I.R.: A study on blended Portland cements containing residual rice husk ash and limestone filler. Constr. Build. Mater. **166**, 873–888 (2018)

Di Bella, C., Griffa, M., Ulrich, T.J., Lura, P.: Early-age elastic properties of cement-based materials as a function of decreasing moisture content. Cem. Concr. Res. **89**, 87–96 (2016)

Di Bella, C., Villani, C., Phares, N., Hausheer, E., Weiss, J.: Chloride transport and service life in internally cured concrete. In: Structures Congress 2012, pp. 686–698 (2012)

Fapohunda, C., Akinbile, B., Shittu, A.: Structure and properties of mortar and concrete with rice husk ash as partial replacement of ordinary Portland cement–a review. Int. J. Sustain. Built Environ. **6**(2), 675–692 (2017)

Kaur, K., Singh, J., Kaur, M.: Compressive strength of rice husk ash based geopolymer: The effect of alkaline activator. Constr. Build. Mater. **169**, 188–192 (2018)

Kovler, K., Jensen, O.M.: Novel techniques for concrete curing. Concr. Int. **27**(09), 39–42 (2005)

Mehta, P.K., Monteiro, J.M.: Concrete Microstructure Properties and Materials, 4th edn. McGraw-Hill Education, New York (2014)

Nduka, D., Ameh, J., Joshua, O., Ojelabi, R.: Awareness and benefits of self-curing concrete in construction projects: builders and civil engineers perceptions. Buildings **8**(8), 109 (2018)

Neville, A.M.: Properties of Concrete, 5th edn. Pearson Educational Limited, Harlow (2012)

Ogunbayo, B.F., Ajao, A.M., Ogundipe, K.E., Joshua, O., Durotoye, T.O., Bamigboye, G.O.: Study of aggregate dormancy and its effects on the properties of aggregates and concrete. Cogent Eng. **5**(1) (2018)

Olawuyi, B.J.: Mechanical behaviour of high-performance concrete with superabsorbent polymers (SAP). Ph.D. thesis, Department of Civil Engineering, Stellenbosch University, Stellenbosch, South Africa (2016)

Olawuyi, B.J., Boshoff, W.P.: Compressive strength of high-performance concrete with absorption capacity of Super-Absorbing-Polymers (SAP). In: Proceedings of the Research and Application in Structural Engineering, Mechanics and Computation, Cape Town, South Africa, pp. 2–4 (2013)

Olawuyi, B.J., Boshoff, W.P.: Influence of SAP content and curing age on air void distribution of high-performance concrete using 3D volume analysis. Constr. Build. Mater. **135**, 580–589 (2017)

Olawuyi, B.J., Boshoff, W.P.: Influence of superabsorbent polymer on the splitting tensile strength and fracture energy of high-performance concrete. In: MATEC Web of Conferences, vol. 199, p. 11004. EDP Sciences (2018)

Orosz, K.: Early age autogenous deformation and cracking of cementitious materials—implications on strengthening of concrete. Ph.D. dissertation, Luleå Tekniska Universitet, Luleå, Sweden (2017)

Persson, B.: Self-desiccation and its importance in concrete technology. Mater. Struct. **30**(5), 293–305 (1997)

Savva, P., Nicolaides, D., Petrou, M.F.: Internal curing for mitigating high temperature concreting effects. Constr. Build. Mater. **179**, 598–604 (2018)

Shetty, M.S.: Concrete Technology - Theory and Practice. S. Chand and Company Limited, New Delhi (2004)

Tu, W., Zhu, Y., Fang, G., Wang, X.M.: Internal curing of alkali-activated fly-ash pastes using superabsorbent polymers. Cem. Concr. Res. **116**, 179–190 (2019)

Wu, L., Farzadnia, N., Shi, C., Zhang, Z., Wang, H.: Autogenous shrinkage of high-performance concrete: a review. Constr. Build. Mater. **149**, 62–75 (2017)

Long-Term Parameters of New Cement Composites

Andina Sprince[✉], Leonids Pakrastins, and Rihards Gailitis

Department of Structural Engineering, Riga Technical University, ST Kipsalas 6A, Riga 1046, Latvia
andina.sprince@rtu.lv

Abstract. Since the beginning of the 20th century, scientists and cement composite technologists are working on developing various types of new structural multi-component cement composites. Several obstacles prevent more widespread use of these newly developed cement composites in construction. One of the main problems is insufficient information about the long-term properties, which are essential in ensuring the safe and long exploitation of structures. The purpose of this research is to determine the long-term properties of several new cement composites: ultra-high strength cement composite with PVA fiber "cocktail" (2% by volume), with micro silica and nano silica additive; ultra-high strength cement composite with 1% montmorillonite mineral nano-size particles; reference composition. Test specimens were prepared and subjected to constant compressive load in permanent room temperature and level of moisture. There were properties such as compressive strength, modulus of elasticity, shrinkage as well as uniaxial creep deformations investigated in the laboratory. Afterward parameters of long-term properties were determined. The obtained results showed that after approximately 90 days of loading the creep coefficient values of new cement composites were 0,5–3; specific creep values were 30–55 microstrain/MPa; creep modulus was 2–90 GPa. The experimental study proves that new elaborated mixes can be successfully used in the production of concrete, thus potentially decreasing the use of cement, which would lead to the reduction of carbon dioxide released into the atmosphere.

Keywords: Creep coefficient · Specific creep · Creep modulus · New cement composite

1 Introduction

Nowadays the construction material industry is developing rapidly, with an ever-increasing tendency towards the use of new materials. Scientists and cement composite technologists are working on developing various types of new structural multi-component cement composites. The findings include mixtures with reduced quantity of cement and smaller aggregate dimensions, different types of fibers being used as disperse reinforcement, introduction of various chemical additives and a lowered water-cement ratio, as well as substituting some of the cement with recycled materials, etc. (Fathifazl et al. 2011; Fehling et al. 2014; Girskas et al. 2016; Kazanskaya and Smirnova 2018; Smirnova 2018; Šinka et al. 2018; etc.). The newly developed cement

W. P. Boshoff et al. (Eds.): SAP 2019, RILEM Bookseries 24, pp. 85–94, 2020.
https://doi.org/10.1007/978-3-030-33342-3_10

composite matrices in general have improved physical properties, e.g., their microstructure—cement paste accounts for the larger part of the volume, and the porosity was reduced, thus leading to cement composites with smaller water absorption and better frost resistance properties (Grinfeld et al. 2014; Lu and Poon 2018; Prisco et al. 2009; etc.). Although the effect on these properties has been conscious there are still several obstacles that prevent more widespread use of these newly developed cement composites in construction. One of the main problems is insufficient information about the long-term properties, which are essential to ensure safe and long exploitation of structures.

2 Materials and Methods

One of the goals of the experiments was to find out whether the new cement composites can be competitive and whether their long-term properties are equivalent to high strength cement composites (HSCC) (the compressive strength 40–120 MPa; w/c ≤ 0.4) (Neville 2002; Naaman and Reinhardt 2003) and ultra high strength cement composites (UHSCC) (the compressive strength 120–400 MPa; w/c ≤ 0.2) (Naaman and Reinhardt 2003; Gilbert and Ranzi 2011).

During the research following cement composites were made and tested (see Table 1), where appeared designations of various cement composites, which were used in the graphs.

Cement composite compositions were made by the design rules of recipes. The cement composite mixtures were prepared using a double shaft laboratory mixer (BHS, 3 kW, 20–100 rpm). For determination of the compressive strength cubes of 100 × 100 × 100 mm, prisms of 40 × 40 × 160 mm and cylinders of Ø47 × 190 mm were used (see Fig. 1). The compressive strength of the specimens was determined in the compression machine "Controls", model No. C56G2, with an accuracy of ±1% and a measurement range of 0–3 kN; the loading speed was 0.8 MPa/s.

Fig. 1. Demolded specimens (RTU lab).

After becoming of a certain age, the specimens were prepared for testing. Before that, specimens were weighted, and measured. For determination of creep strain (in

Table 1. Various cement composites.

Mix designation	Description	*Compressive strength, MPa	Long-term test specimen type	Storage regime	Long-term loading regime	*Specimen age, days
HSC PVA MS	Ultra-high strength cement composite with PVA fiber "cocktail" (2% by weight/mass), + micro silica	112	Cylinder Ø47 × 190 mm, air-dry	t = 23 ± 1 °C, RH = 35 ± 3%	90 days - constant static load corresponding to 25% of the ultimate stress; 30 days - *recovery period	28
HSC PVA NS	Ultra-high strength cement composite with PVA fiber "cocktail" (2% by weight/mass) + micro silica + nano silica	110	Cylinder Ø47 × 190 mm, air-dry	t = 23 ± 1 °C, RH = 35 ± 3%	90 days - constant static load corresponding to 25% of the ultimate stress; 30 days - *recovery period	28
HSC MS REF (reference for HSC PVA NS)	Ultra-high strength cement composite with micro silica + nano silica	128	Cylinder Ø47 × 190 mm, air-dry	t = 23 ± 1 °C, RH = 35 ± 3%	50 days - constant static load corresponding to 25% of the ultimate stress;	28
HSC M	Ultra-high strength cement composite with 1% *montmorillonite mineral nano-size particles	103	Prism 40 × 40 × 160 mm, air-dry	t = 23 ± 1 °C, RH = 35 ± 3%	90 days - constant static load corresponding to 30% of the ultimate stress; 30 days - *recovery period	51

(*continued*)

Table 1. *(continued)*

Mix designation	Description	*Compressive strength, MPa	Long-term test specimen type	Storage regime	Long-term loading regime	*Specimen age, days
HSC R (reference for HSC M)	Ultra-high strength cement composite	100	Prism $40 \times 40 \times 160$ mm, air-dry	$t = 23 \pm 1$ °C, $RH = 35 \pm 3\%$	90 days - constant static load corresponding to 30% of the ultimate stress; 30 days - *recovery period	51

Notes:
*Compressive strength of cube ($100 \times 100 \times 100$ mm), 28 days old specimen.
*Montmorillonite mineral nano-size particles - powder of very fine, especially processed clay particles, applied as an additive and partly replaced cement.
*Recovery period - no load applied
*Specimen age (days) on the long-term testing start date

uniaxial compression) and shrinkage strain, the cylindrical specimens (Ø47 × 190 mm) and prismatic specimens (40 × 40 × 160 mm) were used. For all the creep and shrinkage specimens, aluminum plates (10 × 15 mm) were centrally and symmetrically glued to side surfaces in order to provide the basis for strain gauges. Six aluminum plates were glued to one cylindrical specimen, four – to the prismatic specimen.

The strains were measured using Aistov electrical strain gauges with a scale interval of 1 μm and maximum range of ±5 mm or mechanical clock gauges "ИЧ" with a scale interval of 1/100 mm and maximum range of 10 mm. The manufacturer had calibrated the measuring instruments. The strain gauges were attached in such a way that their "knives" were located on the glued plates; strain gauge base – 50 mm. Strain gauges were attached to the specimens with elastic rubber bands. Specimens with attached strain gauges were put into creep lever test stands (see Fig. 2), two specimens in each test stand. Creep lever test stand allows to use the specimens with dimensions ≤ 70 mm, what is more, characteristic to the dimensions of high and ultra-high strength cement composite structures. With these stands, it is possible to apply constant loading to the specimens and to keep it uniform over a long period. Also, it is not necessary to adjust the stress level during the experiments, the calibration curves are linear, no energy resources are consumed, and it is possible to test cement composites with the maximum aggregate dimension ≤ 5 mm, simultaneously ensuring economical use of materials. The lever arm ratio of the creep testing stand was 1:40. The accuracy of the counterweights was 1/100 kg or 0.01%. Therefore, the accuracy of creep levers is 0.01 × 40 = ±0.4 kg.

Fig. 2. Determination of uniaxial compressive creep strains with the creep lever test stands (RTU lab).

All the creep specimens were loaded with a constant static load, regularly performing strain readings. To determine correct creep behavior, similarly shaped shrinkage specimens were placed in equivalent environmental conditions and their strain changes were monitored (no load applied to the shrinkage specimens). Conclusions were made based on subtracting shrinkage strain values from the compressive creep values. The basic and drying creep components have not been determined separately. The modulus of elasticity was determined from the elastic strains that occurred at the beginning of the creep test. (Reunion Internationale des Laboratoires et Experts des Materiaux [RILEM] 1998; American Concrete Institute [ACI] 2008).

3 Results and Discussion

During the experimental study the values of strength, deformability, long-term properties of various new cement composites were found and parameters for designing of safe structures were determined, which had not been found before.

Figure 3. shows the cubic compressive strength of various high and ultra-high cement composites (testing 100 × 100 × 100 mm cubes). Values range from 80 to 104 MPa at the beginning of the tests and from 107 to 128 MPa at the end of tests. The largest cubic compressive strength was determined for cement composite without unconventional additives.

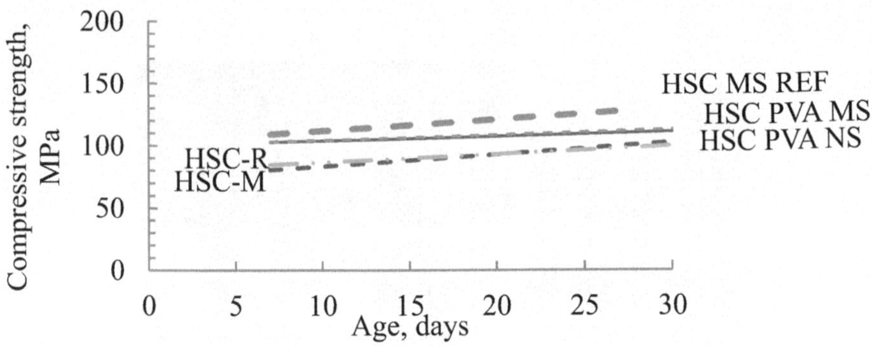

Fig. 3. Compressive strength of different kinds of high and ultra-high strength cement composites.

Creep coefficient (see Fig. 4) shows the proportion of creep strain and elastic strain. Experimental data for various compositions shows that creep coefficient of high and ultra-high strength cement composites was the same as for normal strength cement composites and normal strength cement composites with unconventional additives (Sprince et al. 2018).

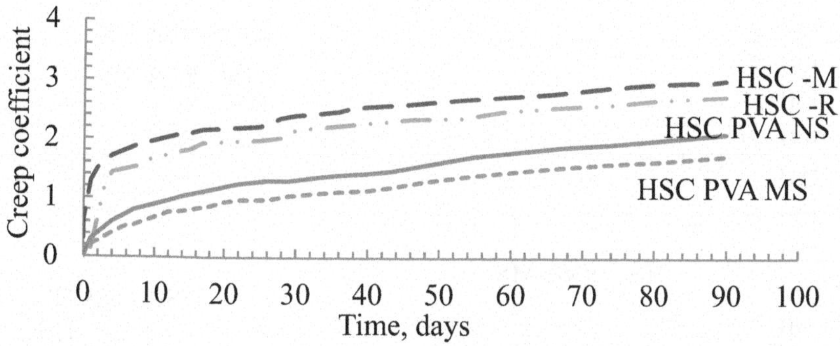

Fig. 4. Creep coefficient in compression of different kinds of high and ultra-high strength cement composites.

The values are within margins of 0.5 to 3, which does not comply with data from literary sources which predict a significant decrease of this coefficient (European Committee for Standardization [CEN] 2004). The smallest creep coefficient values are for ultra-high strength compositions with PVA fibers.

Specific creep (see Fig. 5) is the most objective parameter of long-term loading as it excludes stress effect on long-term strains. Specific creep values are from 30–55 microstrain/MPa. The smallest specific creep values are for ultra-high strength cement compositions.

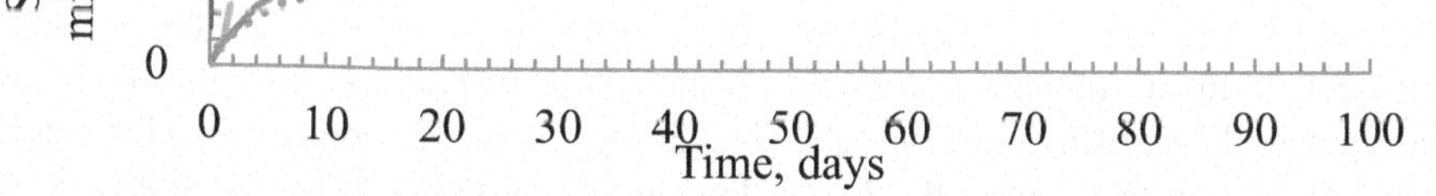

Fig. 5. Specific creep in compression of different kinds of high and ultra-high strength cement composites.

Creep modulus (see Fig. 6) is the proportion of applied stress and creeps strain. This long-term parameter can be used for determining the displacement of long-term loaded structures after a long period. As can be seen, creep modulus tends to decrease within the time which can be explained by the increase in creep strain and total strain. The lower the creep modulus, the less creep in the material.

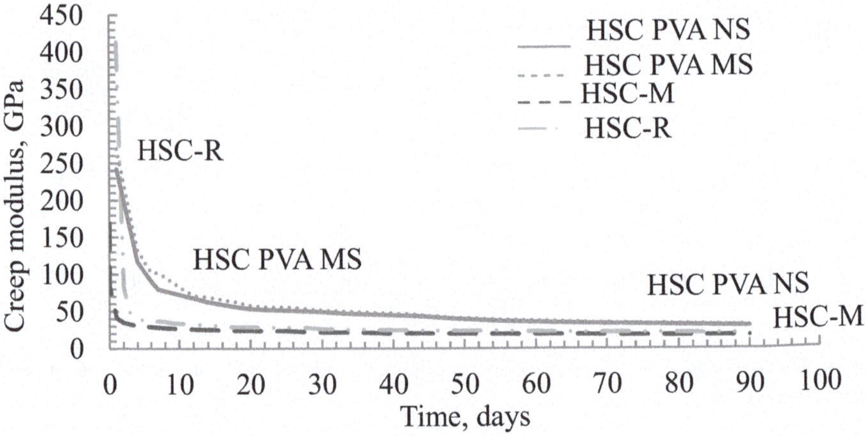

Fig. 6. Creep modulus in compression of different kind of high and ultra-high strength cement composites.

Deformation amount rapidly increases in the first week, then the deformation speed in time decreases and approximately after 60 days significant changes in deformation cannot be observed anymore.

4 Conclusion

Experimentally obtained results for various high and ultra-high strength cement composite tests confirm the hypothesis by various leading researchers (Neville, Brooks, Bazant, Gardner, Lockman, Fanourakis, Gilbert, Ranzi, Baweja, Kim, Wittmann, Rusch and others). The obtained results indicate: the higher the density of cement composite, the stronger it is and strain is lesser (it can be explained by the unfilled gel space relative amount in the hardened cement paste. Gel space ratio is closely related to the amount of cement and water/cement ratio, which is normally decreased to obtain an increase in cement composite strength. Similarly, the hypothesis that, the dryer and more mature the cement composite, the smaller the strain, was confirmed. That can be explained with cement composite chemical processes during its drying. Similarly, it has been experimentally proved that cement composite final strength increases with cement composite age and it is not substantially affected by its subjection to loading. It has been assumed that the difference in specimen's age does not significantly influence the test results.

Similarly, it was experimentally determined that montmorillonite added to cement composites neither significantly improves nor decreases the mechanical and deformability properties.

Nano silica mineral additive does not have a significant effect on parameters of cement composites strength, deformations, long-term properties. The difference in results between cement composites with and without nano silica is within 2–7% margins.

The application of PVA fibers in high and ultra-high strength cement composites does not provide improvements in strength and long-term properties. That can be explained by the fact that distribution of fibers is irregular, random in character, as well as the fact that in mixing process more air is attracted and gaps were formed, and it is technologically harder to work such fibrous cement composite in cement composite molds. Therefore, the cement composite composition is not uniform, it is more porous.

The obtained results show that after approximately 90 days of loading creep coefficients values of new high and ultra-high strength cement composites are the same as for normal strength cement composites 0,5–3; specific creep values are 30–55 microstrain/MPa; creep modulus is 2–90 GPa.

The use of new additives would give an indirect positive effect on the global environment, as, by increasing the use of new unconventional cement compositions and by reducing the dimensions of the cross-sections of structures, the total amount of cement consumption would decrease. By substituting part of the cement with recycled mineral fillers, the use of non-renewable resources and non-biodegradable waste shall also be reduced.

Acknowledgments. "This research is funded by the Latvian Council of Science, project "Long-term properties of innovative cement composites in various stress-strain conditions", project No. lzp-2018/2-0249".

References

ACI 209.2R-08: Guide for Modeling and Calculating Shrinkage and Creep in Hardened Concrete, ACI Committee 209, USA (2008)

BS EN 1992-1-1:2004: Eurocode 2: Design of concrete structures - Part 1-1: General rules and rules for buildings, CEN (2004)

Grinfeld, G.I., Gorshkov, A.S., Vatin, N.I.: Tests result in strength and thermophysical properties of aerated concrete block wall samples with the use of polyurethane adhesive. Adv. Mater. Res. **941–944**, 786–799 (2014)

Fathifazl, G., Razaqpur, A.G., Isgor, B., Abbas, A., Foumier, B., Somon, F.: Creep and drying shrinkage characteristic of concrete produced with coarse recycled concrete aggregate. Cem. Concr. Compos. **74**, 1026–1037 (2011)

Fehling, E., Schmidt, M., Walraven, J., Leutbecher, T., Frolich, S.: Ultra-High Performance Concrete UHPC. Ernst & Sohn, Berlin (2014)

Gilbert, R.I., Ranzi, G.: Time-Dependent Behaviour of Concrete Structures, pp. 3, 5, 9–11, 25–30, 26, 27, 33. Spon Press, London and New York (2011)

Girskas, G., Skripkiunas, G., Šahmenko, G., Korjakins, A.: Durability of concrete containing synthetic zeolite from aluminum fluoride production waste as a supplementary cementitious material. Constr. Build. Mater. **117**, 99–106 (2016)

Kazanskaya, L.F., Smirnova, O.M.: Supersulphated cements with technogenic raw materials. Int. J. Civ. Eng. Technol. **9**(11), 3006–3012 (2018)

Lu, J., Poon, C.S.: Improvement of early age properties for glass-cement mortar by adding nano-silica. Cem. Concr. Compos. **89**, 18–30 (2018)

Naaman, A.E., Reinhardt, H.W.: High-performance fiber reinforced cement composites. In: Proceedings PRO6, France. RILEM Publications S.A.R.L. (2003)

Neville, A.M.: Creep of concrete and behaviour of structures. Concrete International no. 5 (2002)

Prisco, M., Plizzari, G., Vandewalle, L.: Fibre reinforced concrete: new design perspectives. Mater. Struct. **42**, 1261–1281 (2009)
Rilem: TC 107 - CSP: creep and shrinkage prediction models: principles of their formation. Measurement of time-dependent strains of concrete. Mater. Struct. **31**, 507–512 (1998)
Smirnova, O.M.: Evaluation of superplasticizer effect in mineral disperse systems based on quarry dust. Int. J. Civ. Eng. Technol. **9**(8), 1733–1740 (2018)
Sprince, A., Pakrastins, L., Radina, L.: Experimental investigation of new cement composites long-term properties. In: SynerCrete 2018: Interdisciplinary Approaches for Cement-based Materials and Structural Concrete: Synergizing Expertise and Bridging Scales of Space and Time, vol. 1 & 2, pp. 189–194 (2018)
Šinka, M., Van Den Heede, P., De Belie, N., Bajāre, D., Šahmenko, G., Korjakins, A.: Comparative life cycle assessment of magnesium binders as an alternative for hemp concrete. Resour. Conserv. Recycl. **133**, 288–299 (2018)

Shrinkage and Associated Cracking

The Effect of SAP on Volumetric Changes and Microstructural Alterations in PC-GGBS Matrices

Fernando C. R. Almeida[1,2(✉)], Rohollah Rostami[3], and Agnieszka J. Klemm[3]

[1] Department of Civil Engineering, Federal University of São Carlos, São Carlos, SP 13565-905, Brazil
almeida@ufscar.br
[2] Laboratory of Construction Materials, Institute for Technological Research - IPT, São Paulo, SP 05508-901, Brazil
[3] School of Computing, Engineering and Built Environment, Glasgow Caledonian University, Glasgow G4 0BA, UK
{rohollah.rostami,a.klemm}@gcu.ac.uk

Abstract. Despite their growing popularity in construction, cement-based materials containing ground granulated blast-furnace slag (GGBS) may suffer from various deteriorative actions, including development of early cracks caused by shrinkage processes. Moreover, the degree of GGBS reaction (at later stages) can be limited by lack of space already filled by early products of Portland cement (PC) hydration. In attempt to reduce these negative effects, Superabsorbent polymers (SAP) can be used as internal curing agents and facilitate hydration processes. This paper intends to evaluate and correlate shrinkage behaviour and microstructural features of SAP-GGBS mortars during the first 90 days. Six sets of mortars (with 0% and 50% GGBS) modified by two SAPs were analysed. Autogenous, plastic and drying shrinkages were tested, and microstructural characteristics were analysed by the MIP technique. The experimental results showed significant effect of SAPs on shrinkage reduction in GGBS matrices. The paper argues that deposition of later GGBS hydration products in pores below 20 nm, triggered by SAP water, leads to a relative expansion of SAP-PC-GGBS mortars. In this context, the role of SAPs in GGBS-PC matrices is twofold: firstly, a provision of additional water for hydration and, secondly, a provision of the required space for later products formation.

Keywords: Superabsorbent polymers · GGBS · Hydration · Shrinkage · Porosity

1 Introduction

Superabsorbent polymers (SAP) are widely recognised as a new type of admixture, which can provide additional water for internal curing of cementitious materials during setting and hardening processes (Mechtcherine and Reinhardt 2012;

W. P. Boshoff et al. (Eds.): SAP 2019, RILEM Bookseries 24, pp. 97–105, 2020.
https://doi.org/10.1007/978-3-030-33342-3_11

Wyrzykowski et al. 2018). The performance of SAP-modified concrete depends not only on its intrinsic characteristics (shape, size, crosslinking density, chemical structure) but also on the external stimuli surrounding the polymer. In this context, both absorption capacity and sorption kinetics are highly influenced by pH and ionic concentration of a pore solution, given by the type of binder used (Almeida and Klemm 2018a; Snoeck et al. 2015).

Addition of ground granulated blast-furnace slag (GGBS) can affect the rate and extend of cementitious reactions due to its filler effect (Scrivener et al. 2015a, b). GGBS reaction is slower than the reaction of clinker phases from Portland cement (PC), and depends on the chemical composition, fineness, glass content as well as on the composition of interacting solution. Although there is still no consensus on the mechanisms controlling PC-GGBS hydration kinetics, especially beyond one day, it is becoming clear that the amount of space available is a critical factor (Berodier and Scrivener 2015; Scrivener et al. 2015a, b).

Thus, in addition to the prolonged water provision, a collapsed SAP has the potential to provide more room for the deposition of later GGBS hydration products (Almeida and Klemm 2018b). Therefore, this paper aims to evaluate the effect of SAP on volumetric changes and microstructural alterations in PC-GGBS mortars, by correlating shrinkage results and pore size distributions up to 90 days.

2 Experimental Programme

Six mortar compositions have been designed for the experimental programme, with different binders (Portland cement – PC, CEM I 52.5N; and 50% PC, 50% GGBS) and two SAPs. Chemical and physical characteristics of PC and GGBS are listed in Table 1.

Table 1. Chemical and physical characteristics of PC and GGBS.

	SiO_2 (%)	Al_2O_3 (%)	Fe_2O_3 (%)	CaO (%)	MgO (%)	SO_3 (%)	Na_2O (%)	K_2O (%)	LOI (%)	Fineness (m^2/kg)
PC	20.1	4.9	2.7	62.4	2.2	3.2	0.3	0.6	2.8	410
GGBS	34.5	13.1	0.2	38.5	9.7	0.4	0.2	0.6	0.6	390

Two types of SAPs (A and B), provided by BASF Construction Chemicals GmbH, have been used in the proportion of 0.25% by mass of binder (Table 2). Full characterization of SAPs used in this study, including kinetics sorption analysis, can be found in Almeida and Klemm (2018a).

Mortars have been prepared in the proportion of 1:2 (binder:sand) and with total water/binder ratio (w/b) of 0.5. The second stage of research was focused on drying shrinkage analysis. Mortars used in this part had different w/b as shown in Table 3. Fine sand used as fine aggregate contained at least 90% of particles smaller than 0.425 mm (BS EN 13139, 2013). Table 3 shows nomenclature of mortar samples used in this experimental programme.

Table 2. Characterization of SAPs.

Properties	SAP A	SAP B
Type of polymer	Copolymer of acrylamide and acrylic acid	Modified polyacrylamide
Particles size (μm)	30–140	30–140
WAC[a] $_{\text{deionized water}}$ (by tea-bag test)	340 g/g	290 g/g
Time to get max WAC[a] (by tea-bag test)	2880 min	30 min
WAC[a] $_{\text{mixing water}}$ (by flow table)	28 g/g	24 g/g

[a]WAC: water absorption capacity

Table 3. Mixture composition of mortar samples.

Samples name	SAP composition	Binder composition	Binder:Sand proportion	Total w/b ratio[a]	DS total w/b ratio[b]
R0	–	100% PC; 0% GGBS	1:2	0.50	0.52
A0	SAP A	100% PC; 0% GGBS	1:2	0.50	0.58
B0	SAP B	100% PC; 0% GGBS	1:2	0.50	0.58
R50	–	50% PC; 50% GGBS	1:2	0.50	0.52
A50	SAP A	50% PC; 50% GGBS	1:2	0.50	0.58
B50	SAP B	50% PC; 50% GGBS	1:2	0.50	0.58

[a]Total w/b ratio used in samples for tests of plastic shrinkage, autogenous shrinkage and MIP; [b]Total w/b ratio used in samples for tests of drying shrinkage.

Plastic shrinkage (PS) was evaluated by measuring cracking widths of restrained mortars using a steel form insert ASTM C1579-13 (2013). Three restrained prismatic specimens of 50 × 50 × 240 mm^3 were produced for each mortar mix, and stored in a ventilation chamber (T = 40 ± 3 °C and RH = 15 ± 10%) for 24 h (Almeida et al. 2018). After demoulding, cracks on the free surface of specimens were analysed by optical microscopy and measured by the ImageJ software.

Autogenous shrinkage (AS) was tested in triplicate by the sealed corrugated tubes method ASTM C1698-09 (2009) from the final setting time to 90 days. Setting times were determined according to BS EN 196-3 and the results are shown in Almeida and Klemm (2016).

Drying shrinkage (DS) testing was performed on prismatic specimens (75 × 75 × 280 mm^3) according to BS ISO 1920-8 (2009). Samples were demoulded after 24 h and left drying in laboratory environment (T = 21 ± 2 °C and RH = 40 ± 5%) for 90 days. Length was measured daily in the first 28 days, and weekly in the following two months.

Microstructural features were evaluated after unsealed curing in climate chamber (T = 21 ± 2 °C and RH = 40 ± 5%) at 7, 28 and 90 days. Microstructural characteristics (total porosity (%) and pore size distribution (nm)) were analysed by the Mercury Intrusion Porosimetry (MIP) technique.

3 Results and Discussions

Figure 1 shows the representative micrographs of cracked surfaces of each PC-GGBS-SAP mortar subjected to plastic shrinkage (PS) testing. All reference samples (without SAP) displayed cracks visible to the naked eye (greater than 0.1 mm width), while in SAPs mortars, such cracks appeared only for B50 specimen. Due to higher absorption capacity of SAP A (Table 2) more water is available for early cementitious reactions leading to smaller cracking susceptibility (Almeida et al. 2018). However, when compared to the reference specimens both SAPs significantly reduced widths of plastic shrinkage cracking up to 90%.

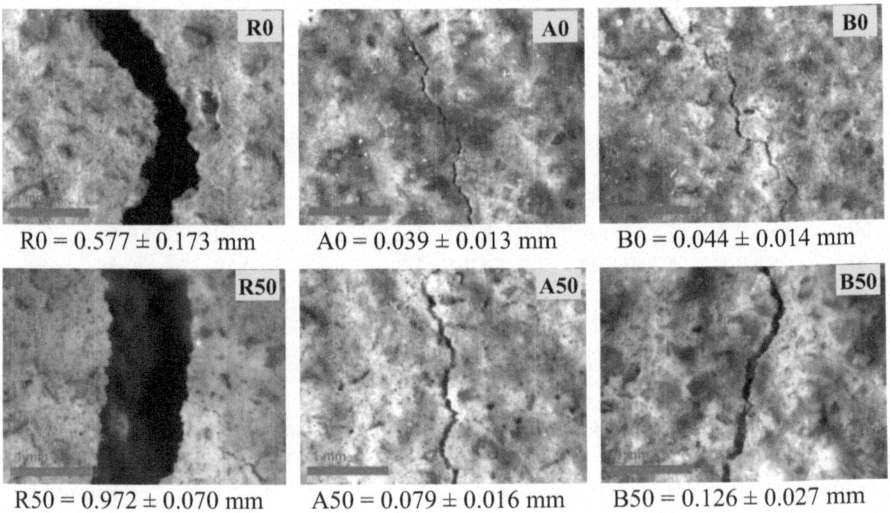

Fig. 1. Average of plastic shrinkage cracking widths formed in each mortar.

Figure 2 shows results of autogenous shrinkage (AS) measurements. Overall, SAPs reduced AS in all mortars, with and without GGBS, when compared to the reference samples. Again, SAP A seemed to be more efficient in reduction of autogenous shrinkage owing to its higher and slower WAC, i.e., more water is available in the system that is gradually and slowly released. As a result, mortars with SAP A had the lowest values of shrinkage in comparison with the other samples at 90 days. Moreover, SAP A had a different behaviour than SAP B, showing an initial swelling in the first week and competing with the early-age shrinkage. As a result, SAP A either counteracts (A0) or decreases (A50) shrinkage. Some studies using pure PC systems showed filling of SAP pores by calcium hydroxide during the first week (Justs et al. 2015).

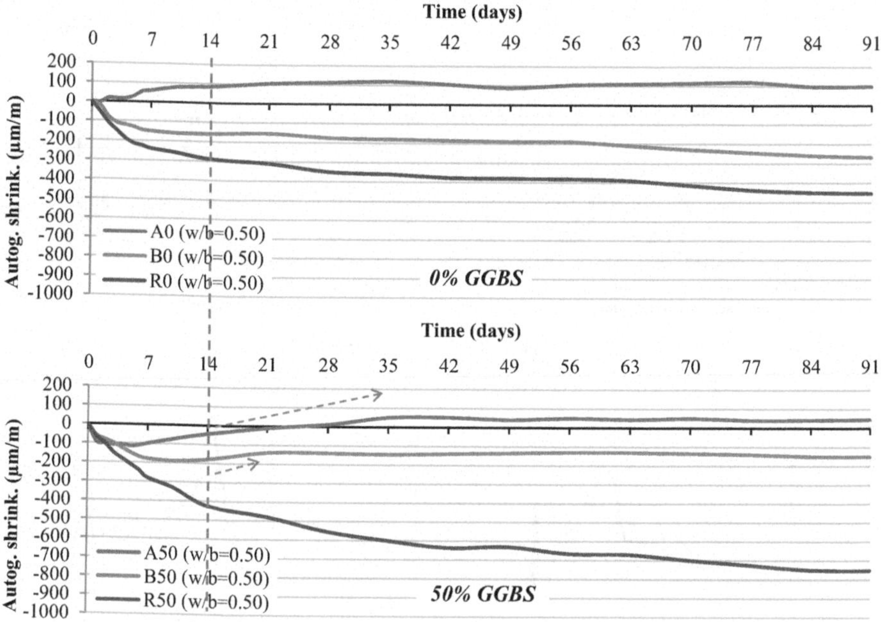

Fig. 2. Autogenous shrinkage development of PC-GGBS mortars with SAPs.

GGBS addition has an impact on performance of SAP mortars over time. The efficiency of SAP in reducing AS is improved by incorporation of slag, since GGBS increases autogenous shrinkage (Lee et al. 2006). Also, plain SAP-mortars slightly shrank (B0) or after initial "dormant period" swelled a little (A0). After that, no volumetric changes took place (A0) or shrinkage proceeded at the same pace as in the reference sample (B0).

However, in PC-GGBS-SAP mortars a slight swelling was noticed after a maximum shrinkage around −200 and −100 mm/m for B50 and A50, respectively. Other studies have also confirmed an expansion in systems containing shrinkage reducing admixtures (Snoeck et al. 2015). This effect of "relative" expansion took place after the second week and was more pronounced for higher GGBS contents (Almeida and Klemm 2018b). This is related to the beginning of GGBS hydration facilitated by the presence of SAP as water supplier in smaller pores.

Figure 3 shows the progressive reduction of pores with diameter under 20 nm, particularly for samples A50 and B50. GGBS has reacted with water and the formation of hydrated products, deposited in that range of pores, led to the "relative" expansion of SAP mortars. When water was consumed and SAP collapsed, no significant external volume changes were recorded, and AS curves had a tendency to flatten out (around the 42nd day).

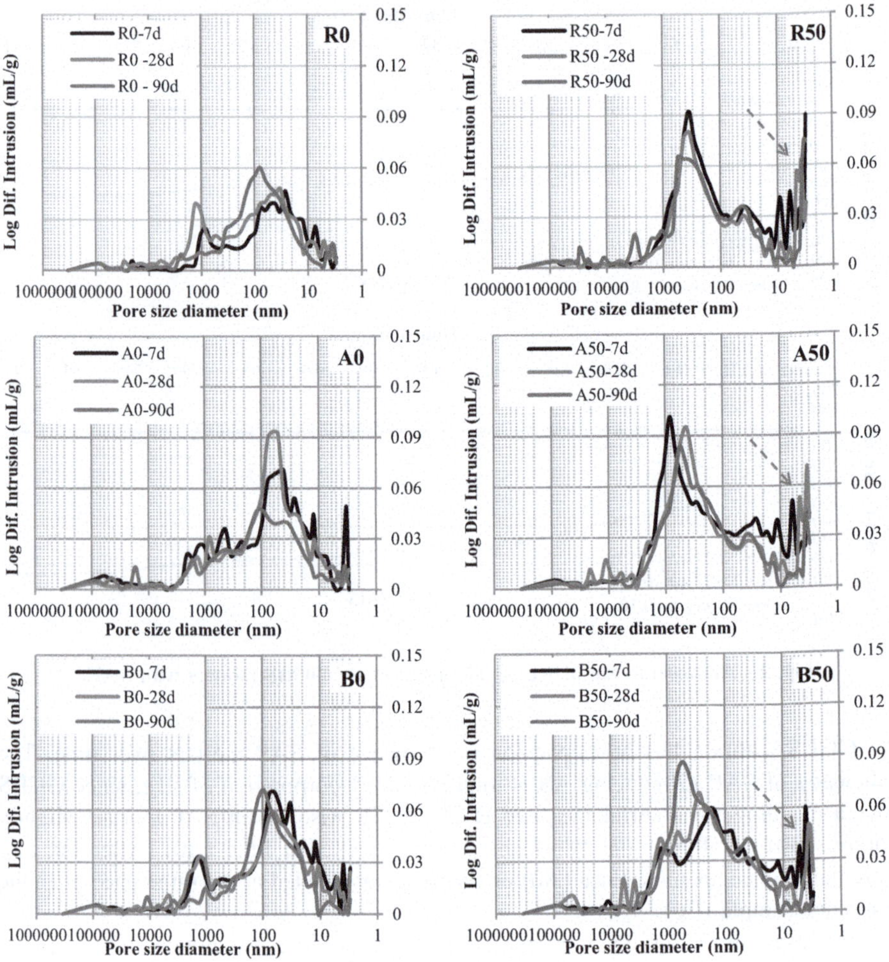

Fig. 3. Pore size distribution (by MIP) of PC-GGBS mortars at different ages (w/b = 0.50 for all samples).

Desorption of SAP is mainly controlled by capillary suction in early-ages. After that, there may be a detachment between SAP and pore wall (from the cement matrix) and/or SAP can be built into the wall (and likely covered with hydration products). Thus, SAP moisture can be released to the paste under humidity gradient (in form of vapour) or can be itself a water supplier for continuous curing (Wang et al. 2016). The water stored by polymer may move through smaller pores since they provide more surface area for water to adhere to. Indeed, there is a preference of GGBS products in precipitating into smaller pores in non-saturated-water condition (Choi et al. 2017).

Moreover, GGBS reaction is directly related to the saturation of portlandite (CH) from PC hydration. According to Lothenbach and Winnefeld (2006), most of CH is formed during the first 1000 h hydration (or 42 days). This is approximately when mortars with slag exhibited maximum relative expansion. The maximum amount of CH means max activation for GGBS hydration, prompting the production of further C-S-H. The more CH, the higher is pH in the vicinity of GGBS grains, and hence the higher is reactivity of slag. These later hydrated products fill extra spaces which were not filled by the cement hydrates due to the lack of clinker (Scrivener et al. 2015).

As smaller pores are being filled with later hydration products, these products start to exert pressure upon pore walls. Due to potential lack of room for C-S-H precipitation, there is a "space competition" between hydrated products and hardened microstructure, and it happens in pores below 20 nm. This outcome seen in Fig. 3 is in line with other authors (Berodier and Scrivener 2015; Scrivener et al. 2015). Consequently, this stress leads to the "relative" expansion of the hardened bulk volume in later ages, particularly for SAP-PC-GGBS matrices (A50 and B50) after the second week (Almeida and Klemm 2018b).

Figure 4 shows a reduction of drying shrinkage (DS) rate in SAP samples as a result of swelling/contraction in the second week. It was more pronounced for mortars with high WAC - SAP A.

Moreover, even with higher w/c ratios for SAP samples, similar or slightly reduced DS was observed in comparison to the reference sample, confirming the positive effect of SAP in mitigating shrinkage.

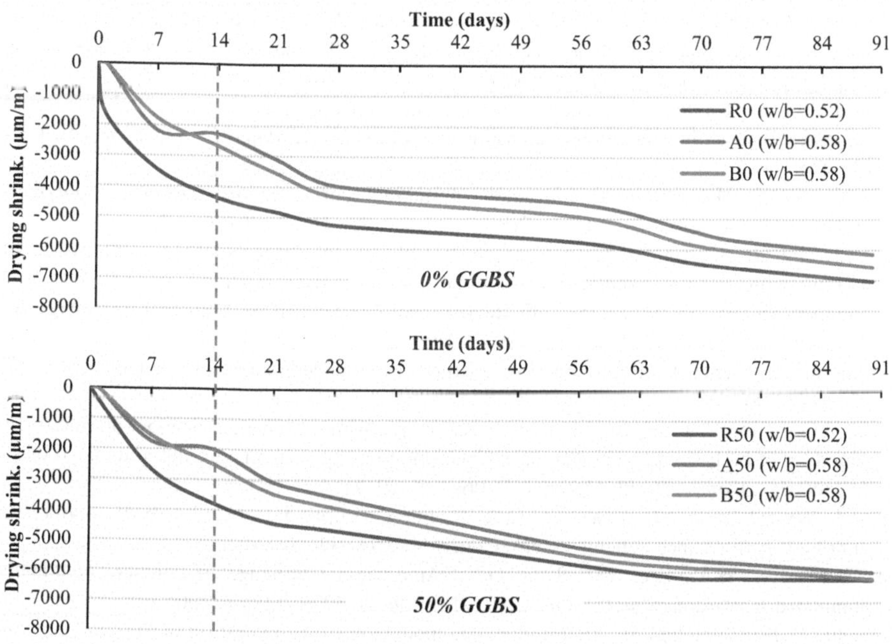

Fig. 4. Drying shrinkage development of PC-GGBS mortars with SAPs.

4 Conclusions

SAP significantly reduces plastic, autogenous and drying shrinkages. Moreover, SAP can help GGBS hydration, contributing to water supply for further reactions in a prolonged time (up to 90 days). Experimental results indicated that the later GGBS reactions can be facilitated not only by SAP water (that is adhered to smaller pores, with high surface area) but also by the potential presence of portlandite from PC hydration (max CH saturation is about day 42). Thus, the later hydration products seem to start to form after the second week into the smaller pores (under than 20 nm) resulting in an increment of internal compressive stress. This, in turn, may lead to a slight "relative" expansion of the bulk volume of SAP-PC-GGBS systems. This effect is more pronounced for SAP A due to its higher capacity to store water and slower desorption kinetics.

Acknowledgements. The authors acknowledge CNPq/Brazil and FAPEMIG for the financial support (CNPq grant numbers: 204097/2014-2 and 409685/2017-8), BASF for SAPs supply, and Prof Valeria Corinaldesi (UNIVPM-Italy) for plastic shrinkage testing support.

References

Almeida, F.C.R., Klemm, A.J.: Effect of superabsorbent polymers (SAP) on fresh state mortars with ground granulated blast-furnace slag (GGBS). In: 5th International Conference on the Durability of Concrete Structures, Shenzhen, pp. 1–7(2016)

Almeida, F.C.R., Klemm, A.J.: Effect of GGBS on water absorption capacity and stability of Superabsorbent polymers partially crosslinked with alkalis. J. Mater. Civ. Eng. **30**(12), 1–11 (2018a)

Almeida, F.C.R., Klemm, A.J.: Efficiency of internal curing by superabsorbent polymers (SAP) in PC-GGBS mortars. Cem. Concr. Compos. **88**, 41–51 (2018b)

Almeida, F.C.R., Klemm, A.J., Corinaldesi, V.: Plastic shrinkage cracking performance of mortars with ground granulated blast-furnace slag (GGBS) modified by superabsorbent polymers (SAP). In: PRO 125 - 4th International Conference on Service Life Design for Infrastructures, pp. 1–12. RILEM, Delft (2018)

Berodier, E., Scrivener, K.: Evolution of pore structure in blended systems. Cem. Concr. Res. **73**, 25–35 (2015)

Choi, Y.C., Kim, J., Choi, S.: Mercury intrusion porosimetry characterization of micropore structures of high-strength cement pastes incorporating high volume ground granulated blast-furnace slag. Constr. Build. Mater. **137**, 96–103 (2017)

Justs, J., Wyrzykowski, M., Bajare, D., Lura, P.: Internal curing by superabsorbent polymers in ultra-high performance concrete. Cem. Concr. Res. **76**, 82–90 (2015)

Lee, K.M., Lee, H.K., Lee, S.H., Kim, G.Y.: Autogenous shrinkage of concrete containing granulated blast-furnace slag. Cem. Concr. Res. **36**(7), 1279–1285 (2006)

Lothenbach, B., Winnefeld, F.: Thermodynamic modelling of the hydration of Portland cement. Cem. Concr. Res. **36**(2), 209–226 (2006)

Mechtcherine, V., Reinhardt, H.-W. (eds.): Application of Superabsorbent Polymers (SAP) in Concrete Construction: State-of-the-Art Report Prepared by Technical Committee 225-SAP. RILEM: Springer (2012)

Scrivener, K.L., Juilland, P., Monteiro, P.J.M.: Advances in understanding hydration of Portland cement. Cem. Concr. Res. **78**, 38–56 (2015a)

Scrivener, K.L., Lothenbach, B., De Belie, N., Gruyaert, E., Skibsted, J., Snellings, R., Vollpracht, A.: TC 238-SCM: hydration and microstructure of concrete with SCMs. Mater. Struct. **48**(4), 835–862 (2015b)

Snoeck, D., Jensen, O.M., De Belie, N.: The influence of superabsorbent polymers on the autogenous shrinkage properties of cement pastes with supplementary cementitious materials. Cem. Concr. Res. **74**, 59–67 (2015)

Wang, F., Yang, J., Hu, S., Li, X., Cheng, H.: Influence of superabsorbent polymers on the surrounding cement paste. Cem. Concr. Res. **81**, 112–121 (2016)

Wyrzykowski, M., Igarashi, S.-I., Lura, P., Mechtcherine, V.: Recommendation of RILEM TC 260-RSC: using superabsorbent polymers (SAP) to mitigate autogenous shrinkage. Mater. Struct. **51**(135), 1–7 (2018)

Exploring Different Choices of "Time Zero" in the Autogenous Shrinkage Deformation of Cement Pastes Containing Superabsorbent Polymers

José Roberto Tenório Filho[1,2(✉)],
Maria Adelaide Pereira Gomes de Araújo[1,3], Didier Snoeck[1],
Els Mannekens[4], and Nele De Belie[1]

[1] Magnel Laboratory for Concrete Research, Department of Structural Engineering, Faculty of Engineering and Architecture, Ghent University, Technologiepark Zwijnaarde 60, 9052 Ghent, Belgium
{joseroberto.tenoriofilho, adelaide.araujo, didier.snoeck, nele.debelie}@ugent.be
[2] SIM vzw, Technologiepark 48, Zwijnaarde, 9052 Ghent, Belgium
[3] Polymer Chemistry and Biomaterials Group, Department of Organic and Macromolecular Chemistry, Ghent University, Ghent, Belgium
[4] Chemstream bvba, Drie Eikenstraat 661, 2650 Edegem, Belgium
els.mannekens@chemstream.be

Abstract. Shrinkage in concrete structures has been the focus of many studies. Lately a lot of attention has been given to autogenous shrinkage. Although it may not be prominent in ordinary concrete structures, in systems with very low water-to-cement/binder ratio (ultra-high performance concrete for example) it can become a serious issue associated with the cracking of the structure at early age. This type of shrinkage develops due to a reduction in the internal relative humidity of the material and it is also associated to the development of capillary pressure in the pore system due to receding menisci. A big challenge in studying autogenous shrinkage is determining the "time-zero". Given a lack of consensus in literature, this study aimed to investigate the influence of different estimations of time-zero: the final setting time determined by both an electronic Vicat apparatus and ultrasonic measurements; the "knee-point" in the shrinkage curve; and the capillary pressure build-up. Cement pastes with and without superabsorbent polymers (SAPs) were produced with Portland cement CEM III-B 42.5 N and superplasticizer (Glenium 51, 35% conc.). SAPs have proven to be quite effective in the mitigation of autogenous shrinkage as they can act as water reservoirs for the system. Among all methods, the capillary pressure was very suitable for all mixtures. For those containing SAPs no difference was found in picking the time-zero with any method. For the one without SAPs and lower w/c the choice of time-zero based on the setting time led to a different magnitude of autogenous shrinkage deformation in comparison to the other methods, which could be interpreted as an underestimation of the autogenous shrinkage deformation.

Keywords: Autogenous shrinkage · Time-zero · Hydrogels · Superabsorbent polymers

W. P. Boshoff et al. (Eds.): SAP 2019, RILEM Bookseries 24, pp. 106–113, 2020.
https://doi.org/10.1007/978-3-030-33342-3_12

1 Introduction

In cementitious materials after contact of water with cement and during hydration, many changes take place in the material structure. From that moment until the cementitious material reaches its final setting chemical and physical processes result in expansion and shrinkage that can initiate cracking in the hardening material. Shrinkage in concrete structures has been the focus of many studies, and lately a lot of attention has been given to autogenous shrinkage. Although the autogenous shrinkage may not be prominent in ordinary concrete structures, in systems with very low water-to-cement/binder ratio (ultra-high performance concrete for example) (lower than 0.42 (Jensen and Hansen 2001)) it can become a serious issue associated with the cracking of the structure at early age.

This type of shrinkage is known to be related to a reduction in the internal relative humidity of the material (consequence of chemical shrinkage) and associated to the development of capillary pressure in the pore system due to receding menisci (Lura 2003; Bentz and Jensen 2004; Jensen and Hansen 2001). The use of internal curing agents has proven to be quite effective in the mitigation of this kind of shrinkage and special attention has been given to the use of hydrogels as they can act as water reservoirs for the system, keeping its levels of internal relative humidity high for a considerable time frame (Snoeck 2015; Mechtcherine *et al.* 2009; Jensen 2008).

One of the main challenges in the study of the autogenous shrinkage in cementitious materials is the determination of time-zero. Some authors refer to it as the time when the material presents a skeleton structure that can transfer the tensile stresses along the material and suggest the final setting time as the time-zero (Hammer and Bjøntegaard 2006). Others consider it as the moment when there is a drop in the internal relative humidity of the material. The drop in relative humidity associated to the self-desiccation results in an increase in the capillary pressure, which was studied by Miao *et al.* (2007) as an indication for the time-zero. Darquennes *et al.* (2011) pointed out that the moment of change in the curvature of the shrinkage curve (a knee-point) could indicate the point from which the chemical shrinkage is no longer capable of promoting relevant changes in the volume of the material given the existence of solid skeleton restraining such deformations and for that could also be considered as a logical choice for the time-zero.

This study was based on a comparison of the deformation due to autogenous shrinkage in cement pastes with and without superabsorbent polymers as internal curing agents. Given the lack of consensus in literature, the time-zero was taken as the final setting time determined by both an electronic Vicat apparatus and ultrasonic measurements; the so-called "knee-point" observed in the shrinkage deformation curve of the pastes; and by the build-up in the capillary pressure.

2 Materials and Methods

2.1 Mixture Compositions

All tests were performed on cement pastes produced with cement type CEM III-B 42.5 N – LH/SR; a polycarboxylate superplasticizer (at a constant dosage of 0.3 m% in relation to the cement mass; Glenium 51, 35% conc.); and two superabsorbent polymers hereby identified as SAP1 and SAP2.

SAP1, provided by BASF, is a copolymer of acrylamide and sodium acrylate produced by bulk polymerization and subsequently ground to a mean particle size (D_{50}) of 40 μm and absorption capacity of 27 g of water per g of SAP in cement paste. SAP2, produced by ChemStream, consists of the co-monomer NaAMPS (2-acrylamido-2-methyl-1-propanesulfonic acid sodium salt), diluted with a non-charged or neutral monomer ACMO (acryloyl morpholino acrylate), and has a mean particle size (D_{50}) of 100 μm and absorption capacity of 50 g of water per g of SAP in cement paste. In both cases the absorption capacity was measured by means of comparing flow table values (EN 1015-3 (CEN, 1999) 10 min after the first contact of the dry SAPs with water in the paste. Information on the composition of the pastes can be found in Table 1. The mixing procedure adopted for all the mixtures followed the standard NBN EN 196-1 with the superplasticizer added after the water.

Table 1. Composition of the cement pastes

Mixture	Effective w/c	Total w/c	Amount of SAPs [m%][a]	Amount of superplasticizer [m%][a]
REF0.30	0.3	0.3	0	0.3
REF0.354	0.354	0.354	0	0.3
SAP1	0.3	0.354	0.2	0.3
SAP2	0.3	0.4	0.2	0.3

[a]m% versus cement

The 0.30 w/c ratio was chosen, as this REF mixture would show pronounced autogenous shrinkage. The second reference, REF0.354 was included based on the theory of Powers and Brownyard (1948) which was adapted by Jansen and Hansen (2001) for the case of internal curing. According to their study, and additional amount of water corresponding to w/c = 0.054 stored in the SAPs is enough for an effective internal curing in mixtures produced with ordinary Portland cement. The mixtures with SAP1 and SAP2 contain the minimum amount of SAP and additional water that was found to effectively mitigate the autogenous shrinkage.

The experimental program was based on the measurement of the autogenous shrinkage deformation of the mixtures; the hardening of the mixtures studied by means of Vicat test and ultrasonic pulse velocity (UPV); the build-up in the capillary pressure.

2.2 Measuring the Autogenous Shrinkage

The autogenous shrinkage was assessed according to the standard ASTM C1698-09. The test consists of the measurement of the deformation of specimens in corrugated tubes with a nominal length of 425 ± 5 mm and a diameter of 29 ± 0.5 mm. The specimens were placed on metallic supports with one linear variable differential transducer (LVDT) with a range of 5 mm on one end. The other end was restrained in movement. The measurements were performed continuously every 10 min for 7 days in a room with controlled humidity (60 ± 5%) and temperature (20 ± 1 °C). The knee-point in the curve was determined as the point with a (local) absolute maximum value of strain in time, immediately followed by a reduction in the strain values. After this point a change in the deformation rate was observed for all mixtures, thus being considered as the point of solid-fluid transition.

2.3 The Hardening of the Mixtures

The final setting time of the mixtures was studied by means of an electronic Vicat apparatus in compliance with EN 196-3. The UPV was determined with a FreshCon equipment (Reinhardt and Grosse 2004) using compressive pulse waves. The measurements were automatically performed each five minutes during 48 h with an amplifying voltage of 450 V. The final setting time was then approximately determined based on the first derivative of the wave velocity in time. The maximum value of the derivative was chosen, representing the point where the rate of increase in the wave velocity switches to a slower pace.

2.4 The Build-Up in the Capillary Pressure

A small pressure transducer (RVAP015GU, Sensortronics) monitored the development of the capillary pressure, from the voltage variation due to the water flow, every 10 min for 24 h. The test apparatus (Fig. 1) consists of a plastic cup (400 mL) where a plastic tube is inserted with a piece of sponge at one end and the pressure transducer on the other. Right after a continuous flow of water could be observed in the narrow plastic tube, the transducer was inserted at the outer end of the tube and the measurement of voltage was continuously recorded.

Fig. 1. Set-up for capillary pressure test

3 Results and Discussion

From Figs. 2, 3, 4 and 5 the shrinkage strain is compared with the development of capillary pressure for each mixture. The time corresponding to the knee-point of the shrinkage curve, final setting with both Vicat and UPV is also indicated.

In the reference with a w/c of 0.30 (Fig. 2) the knee-point coincides with the breakdown in the pressure and both occur first. The Vicat and the FreshCon thresholds show points very close in time but with a later occurrence in comparison with the others. The lower amount of water in the system (in comparison with all mixtures studied) causes the matrix to reach the fluid-solid transition earlier and that is immediately followed by the build-up/breakdown in the capillary pressure.

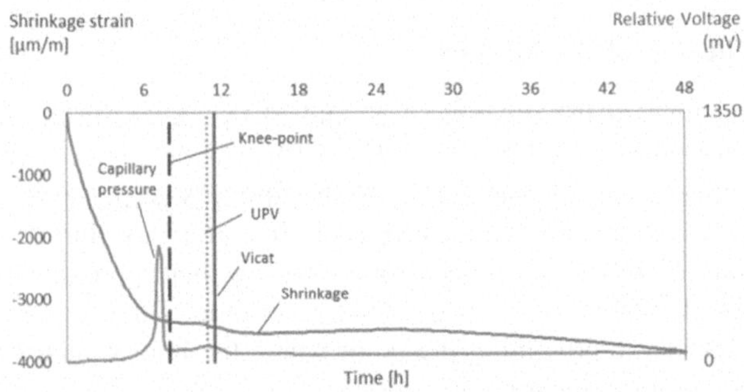

Fig. 2. Shrinkage strain and development of capillary pressure for the mixture REF 0.30

In the reference with w/c of 0.354 (Fig. 3) the same sequence for knee-point, FreshCon and Vicat is found. The capillary pressure shows a breakdown that is delayed in comparison to the rest but coincident with the final setting determined with the Vicat apparatus.

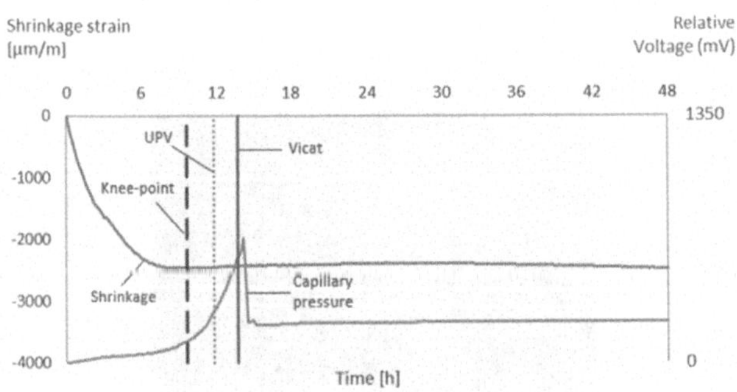

Fig. 3. Shrinkage strain and development of capillary pressure for the mixture REF 0.354

Due to the higher amount of water in REF0.354 in comparison with REF0.30, the pores formed after the transition take longer to be empty so the time when the radius of the water menisci changes and the moment of breakdown in the pressure are both delayed, occurring around the final setting. According to Bentz and Jensen (2004), the formation of a stable solid skeleton is an essential condition for the self-desiccation, but it does not necessarily mean that this phenomenon will take place immediately after the formation of the skeleton when it comes to mixtures with w/c around or higher than 0.42.

In the mixtures containing the SAP1 (Fig. 4) and SAP2 (Fig. 5), the time-zero defined by all the different methods are very close meaning that the final setting might occur right after the transition point and approximately at that moment of pressure breakdown. In relation to the action of the SAPs as internal curing agents the water of internal curing causes a delay in the moment when the pressure breakdown occurs, in relation to the reference with a w/c of 0.3.

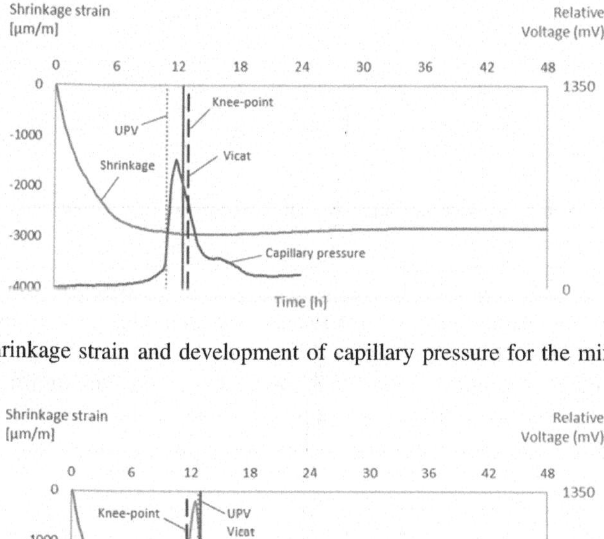

Fig. 4. Shrinkage strain and development of capillary pressure for the mixture SAP1

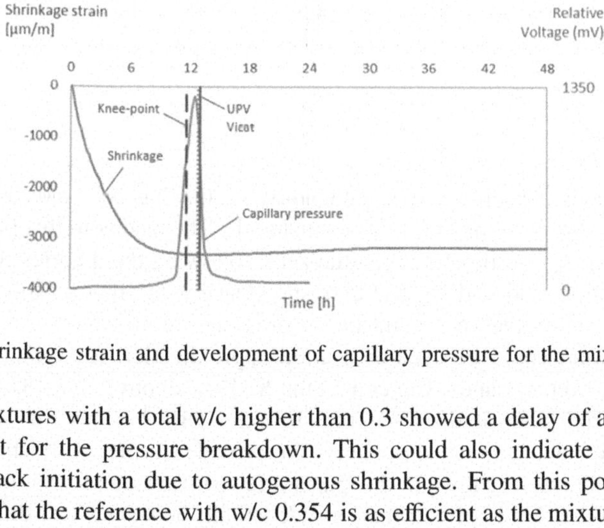

Fig. 5. Shrinkage strain and development of capillary pressure for the mixture SAP2

All the mixtures with a total w/c higher than 0.3 showed a delay of about six hours in the moment for the pressure breakdown. This could also indicate a delay in the moment of crack initiation due to autogenous shrinkage. From this point of view, it might appear that the reference with w/c 0.354 is as efficient as the mixtures with SAPs in terms of controlling the self-desiccation, which is not the case and will be shown in the next part of the discussion.

Picking the time-zero based on different methodologies can influence the correct analysis of the autogenous shrinkage depending on the composition of the mixture. Figure 6 shows the autogenous shrinkage strain of all pastes during the first seven days of age with different choices for time-zero. For the REF with a w/c of 0.3 the time when the maximum voltage was recorded in the capillary pressure test was compared to the final setting time determined by the Vicat. For REF0.354 the knee-point versus the maximum voltage was plotted. Given the proximity of the results in the mixtures with SAPs only the moment of maximum voltage was chosen.

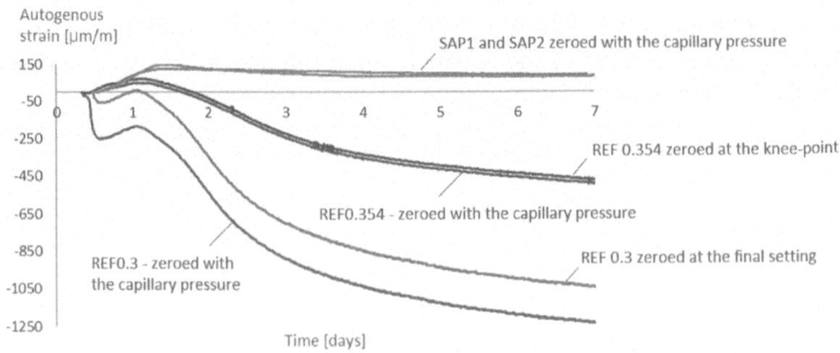

Fig. 6. Autogenous deformation for all mixtures with different zero-times

In the REF0.3 the choice of time-zero based on the final setting would lead to an underestimation of the autogenous shrinkage strain with a difference around 200 μm/m. Regarding the REF0.354 even though a difference in time was obtained between the occurrence of the knee-point and the breakdown of pressure, no significant difference was found in the autogenous shrinkage strain. The same can be extended to the mixtures with SAPs, where all the test methods tend to converge to a same point in time.

4 Conclusion

The build-up of capillary pressure is a phenomenon that has been already associated to the autogenous shrinkage in cementitious materials. The results in this paper showed it can be quite precise in estimating the time-zero, following the theories connecting it to the self-desiccation, and with a good correspondence with other existing methods and definitions, especially the knee-point in the shrinkage strain curve.

The knee-point could be used as a good approximation for the time-zero in the mixtures with lower w/c and those containing SAPs. Adopting this method could save time and material since it does not require extra tests besides the already needed measurement of autogenous shrinkage. To provide further validation and wider application of the results shown in this paper a few suggestions for future investigation can be given: other kinds of cement and w/c should be tested; test different types of SAPs to verify a possible influence of their chemistry composition in terms of mitigating autogenous shrinkage.

Acknowledgments. The work has been financed by SIM program SHE (Engineered Self-Healing Materials) within the ICON project iSAP (Innovative SuperAbsorbent Polymers for crack mitigation and increased service life of concrete structures).

References

Bentz, D.P., Jensen, O.M.: Mitigation strategies for autogenous shrinkage cracking. Cem. Concr. Compos. **26**, 677–685 (2004)

Darquennes, A., Staquet, S., Delplancke-Ogletree, M., Espion, B.: Effect of autogenous deformation on the cracking risk of slag cement concretes. Cem. Concr. Compos. **33**, 368–379 (2011)

Hammer, T.A., Bjøntegaard, Ø.: Testing of autogenous deformation (AD) and thermal dilation (TD) of early age mortar and concrete-recommended test procedure. In: Proceedings of the International RILEM Conference on Volume Changes of Hardening Concrete: Testing and Mitigation, Lyngby, Denmark, pp. 341–346 (2006)

Jensen, O.M., Hansen, P.F.: Autogenous deformation and RH-change in perspective. Cem. Concr. Res. **31**(12), 1859–1865 (2001)

Jensen, O.M.: Use of superabsorbent polymers in construction materials. In: International Conference on Microstructure Related Durability of Cementitious Composites, Nanjing, pp. 757–764 (2008)

Lura, P.: Autogenous Deformation and Internal Curing of Concrete. Ph.D. thesis Delft University of Technology, The Netherlands, 15 (2003)

Mechtcherine, V., Dudziak, L., Hempel, S.: Mitigating early age shrinkage of ultra-high-performance concrete by using super absorbent polymers (SAP). Paper presented at the Creep, Shrinkage and Durability Mechanics of Concrete and Concrete Structures, Ise-Shima, pp. 847–853. Taylor & Francis (2009)

Miao, C.-W., Qian, T., Wei, S., Jia-Ping, L.: Water consumption of the early-age paste and the determination of "time-zero" of self-desiccation shrinkage. Cem. Concr. Res. **37**(11), 1496–1501 (2007)

Powers, T.C., Brownyard, T.L.: Studies of the Physical Properties of Hardened Portland Cement Paste, vol. 22, p. 892. Portland Cement Association, Research Laboratories, Cornell (1948)

Reinhardt, H.W., Grosse, C.U.: Continuous monitoring of setting and hardening of mortar and concrete. Constr. Build. Mater. **18**(3), 145–154 (2004)

Snoeck, D.: Self-Healing and microstructure of cementitious materials with microfibres and superabsorbent polymers. Doctorate thesis – Faculty of Architecture and Engineering/Ghent University, Belgium (2015)

Comparative Study Between Strain Gages for Determination of Autogenous Shrinkage

Paulo Francinete Jr.[1(✉)], Eugênia Fonseca da Silva[2], and Anne Neiry de Mendonça Lopes[3]

[1] Instituto Federal de Goiás, 75 street, n. 46, Downtown, Goiânia, Goiás 74055-110, Brazil
paulo.junior@ifg.edu.br

[2] Universidade de Brasília – PECC/UnB, Campus Darcy Ribeiro, Universidade de Brasília, Brasília - DF 70910-900, Brazil
eugenia@unb.br

[3] Furnas Centrais Elétricas – GST, BR 153 Highway, countryside, Aparecida de Goiânia, Goiás 74923-650, Brazil
anne@furnas.com.br

Abstract. The Carlson-type strain gages are the most suitable for the measurement of autogenous shrinkage. However, your high cost is usually a limiting factor for your employment. A more economical alternative would be the use of Self-Temperature-Compensation Gages to be embedded in concrete. The main objective of this work was to verify if Self Temperature-Compensation Gages can be used instead of the Carlson type for the determination of the autogenous shrinkage in high performance concrete. It was carried out measurements of the autogenous shrinkage, for the same concrete mix, using the two types of extensometers. The difference in the means of the autogenous shrinkage results between these two types of extensometers was typically less than 10×10^{-6} m/m, showing that the Self-Temperature-Compensation Gage is an alternative to replace the Carlson type strain gage for the measurement of autogenous shrinkage of concrete.

Keywords: High strength concrete · Autogenous shrinkage · Carlson

1 Introduction

The advances in concrete technology have allowed the production of concrete with better performances and high resistances. The High Strength Concrete (HSC) presents a low water/cement ratio and considerable consumption of cement materials (Portland cement and pozzolanic additions), resulting in a very dense microstructure with fine pores (Silva 2007). These characteristics make the autogenous shrinkage in the HSC significant, and can cause, specifically in the early ages, cracking of the concrete (Lopes 2011). The occurrence of autogenous shrinkage in HSC is unavoidable, as it is caused by the phenomenon of self-drying, which is the reduction of relative humidity during the cement paste hardening due to capllary water consumption in the cement hydration process (Tazawa 1999). This fact, together with the damages that this

W. P. Boshoff et al. (Eds.): SAP 2019, RILEM Bookseries 24, pp. 114–122, 2020.
https://doi.org/10.1007/978-3-030-33342-3_13

retraction can cause to the HSC, shows the importance of better understanding it and deepening the studies on the subject, in order to control it to acceptable levels.

There are several methodologies for the determination of the autogenous shrinkage by linear deformation measurements. The Japanese Institute of Concrete (Tazawa and Miyazawa 1999) and the dilatometer, developed by Jensen and Hansen (1995) and later standardized by the American Society for Testing Materials (2014), stand out.

The methodology proposed by the Japanese Institute of Concrete has been used, for the measurement of deformations, dial indicators (Silva et al. 2014; Manzano 2016; Cunha et al. 2017) or strain gages to be embedded in concrete specimens (Aïtcin 1998; Hanehara et al. 1998; Kojima et al. 2001; Lopes 2011). The use of embedded strain gages has the advantage of allowing deformation measurements to start as soon as the specimens are molded and prepared.

One of the most suitable extensometers for the measurement of autogenous deformation is the Carlson type strain gages that use two distinct electromechanical principles: that of the variation of the tension in the wire, that causes variation of the electric resistance, and that of the change in the temperature in the wire which has the same effect. Thus, the Carlson-type strain gage also functions as a thermometer, allowing deformations to be determined due to the effects of temperature variation (Andrade 1997). However, their high cost may be a limiting factor to their employment. In this case, a more economical alternative would be the use of Self Temperature-Compensation Gages.

The basic principle of operation of Self Temperature-Compensation Gages is like that of Carlson strain gages (Andrade 1997). But these extensometers do not function as thermometers and, therefore, do not allow the determination of thermal deformation. However, they are free from the effects of temperature variations. It means that it is a meter in which the temperature resistance coefficient of the wire or blade is controlled. Thus, if the adequate meter is used for the studied material, the expectation of error should be within $\pm 1{,}8 \times 10^{-6}/^{\circ}C$, which means high accuracy.

The Self Temperature-Compensation Gages can be supplied in six types, depending on the application: wood, construction steel and concrete, stainless, aluminum, magnesium alloy and plastics. Table 1 presents the coefficient of linear expansion for the temperature-controlled extensometer as a function of the type of compatible material.

Table 1. Coefficient of thermal expansion for the temperature-controlled extensometer as a function of material type (Andrade 1997).

Material	Coefficient of thermal expansion ($\times 10^{-6}/^{\circ}C$)
Wood	5,0
Steel, concrete	10,8
Stainless	16,2
Aluminum alloy	23,4
Magnesium alloy	27,0
Polymer	65,0

The main objective of this study is to verify if the Self Temperature-Compensation Gages model can be used in substitution of the Carlson type strain gage to determine the autogenous shrinkage in HSC.

2 Experimental Program

The hydraulic cement used in the concrete production was the Brazilian Portland cement of high early strength, CP V ARI (ABNT 1991). Two mineral additions were used in the production of the concrete: (1) silica fume non-densified and (2) commercially available nano-silica in aqueous solution of colloidal silica with solids content of 30%. Table 2 shows the properties of Portland cement and silica fume used in the manufacture of mixtures. Table 3 presents other relevant properties of cement.

Table 2. Chemical composition of Portland cement and silica fume used (%)

Material	Loss on ignition	Insoluble residue	SO_3	$CaSO_4$	MgO	SiO_2	Fe_2O_3	Al_2O_3	CaO	Free CaO	Na_2O	K_2O
Cement	1,82	0,79	3,28	5,57	4,36	24,41	3,02	7,09	53,74	2,16	0,29	0,77
Silica fume	3,22	–	–	–	0,49	93,55	0,16	0,15	0,37	–	0,26	0,85

Table 3. Characterization of the physical and mechanical properties of Brazilian Portland Cement CP V ARI

Properties		Results	Limits by NBR 5733	Test method
Cement Fineness	Residue on the sieve 75 μm (%)	0,4	≤ 6,0	NBR 11579/2012
	Specific surface by Blaine (cm^2/g)	5440	≥ 3000	NBR 16372/2015
Normal consistency paste (%)		32,4	–	NBR NM 43/2003
Setting time	Initial (h:min)	02:30	≥ 1:00	NBR NM 65/2003
	Final (h:min)	03:10	≤ 10:00	
Autoclave expansion (%)		0,03	–	ASTM C 151
Compressive strength (MPa)	1 day	32,1	≥ 14,0	NBR 7215/1996
	3 days	37,2	≥ 24,0	
	7 days	41,8	≥ 34,0	

The used fine aggregate was washed natural sand of river, with fineness modulus of 2.67 and absorption of 0.3%. The coarse aggregate used was crushed stone from rock of the lithological type gneiss, with fineness modulus of 5.95 and absorption of 0.3%.

The chemical admixture used was a high efficiency (superplasticizer) type of water reducer based on polycarboxylate.

A concrete with slump of 200 ± 10 mm was produced as shown in Table 4. The preparation and production of the specimens is described in Francinete Jr. et al (2018).

Table 4. Mixture proportion of Concrete

kg/kg of cement					% Cement weight	
Cement	Silica fume	Sand	crushed rock	Water	Nano-sílica	Superplasticizer
1	0,10	1,914	1,686	0,30	1	2,69

2.1 Methods

In the present study, the concrete autogenous shrinkage was determined by means of the embedded strain gage method, according to IT.MC 201 (Furnas 2015). Measurements of the autogenous shrinkage were carried out on specimens of the same dimensions and molded with the same concrete. The variables were only the strain gages types used.

The Carlson M-4 strain gage, whose characteristics are presented in Table 5, and the Self Temperature-Compensation Gages of electrical resistance equal to 120 Ω, were tested.

Table 5. Characteristics of the Carlson M-4 strain gage (Carlson 1995).

Feature	*Carlson* M-4 strain gage type
Range (micro-strain)	3900
Resolution (micro-strain)	5,8
Resolution temperature (°C)	0,05
Gauge length (cm)	10,2
Weight (gram)	86

To facilitate the understanding and identification of the two types of extensometers, the Carlson M-4 type only will be called Carlson and the Self Temperature-Compensation Gages type will be called the SC120. Six test specimens, two with Carlson type (Carlson 01 and Carlson 02) and four with Self Temperature-Compensation Gages (SC120-01, SC120-02, SC120-03 and SC120-04) were molded. In November of 2015, the Carlson strain gage type cost US$ 335 while the Self Temperature-Compensation Gage was purchased for US$ 25. Figure 1 shows the two models of strain gages that were studied.

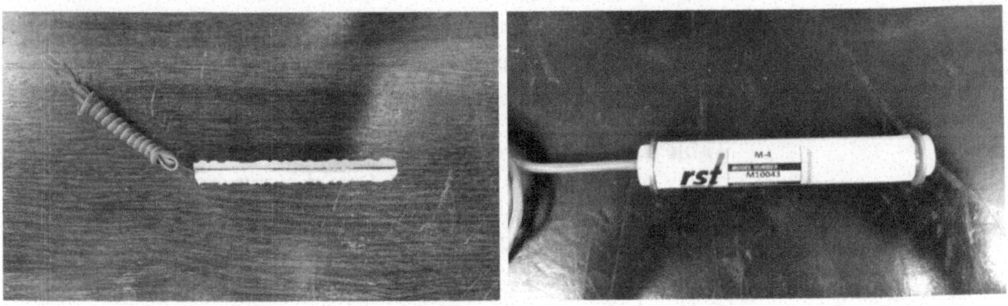

Fig. 1. Self-compensating temperature gage (left) and Carlson type (right).

3 Results and Discussion

The loss of water mass was determined for each specimen, in order to verify if the shrinkage was an eminently autogenous phenomenon or if the specimen was considerably dried. The loss of water mass of the specimens, in relation to the total mass, at the end of the autogenous shrinkage test (28 days) was 0.04%, on average. Therefore, it can be affirmed that there was a good sealing of the specimens and that the measured was really about autogenous shrinkage.

Table 6 shows the results obtained for the autogenous shrinkage over time of all specimens tested. It presents also the arithmetical average of the results, the standard deviation (SD) and coefficient of variation (CV) for each type of extensometer.

It can be observed in Table 6 that the behavior of the autogenous shrinkage over time are quite similar for all specimens, regardless of the type of strain gage used, showing a low dispersion of the results. It is observed that within each type of extensometer the greatest discrepancies of the results occurred at 1 day of age, and the Self Temperature-Compensation Gages showed a CV at this age equal to 16%, while the Carlson type strain gage, at the same age, a CV equal to 25%. At the other ages, the CV was always lower than 14%, and in those ages over 3 days this coefficient was always equal to or less than 3%. The higher values of the CV, shown in the results of the Carlson strain gage, are due to the reduced number of specimens with this type of extensometer (two), it is believed that if the number of specimens with this type of extensometer was higher the CV values would be lower. However, even considering the greater dispersion of the results in the early ages, it can be affirmed that both extensometers have a good repeatability of the results.

It is also observed that the largest difference between the average of the results of each type of extensometer was 40×10^{-6} m/m, also observed at 1 day of age. From 3 days of age this difference was typically less than 11×10^{-6} m/m. To verify if these differences were significant, a variance analysis (ANOVA) was performed. The use of variance analysis in group comparison is based on the relationship between the variability of the results within the groups and the variability of the averages between the groups, and on the Fischer (F) distribution, with the level of significance α (Gomes 1982; Montgomery and Runger 2003). In the present study, a significance level of 5% ($\alpha = 0.05$) was used in the statistical test performed with the *Statistica 10.0®* software. Table 7 shows the results of the variance analysis.

Table 6. Autogenous shrinkage (μm/m) according to the age and strain gage type

Age (days)	Self temperature-compensation gages							Carlson				
	01	02	03	04	Average	sd	CV (%)	01	02	Average	sd	CV (%)
0	0	0	0	0	0	0	0	0	0	0	0	0
1	−75	−83	−107	−86	−88	14	16	−151	−105	−128	32	25
2	−142	−150	−178	−165	−159	16	10	−160	−130	−145	21	14
3	−180	−192	−207	−196	−194	11	6	−185	−182	−183	2	1
6	−238	−246	−254	−245	−246	7	3	−235	−235	−235	0	0
7	−250	−260	−260	−250	−255	6	2	−250	−250	−250	0	0
8	−259	−265	–	–	−262	4	2	−249	−254	−252	3	1
9	−266	−274	−277	−265	−271	6	2	−260	−264	−262	3	1
10	−274	−275	−280	−269	−275	5	2	−264	−269	−267	3	1
11	−283	−286	–	–	−285	2	1	−275	−279	−277	3	1
13	−294	−295	−291	−282	−291	6	2	−290	−285	−288	4	1
14	−300	−299	−295	−287	−295	6	2	−296	−299	−297	3	1
15	−305	−303	−300	−292	−300	6	2	−300	−299	−300	1	0
16	−306	−308	−300	−298	−303	5	2	−299	−303	−301	3	1
17	−321	−310	−310	−301	−311	8	3	−299	−307	−303	6	2
20	−323	−318	–	–	−321	4	1	−313	−317	−315	2	1
21	−326	−323	−309	−311	−317	9	3	−317	−321	−319	3	1
22	−327	−323	−322	−315	−322	5	2	−322	−325	−323	2	1
23	−330	−327	−330	−318	−326	6	2	−321	−325	−323	2	1
24	−333	−329	−330	−319	−328	6	2	−321	−329	−325	6	2
27	−344	−337	−346	−328	−339	8	2	−338	−345	−341	6	2
28	−351	−343	−342	−334	−343	7	2	−343	−346	−344	2	1

The results presented in Table 7 show that more than 98% of the total variation of the data is explained by the criterion adopted, since $R^2_{mod} = 0.9887$. It is observed in the column of the values of F that the variable strain gage type did not have a significant effect on the results of autogenous shrinkage. Therefore, it can be affirmed that there are no differences between the autogenous shrinkage results determined with the Carlson-type strain gage and with the Self Temperature-Compensation Gage type, confirming the efficiency of the self-compensating temperature system and showing that it can be employed replacing the Carlson strain gage type.

The age variable, as expected, was the most important effect (F = 340.6). As the aim of the study was to compare the two types of extensometers, the behavior of autogenous shrinkage with age will not be discussed. However, considering that the interaction between the type of extensometer and age was also significant, a multiple comparison of averages was done by the Duncan method (Gomes 1982; Montgomery and Runger 2003), to determine if and which groups of values associated with

Table 7. Variance analysis

$R^2_{mod} = 0,9887$						
Variable	Degress of freedom	Sum of Squares	Mean squares	F	$F_{0,05}$	Significance
Strain gage type (A)	1	54	54	0,8	3,964	No
Age (B)	20	457164	22858	340,6	1,707	Yes
AB	20	3148	157	2,3	1,707	Yes
Erro	78	5235	67			
Total	119					

interaction between the types of extensometers and ages differ from each other. The conclusion was that, in all ages studied, the results of the two types of extensometer are associated with each other in a single group, except at 1 day of age, where the results of each type of extensometer form two groups that differ significantly from each other, as shown in Fig. 2.

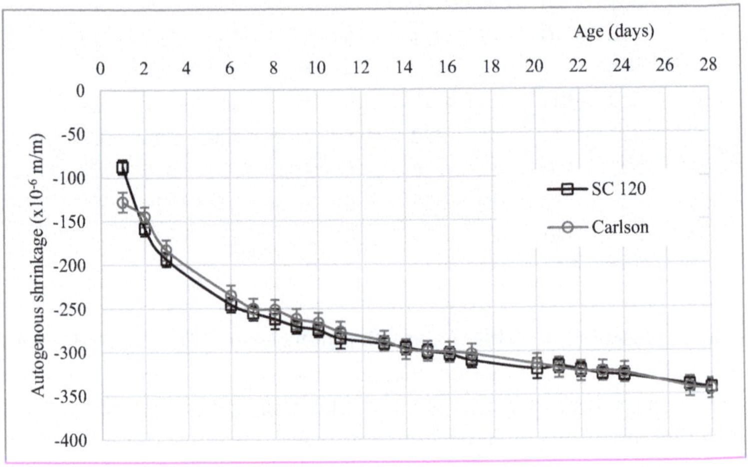

Fig. 2. Interaction of the average results of the autogenous shrinkage for each strain gage type, with a 95% confidence interval.

It should be noted that the curves show very similar behavior and that the difference between the averages of the results of each type of extensometer are not significant, showing that, under the conditions of the present study, the autogenous shrinkage can be measured using the Self Temperature-Compensation Gage.

Figure 3 shows there is a linear relationship between the autogenous shrinkage measurements obtained by the two strain gages types studied.

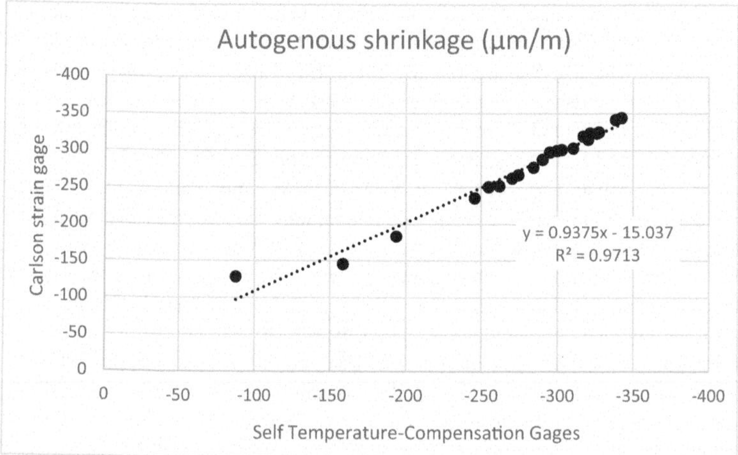

Fig. 3. Linear relationship between autogenous shrinkage measurements obtained with the Carlson type and the Self Temperature-Compensation Gage

4 Conclusion

It can be concluded that Self Temperature-Compensation Gages in concrete are a viable alternative to the Carlson strain gage type for measuring the concrete autogenous shrinkage. Statistical analysis showed that, in all ages studied, the results of the two types of extensometers are associated with each other in a single group, except at 1 day of age. The continuity of the study with a larger number of specimens and other concrete mixtures is necessary for the validation of the results.

The possibility of using the Self Temperature-Compensation Gages from the practical point of view reduces the financial cost and/ or allows the instrumentation of a greater quantity of structural parts or elements, both in laboratories and in concrete structures, since the cost of the Self Temperature-Compensation Gages is about 1/10 the cost of the Carlson strain gage. type.

Acknowledgments. The authors thank the University of Brasilia, Furnas Central Hydropower S.A., the Federal Institute of Goias (IFG), the Coordination for the Improvement of Higher Education Personnel (Capes) and Natália Carvalho de Camargo.

References

ABNT: NBR 5733: Cimento Portland de alta resistência inicial, Brasil (1991)
Aïtcin, P.C.: Autogenous shrinkage measurement. In: Proceedings of the International Workshop on Autogenous Shrinkage of Concrete, Hiroshima, Japan, pp. 245–256 (1998)
American Society for Testing and Materials: ASTM C1698: Standard Test Method for Autogenous Strain of Cement Paste and Mortar, Philadelphia, USA, 8p. (2014)
Andrade, W.P.: Concretos: massa, estrutural, projetado e compactado com rolo: ensaios e propriedades, 1st edn. Editora Pini, São Paulo, Brasil (1997)

Carlson, R.W.: Carlson Strain Meters and Other Instruments for Embedment in Concrete Structures, 5th edn. RST Instruments Ltd., Canada (1995). Edited by Robert M. Taylor
Cunha, T.A., Francinete, P., Agostinho, L.B., Silva, E.F, Lopes, A.N.: Study of the autogenous shrinkage in microconcretes containing superabsorbent polymer and nano-silica. In: 2nd International RILEM/COST Conferece on Early Age Cracking and Serviceability in Cement-Based Materials and Structures, Brussels, Belgium, pp. 251–256 (2017)
Francinete Jr., P., Silva, E.F., Lopes, A.N.M.: Estudo comparativo entre o emprego dos extensômetros do tipo Carlson e do tipo de temperatura auto compensável para determinação da retração autógena. Revista Matéria, v. 23, n. 3. UFRJ, Rio de Janeiro, Brasil (2018)
Furnas: IT.MC.201 Concreto – Determinação da variação autógena. Instrução de Trabalho – Métodos Construtivos. Furnas Centrais Elétricas S.A. Gerência de Pesquisa, Serviços e Inovação Tecnológica, Aparecida de Goiânia, Brasil (2015)
Gomes, F.P.: Curso de estatística experimental. USP, ed. 10, Piracicaba, Brasil (1982)
Hanehara, S., Hirao, H., Uchikawa, H.: Relationships between autogenous shrinkage, the microstructure and humidity changes at inner part of hardened cement paste at early ages. In: Proceedings of the International Workshop on Autogenous Shrinkage of Concrete, Hiroshima, Japan, pp. 89–100 (1998)
Jensen, O.M., Hansen, P.F.: A dilatometer for measuring autogenous deformation in hardening Portland cement paste. Mater. Struct. **28**(7), 406–409 (1995)
Kojima, T., Takagi, N., Horikawa, S.: Study on shrinkage characteristics of high-strength silica fume concrete. In: Proceedings of Seventh CANMET/ACI International Conference on Fly Ash, Silica Fume, Slag and Natural Pozzolans in Concrete, Chennai, India, pp. 719–735 (2001)
Lopes, A.N.M.: Mitigação da retração autógena em concretos de alta resistência contendo aditivo redutor de retração e seus efeitos na macro e microestrutura. Tese de D.Sc., Universidade Federal do Rio Grande do Sul, Escola de Engenharia, Curso de Pós-graduação em Engenharia Civil, Porto Alegre, Brasil (2011)
Manzano, M.A.R.: Estudo Experimental de Materiais Cimentícios de Alta Resistência modificados com Polímeros Superabsorventes (PSAs) como Agentes de Cura Interna. Tese de D.Sc., Universidade de Brasília. Faculdade de Tecnologia, Departamento de Engenharia Civil e Ambiental, Brasília, Brasil (2016)
Montgomery, D.C., Runger, G.C.: Applied Statistics and Probability for Engineers, 3rd edn. Wiley, Hoboken (2003)
Silva, E.F.: Variações dimensionais em concretos de alto desempenho contendo aditivo redutor de retração. Tese de D.Sc., COPPE/UFRJ, Rio de Janeiro, Brasil (2007)
Silva, E.F., Manzano, M.A.R., Lopes, A.N.M., Toledo Filho, R.D.: Effect of SAP on the autogenous shrinkage and compressive strength of high-strength fine-grained concrete. In: International RILEM Conference on Application of Superabsorbent Polymers and Other New Admixtures in Concrete Construction, Dresden, pp. 211–219 (2014)
Tazawa, E.: Technical committee on autogenous shrinkage of concrete – committee report. In: Tazawa, E.-C. (ed.) Autogenous Shrinkage of Concrete – Proceedings of the International Workshop organized by Japan Concrete Institute. E & FN Spon, London (1999)
Tazawa, E., Miyazawa, S.: Effect of constituents and curing condition on autogenous shrinkage of concrete. In: Tazawa, E.-C. (ed.) Autogenous Shrinkage of Concrete – Proceedings of the International Workshop organized by Japan Concrete Institute, pp. 269–280. E & FN Spon, London (1999)

Dewatering Kinetics from Fresh Cement Pastes Enriched with Superabsorbent Polymer (SAP) Samples at Ambient and Elevated Temperatures Visualised and Quantified by Neutron Radiography Imaging

Christof Schröfl[1(✉)], Viktor Mechtcherine[1], and David Mannes[2]

[1] Institute of Construction Materials, Department of Civil and Environmental Engineering, Technische Universität Dresden, 01062 Dresden, Switzerland
{christof.schroefl, viktor.mechtcherine}@tu-dresden.de
[2] Laboratory for Neutron Scattering and Imaging, Paul Scherrer Institute-PSI, 5232 Villigen, Switzerland
david.mannes@psi.ch

Abstract. Two superabsorbent polymer (SAP) samples were studied with respect to their sorption kinetics in freshly prepared cement pastes that were exposed to evaporation. SAP 1 was a self-releasing type material when tested in extracted cement pore solution, whereas SAP 2 was a retentive type. These polymers had proven efficient to mitigate plastic shrinkage and its related cracking propensity, however, to different degrees. The observed difference in the performance was elucidated using neutron radiography imaging. Cement pastes with water-to-cement ratios of 0.25 and 0.50 were investigated which were exposed to two different climates: the one mimicking ordinary central European climate (20 °C, intermediate relative humidity) and the other one representing hot and dry weather conditions (40 °C, very low relative humidity).

It was found that SAP 1 released its absorbed aqueous solution right from the beginning, nearly independent of w/c and climatic conditions. This corresponded to its individual ab- and desorption behaviour in extracted pore solution. Contrarily, the inherently retentive SAP 2 desorbed water upon demand that arouse in the paste. While the cracking propensity of the pastes generally largely decreased in the presence of SAP, SAP 2 was found to be more efficient than SAP 1 in this aspect, which can be traced back to the temporally delayed supply of water by SAP 2 upon demand. Obviously, it did not matter much by which physico-chemical mechanism the water was extracted from the SAP particles, i.e. low w/c or hot and dry climate.

Keywords: Fresh cement paste · Neutron radiography imaging · Plastic shrinkage · Water evaporation

W. P. Boshoff et al. (Eds.): SAP 2019, RILEM Bookseries 24, pp. 123–131, 2020.
https://doi.org/10.1007/978-3-030-33342-3_14

1 Introduction

Superabsorbent polymers (SAP) have intensely been studied as concrete admixtures to mitigate autogenous shrinkage of high-performance concrete, e.g. (Jensen and Hansen 2001, Mechtcherine et al. 2014). They have as well proven efficient in reducing the cracking propensity of freshly placed concretes due to plastic shrinkage prior to hardening and formation of the rigid framework of the binder matrix (Serpukhov and Mechtcherine 2015, Snoeck et al. 2018, TC 260 2019). Similar to the mode of action as internal curing agents on the long term, they deliver their stored liquid to the surrounding matrix and provide sufficient moisture throughout the concrete volume (Wyrzykowski et al. 2011, Schröfl et al. 2015, Serpukhov and Mechtcherine 2015). Such working mechanism accords to that of pre-saturated lightweight aggregates (Wyrzykowski et al. 2015).

To consolidate experience with SAP in the context of mitigating plastic shrinkage, the RILEM TC 260–RSC has recently performed an inter-laboratory study that took into account two SAP samples with pronouncedly different inherent sorption properties in cement pore solution (TC 260 2019). While SAP 1 individually released the liquid intaken within just about two hours, SAP 2 was retentive on the long term (Schröfl et al. 2015, Mechtcherine et al. 2018). It was found that SAP 1 did have a positive effect, but it was limited. Contrarily, SAP 2 was clearly more beneficial. Scope of the study at hand is to connect these SAP-inherent sorption properties with the effects of the polymers inside cement-based materials. Neutron radiography imaging was selected as the experimental method, since it had proven successful in studying such water-migration kinetics previously (Schröfl et al. 2015, Wyrzykowski et al. 2015).

Two distinct climates were regarded. One mimicked ordinary central European climate with 20 °C and intermediate relative humidity. The other one represented harsh, hot and dry weather at 40 °C and very low relative humidity which is renown to provoke high evaporation rates and, consequently, intense plastic shrinkage and cracking propensity.

2 Experimental

Table 1 summarises mixture compositions, climates and imaging durations.

A climate chamber was mounted at measurement position #2 in the beamline NEUTRA at Paul-Scherrer-Institute (Mannes et al. 2017). Its interior was adjusted to the target climate. At 20 °C, the relative humidity varied between 35% and 50% for metrological reasons. There was no regulated wind streaming through the box but just shallow air stream to steer temperature and relative humidity.

The devices and raw materials for imaging at 20 °C were kept at ambient laboratory temperature (approximately 23 °C, not controlled). Those for imaging at 40 °C were pre-conditioned at that temperature until they had acquired it. All pastes, however, were prepared at the ambient temperature, whereby it was accepted that the hot materials for the 40 °C sequence cooled down a bit. Pastes of a normal Portland cement CEM I 42.5 R according to DIN EN 197-1, local tap water, and – where applicable – the SAP and a polycarboxylate-based high-range water-reducing admixture (HRWRA, aqueous solution, Glenium 51, BASF) were mixed manually by intense agitation with a spoon in a casserole. The weight of cement was 70 grams, to which the other amounts were adjusted. SAP dosage was 0.3% by weight of cement (bwoc) in accordance to previous studies (Schröfl et al. 2015, TC 260 2019). The SAP samples stemmed from the same batches as those used in (Schröfl et al. 2015, Mechtcherine et al. 2018, TC 260 2019) and were denominated accordingly: SAP 1 – inherently desorbing; SAP 2 – retentive upon absorption of cement pore solution. The HRWRA was used in the pastes of w/c (water-to-cement ratio) 0.25 only. Its dosage was 1.0% bwoc.

Table 1. Mixture compositions, climate, and imaging durations.

Specimen	w/c	SAP type	HRWRA	Temperature/°C	Relative humidity/%	Imaging time/h
Ref 0.25 20	0.25	none	yes	20	35…50	26
Ref 0.25 40	0.25	none	yes	40	2…4	21
Ref 0.50 20	0.50	none	no	20	35…50	26
Ref 0.50 40	0.50	none	no	40	2…4	21
1 – 0.25 20	0.25	1	yes	20	35…50	26
1 – 0.25 40	0.25	1	yes	40	2…4	20
1 – 0.50 20	0.50	1	no	20	35…50	26
1 – 0.50 40	0.50	1	no	40	2…4	20
2 – 0.25 20	0.25	2	yes	20	35…50	26
2 – 0.25 40	0.25	2	yes	40	2…4	20
2 – 0.50 20	0.50	2	no	20	35…50	26
2 – 0.50 40	0.50	2	no	40	2…4	20

Two pastes were prepared at the same time. After formation of a visually homogeneous paste, i.e. 2–3 min of mixing, an appropriate portion of each paste was filled in a compartment of a twin aluminium container (Fig. 1). The top face of the container was open to allow for evaporation. Immediately after these steps of preparation, the container was inserted in the climate box (Fig. 2), and imaging was initiated.

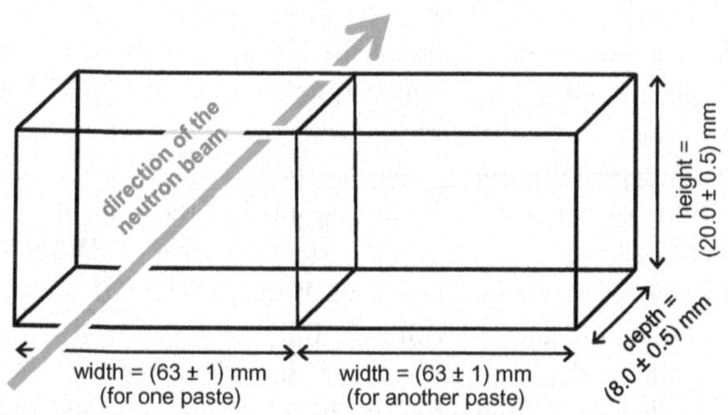

Fig. 1. Dimensions of the specimens and their alignment in the beamline.

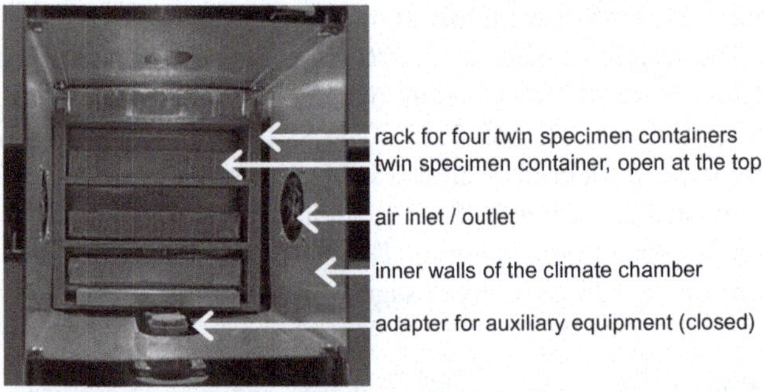

Fig. 2. View into the opened climate chamber in the direction of the neutron beam after the imaging sequence.

The time stamps of the images were re-adjusted to the start of mixing. Time-resolved imaging accorded to studies related to sorption kinetics of SAP within cement-based pastes under autogenous conditions. Details on this, as well as on image processing, especially the physical and computational strategies to derive the water contents from the grey scales, has been described in, e.g., (Schröfl et al. 2015) at length.

3 Results and Discussion

Figures 3, 4 and 5 exemplarily show selected images and calculated water contents. Visual representation is limited to 40 °C and w/c 0.50, but all specimens are discussed.

With reference to the last image of each time-resolved imaging sequence, relative water contents were calculated from the grey scale intensities. Naturally, since the final image is used as the reference, only relative water contents with respect to this stage could be obtained. The darker the shade of grey, the more water is present at the distinct location at the distinct time. Four stripes were placed over the specimen height at most suitable places. In SAP-doped samples, stripes purposefully passing the dark spots of SAP moieties allowed for quantifying the desorption from the SAP in time. In cracked specimens, water-filled (i.e. dark) streaks through slightly brighter matrix indicated the crack appearing, which later on dried quickly. While the top side of the specimens is in fact at the top side of each image, the charts of the extracted water contents show the top side of the image at the left side of the abscissa axis.

The loss of water at 40 °C was much quicker than at 20 °C for every mixture composition. Furthermore, any of the pastes with w/c 0.50 evaporated more water than those with w/c 0.25, especially during the first hours.

All height profiles of water loss at some point in time indicate more or less pronounced enrichment of humidity at the top surface in form of darker shade of grey as compared to depths of below about 2 to 3 mm. With w/c 0.50, this may be the result of bleeding. Contrarily, in the pastes with w/c 0.25, it may be a consequence of the formation of a so-called "elephant skin". Such thin, densified top layer contains finer particles than the bulk below (Wetzel et al. 2015). Obviously, the transition of water molecules to the gas phase is impaired in this zone as compared to evaporation from a

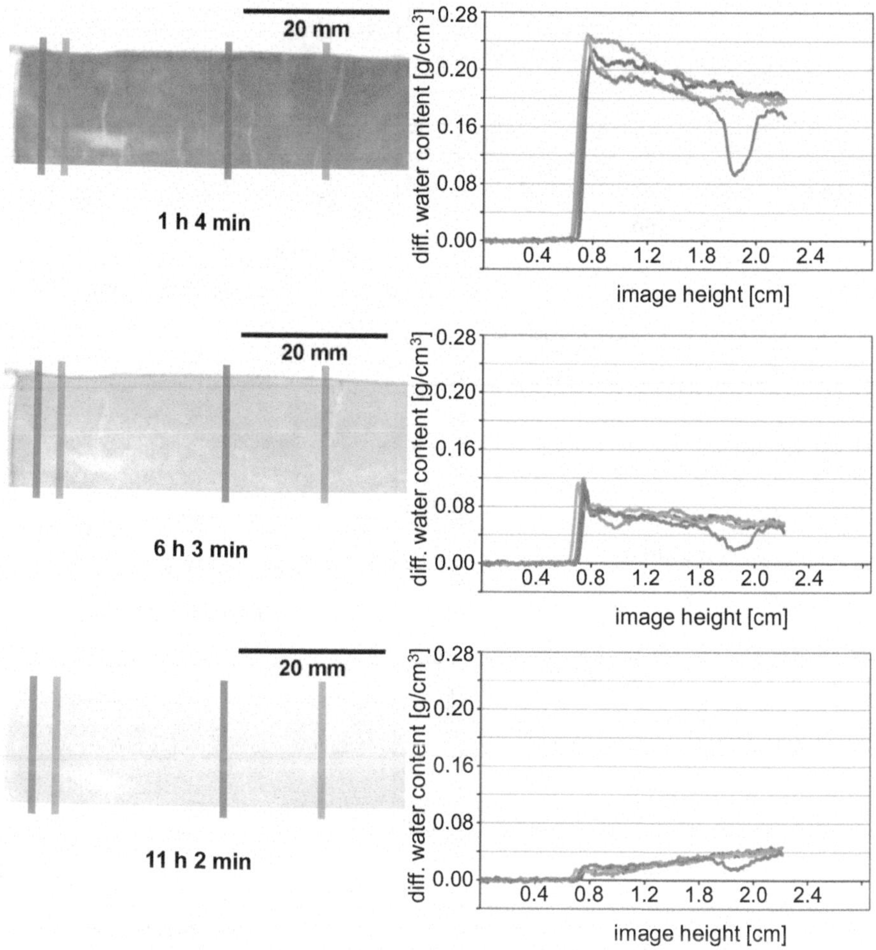

Fig. 3. Differential neutron radiography images with indication of time, stripe positions for quantification and extracted water contents: Ref., w/c 0.50, 40 °C

less dense surface layer. The supply of water from below to the surface, however, remains unaffected and it is sufficient to feed the upper region. This effect is visible for the first few hours only. Afterwards, a relatively steady drying gradient from the top to the bottom is formed.

Irrespective of the w/c and climate, the swollen SAP 1 particles release their water right from the beginning. This corresponds to both the individual sorption characteristics of the particles upon contact with cement pore solution and their behaviour in cement-based pastes under sealed (i.e. autogenous) conditions (Schröfl et al. 2015). At 20 °C, desorption practically ended after 12–15 h. At 40 °C for the paste with w/c 0.25 the release was finished after as short as 5–7 h and 8–9 h for that with w/c 0.50, respectively. Contrarily, SAP 2 released its water significantly slower and continued until approximately 18 h at 20 °C for the paste with w/c 0.25 and 20 to 21 h for the

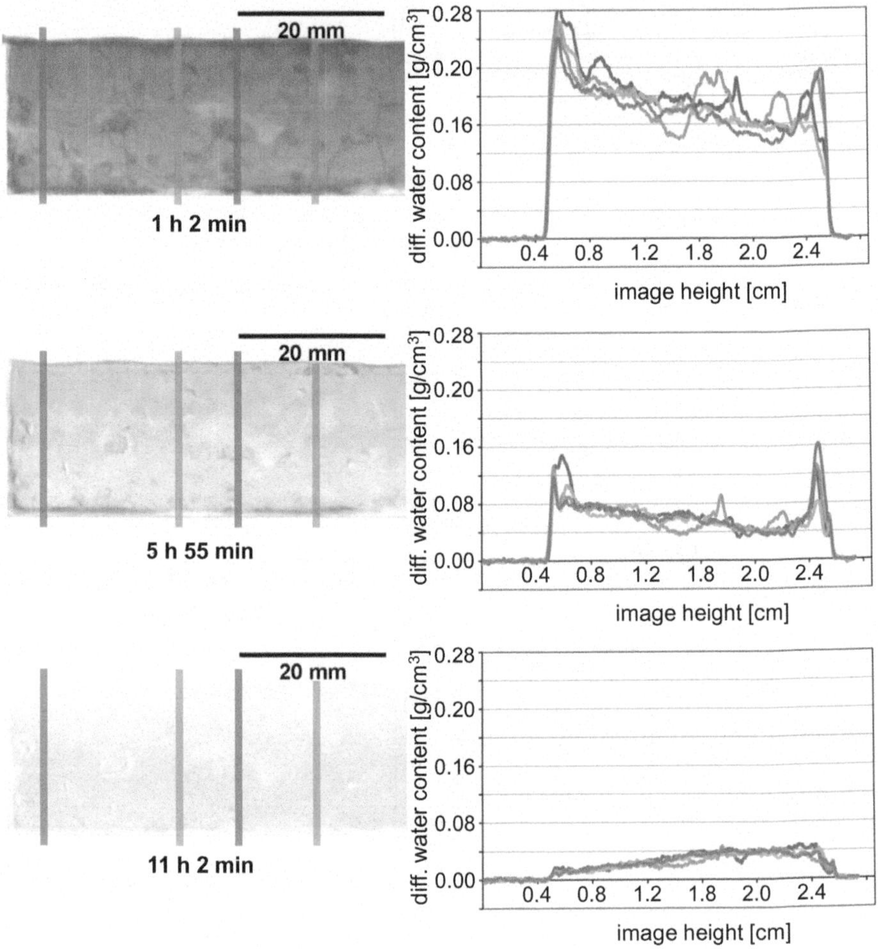

Fig. 4. Differential neutron radiography images with indication of time, stripe positions for quantification and extracted water contents: SAP 1, w/c 0.50, 40 °C

paste with w/c 0.50, respectively. At 40 °C, the desorption from SAP 2 lasted until approximately 14 h for both types of cement paste.

The SAP-free paste with w/c 0.50 at 40 °C featured several cracks, whereas that with w/c 0.25 and both references at 20 °C had less cracks. All cracks dried out quickly. At 40°C, both SAP 1 and SAP 2 prevented cracking completely. Interestingly, at 20 °C, neither SAP 2 with w/c 0.25 nor SAP 1 added to the pastes with both w/c values could completely inhibit cracking, whereas the paste with SAP 2 and w/c 0.50 remained crack-free. These findings indicate that while both SAP materials reduce the cracking propensity of cement paste, SAP 2 is more efficient in this respect.

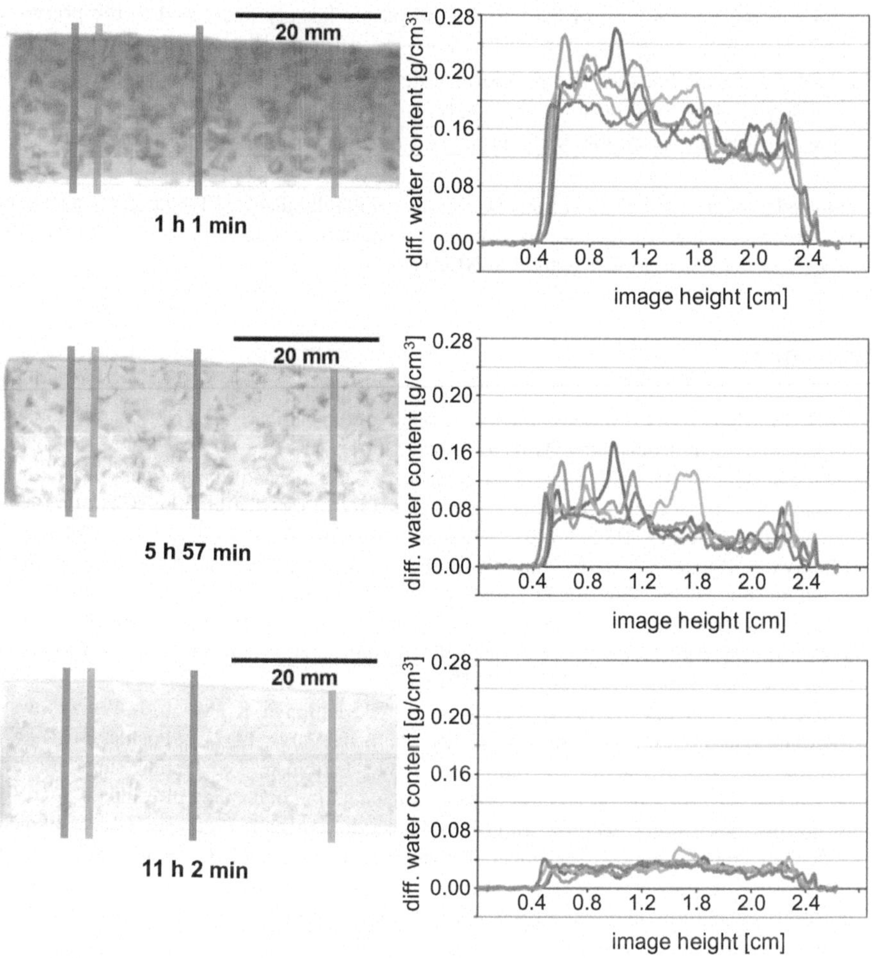

Fig. 5. Differential neutron radiography images with indication of time, stripe positions for quantification and extracted water contents: SAP 2, w/c 0.50, 40 °C

4 Conclusion

Similar to its individual sorption behaviour in extracted cement pore solution, SAP 1 released its absorbed aqueous solution right from the beginning inside the cement pastes. Such performance was nearly independent of the w/c and the climatic conditions. Contrarily, the inherently retentive SAP 2 desorbed water upon an imposed demand which could stem from either a low w/c or pronouncedly elevated evaporation from the surface of the paste due to harsh climatic conditions.

In general, the cracking propensity of the pastes largely decreased in the presence of SAP. However, SAP 2 was found to be more efficient than SAP 1 in this aspect. This can be traced back to the temporally delayed supply of water by SAP 2 upon demand. Obviously, it did not matter much by which physico-chemical mechanism the water was extracted from the SAP particles, i.e. low w/c or hot and dry climate.

Acknowledgements. The authors from TU Dresden gratefully thank SNF Floerger (Andrézieux Cedex, France) for the generous supply with the superabsorbent polymer samples. They as well thank PSI for the access to and support at NEUTRA.

References

Jensen, O.M., Hansen, P.F.: Water-entrained cement-based materials: I. principles and theoretical background. Cem. Concr. Res. **31**(4), 647–654 (2001)

Mannes, D., Schmid, F., Wehmann, T., Lehmann, E.: Design and applications of a climatic chamber for in-situ neutron imaging experiments. Phys. Procedia **88**, 200–207 (2017)

Mechtcherine, V., Gorges, M., Schröfl, C., Ribeiro, A.B., Cusson, D., Custódio, J., da Silva, E.F., Ichimiya, K., Igarashi, S.-I., Klemm, A., Kovler, K., de Lopes, A.N.M., Lura, P., Nguyen, V. T., Reinhardt, H.-W., Filho, R.D.T., Weiss, J., Wyrzykowski, M., Ye, G., Zhutovsky, S.: Effect of internal curing by using superabsorbent polymers (SAP) on autogenous shrinkage and other properties of a high-performance fine-grained concrete: results of a RILEM round-robin test. Mat. Struct. **47**(3), 541–562 (2014)

Mechtcherine, V., Snoeck, D., Belie, N.D., Klemm, A.J., Ichimiya, K., Moon, J., Wyrzykowski, M., Lura, P., Toropovs, N., Assmann, A., Igarashi, S.-I., Varga, I.D.L., Almeida, F.C.R., Erk, K., Ribeiro, A.B., Custódio, J., Reinhardt, H.W., Falikman, V.: Testing superabsorbent polymers (SAP) sorption properties prior to implementation in concrete: Results of a RILEM round-robin test. Mat. Struct. **51**, 28 (2018). https://doi.org/10.1617/s11527-018-1149-4

Schröfl, C., Mechtcherine, V., Vontobel, P., Hovind, J., Lehmann, E.: Sorption kinetics of superabsorbent polymers (SAP) in fresh Portland cement-based pastes visualized and quantified by neutron radiography and correlated to the progress of cement hydration. Cem. Concr. Res. **75**, 1–13 (2015)

Serpukhov, I., Mechtcherine, V.: Early-age shrinkage of ordinary concrete and a strain-hardening cement-based composite (SHCC) in the conditions of hot weather curing. In: Hellmich, C., Pichler, B., Kollegger, J. (eds.) Mechanics and Physics of Creep, Shrinkage and Durability of Concrete and Concrete Structures (Proceedings of CONCREEP 10), pp. 1504–1513. ASCE, Reston, VA (2015)

Snoeck, D., Pel, L., Belie, N.D.: Superabsorbent polymers to mitigate plastic drying shrinkage in a cement paste studied by NMR. Cem. Concr. Compos. **93**, 54–62 (2018)

TC 260: (RILEM Technical Committee 260 – Recommendations for use of superabsorbent polymers in concrete construction). The effect of superabsorbent polymers on the mitigation of plastic shrinkage cracking of conventional concrete, results of a RILEM inter-laboratory test, manuscript in preparation by sub-committee 2 (2019)

Wetzel, A., Glotzbach, C., Maryamh, K., Middendorf, B.: Microstructural investigations on the skinning of ultra-high performance concrete. Cem. Concr. Compos. **57**, 27–33 (2015)

Wyrzykowski, Mateusz, Lura, Pietro, Pesavento, Francesco, Gawin, Dariusz: Modeling of internal curing in maturing mortar. Cem. Concr. Res. **41**(12), 1349–1356 (2011)

Wyrzykowski, M., Trtik, P., Münch, B., Weiss, J., Vontobel, P., Lura, P.: Plastic shrinkage of mortars with shrinkage reducing admixture and lightweight aggregates studied by neutron tomography. Cem. Concr. Res. **73**, 238–245 (2015)

Kinetics of SAPs During Hardening, Drying and Healing in Cementitious Materials Studied by NMR

Didier Snoeck[1(✉)], Leo Pel[2], and Nele De Belie[1]

[1] Magnel Laboratory for Concrete Research, Department of Structural Engineering, Faculty of Engineering and Architecture, Ghent University, Tech Lane Ghent Science Park, Campus A, Technologiepark Zwijnaarde 60, 9052 Ghent, Belgium
didier.snoeck@ugent.be
[2] Transport in Permeable Media, Department of Applied Physics, Eindhoven University of Technology, Eindhoven, The Netherlands

Abstract. At early ages, plastic shrinkage can arise when a cement paste is subjected to harsh drying conditions during hardening. Furthermore, when using a low water-to-binder ratio, the cementitious material may show autogenous shrinkage during setting. Super Absorbent Polymers (SAPs) are a promising admixture to mitigate shrinkage in cement pastes. By introducing internal curing by means of the stored mixing water in the SAPs, the plastic shrinkage can be partially mitigated, next to the mitigation of autogenous shrinkage during setting of the cement paste. The kinetics of water release by the SAPs towards the cementitious matrix are an important factor. To effectively and non-destructively monitor the effects induced by the SAPs during the plastic and hardening period as a function of time, Nuclear Magnetic Resonance (NMR) was applied. Using NMR, a clear distinction could be made in terms of the free water signal and the entrained water signal for SAP particles. The SAPs are able to protect the cement paste internally from the harsh ambient drying conditions by sustaining the internal relative humidity. The plastic settlement was reduced and there was less plastic shrinkage measured. By mitigating shrinkage, shrinkage cracking can be partially prevented. However, upon acting mechanical stresses, the cementitious material may crack nevertheless. SAPs are also interesting to first seal a crack due to their swelling capacity and to heal cracks afterwards by promoting autogenous healing. This healing was also monitored by NMR as a function of healing cycles and the amount of healing products formed were estimated based on the water signals obtained by NMR. Part of the water going into the crack was used to trigger further hydration of unhydrated cement particles. Healing in wet/dry cycles was stimulated by means of SAPs and healing at high relative humidity conditions only occurred in samples containing SAPs.

Keywords: Internal curing · Shrinkage · Autogenous healing · Hydrogel · Water state

W. P. Boshoff et al. (Eds.): SAP 2019, RILEM Bookseries 24, pp. 132–139, 2020.
https://doi.org/10.1007/978-3-030-33342-3_15

1 Introduction

In the last decades, SuperAbsorbent Polymers (SAPs) are studied as a new and interesting concrete admixture (Mechtcherine and Reinhardt 2012). Not only interesting to increase the freeze-thawing resistance (Mechtcherine et al. 2017) they also mitigate shrinkage (Jensen and Hansen 2001). Shrinkage may occur during the first hours after casting while the cement paste is still in a plastic state, due to the evaporation of water near the surface leading to high capillary pressures and tension among the solid particles. If the specimen deformation is restrained, the specimen may crack. By using SAPs, the capillary pressure and plastic deformations can be reduced (Dudziak and Mechtcherine 2010).

Furthermore, due to their interesting swelling ability, they are also used for sealing purposes in occurring cracks (Lee et al. 2010; Snoeck 2018). The absorbed water may afterwards be used to trigger further hydration and calcium carbonate crystallization (Snoeck et al. 2014; Snoeck 2015).

Water and SAPs are intertwined so it would be interesting to study the water kinetics in and from SAPs used for above-mentioned various purposes. However, it is difficult to study the water kinetics of entrained water in SAPs without destructing the fragile microstructure. Nuclear Magnetic Resonance (NMR), on the contrary, is a non-destructive technique to study the water amounts (^1H) in a material. It is a powerful technique to study the cement hydration and the fluid transport in porous materials. Some studies have investigated the effect of SAPs by means of NMR tests. Friedemann et al. showed that water is available during post-internal curing of cement pastes, when most of the internal curing already took place (Friedemann et al. 2006). Nestle et al. used NMR to study the water balance in a cement paste with SAPs (Nestle et al. 2009). Recent studies evaluated the influence on internal curing (Snoeck et al. 2017) and plastic shrinkage (Snoeck et al. 2018a) and will be discussed in this paper. In terms of autogenous healing, one paper investigated the formation of hydration products by means of NMR (Huang et al. 2016). They found a water migration from cracks into the bulk paste during the healing process and were able to quantify the amount of further hydration.

This paper describes the curing by the SAPs during plastic shrinkage and autogenous shrinkage, monitored by means of NMR tests. The different water signals (free water and entrained water) obtained with NMR were compared, in order to draw conclusions on the internal curing effect as a function of time. On hardened specimens, NMR was used to study the effect of SAPs on autogenous healing and to quantify the amount of hydration products formed. The specimens were healed in conditions at high relative humidity and by means of wet/dry cycling.

2 Materials and Methods

Mixture Compositions. For the shrinkage measurements, three mixtures were studied. These were two reference cement pastes with a water-to-binder ratio (W/B) of 0.3 (**R0.30**) and 0.354 (**R0.354**), respectively. The cement used was a CEM I 52.5 N

Portland type. A polycarboxylate type superplasticizer (Glenium 51, conc. 35%, BASF) was added in an amount of 0.5 m% and 0.3 m% (mass-% of cement weight) to ensure practical workability for the R0.30 and R0.354 mixture. Based on the model of Powers and Brownyard, the entrained water amount in the SAPs for internal curing of R0.30 amounts to 0.054 (Jensen and Hansen 2001), leading to a total water-to-binder ratio of 0.354. The effective water-to-binder ratio in a SAP system is thus 0.30. A commercially available bulk-polymerized SAP (BASF, Germany) was studied. This SAP is a copolymer of acrylamide and sodium acrylate (particle size 100.0 ± 21.5 μm (n = 100)) (**A_e**). The SAP is able to absorb 305 ± 4 g/g SAP in demineralized water and 61 ± 1 g/g SAP in cement filtrate solution, as determined by a filtration test (Schröfl et al. 2017; Mechtcherine et al. 2018; Snoeck et al. 2018b). An amount of 0.22 m% (mass-% of binder weight) was added to absorb the entrained water. The same amount of superplasticizer (0.5 m%) was added as compared to the R0.30 reference system. The storage and mixing procedure are described elsewhere (Snoeck et al. 2017).

For the healing mixtures, the reference mortar mixture (**REF**) contained 608 kg/m^3 CEM I 52.5 N (Holcim, Belgium), 608 kg/m^3 Class F fly ash (OBBC, Belgium), 426 kg/m^3 silica sand (D_{50} = 170 μm; Sibelco, Belgium), 365 kg/m^3 water, 12 kg/m^3 polycarboxylate superplasticizer (Glenium 51, conc. 35%; BASF, Germany), and 26 kg/m^3 Polyvinyl-Alcohol (PVA) fibers (2 v%; 15 dtex; 8 mm cutting length; 12 cN/dtex tenacity; Kuraray, Japan). In the SAP mixture (**05A**), 0.5 m% of SAP was added on top of the reference mixture, with additional water of 84 kg/m^3. The storage and mixing procedure are described elsewhere (Snoeck 2015).

Nuclear Magnetic Resonance (NMR) Setup. Sample containers (cylinder ø 27 mm × 100 mm high) were filled with fresh cement paste for approximately 45 mm for the plastic shrinkage, and completely filled and sealed for the autogenous shrinkage test. Dry air was blown over the top of the specimen in such a way that the evaporation rate was 1.04 ± 0.02 kg/h/m^2, as in that way plastic shrinkage will likely develop (ASTM Standard C1579–06). No dry air flow was used for the autogenous shrinkage test. In case of the healing tests, a sample with the same diameter and a height of 40 mm was pre-cracked perpendicular to the axis of the cylinder by means of a Brazilian splitting test at an age of 28 days. This sample was inserted in the sample container for measurement in the NMR setup. The healing in different conditions was monitored by subsequent mass scale and NMR measurements. The mass scale measurements served as calibration of the NMR water intensity signal. The healing conditions were storage at a relative humidity (RH) of 60%, at more than 90% and in wet/dry cycling. The latter is storing the sample for 1 h in water and 23 h at a RH of 60%. Specimens were investigated after 1, 3, 7, 14 and 28 healing cycles.

An external magnetic field B_{main} of 0.8 T was applied corresponding to a frequency of 34 MHz. This field was provided by a water-cooled iron-cored electromagnet with poles 50 mm apart from each other. A coil was placed around the sample for creating and receiving the radiofrequency fields. A Faraday shield was added between the coil and the sample to suppress the effect of the changes of the dielectric permittivity by variations of the moisture content, hence making the NMR measurements quantitative. The intensity of the signal was recorded along the height of the specimen by moving the specimen by means of a step motor (in steps of 0.84 mm). The spin-echo signal S was recorded after an echo time T_E equal to 180 μs. Carr-Purcell-Meiboom-Gill (CPMG) sequences were measured along the height and the overall total water signal intensity was recorded. Fast Laplace Inversion (FLI) was used to obtain the relaxation times T_2 as a spectrum. All peaks were normalized to the first signal to obtain the percentage. The weighed percentage of each peak was then multiplied with the normalized signal, resulting in the relative ratio of each studied peak as a function of time.

3 Results

The most ideal swollen SAP size for mitigating autogenous shrinkage is approximately 100 μm (Jensen and Hansen 2001; Snoeck et al. 2015) and this range was studied in this paper. First, its effect on possible mitigation of plastic shrinkage was investigated. Figure 1 shows the water signal intensity along the height of the hydrating specimen when being put in harsh drying conditions with an evaporation rate of approximately 1 kg/h/m^2, prone to showing plastic shrinkage. The curves progress from high values to lower values as water is used for hydration. When looking at the bottom of the specimen (left-hand side of the curves) there is a first quick decrease of water during setting, reaching the final setting where the lines are closer together at around 6 a.u. of water signal intensity. After this time, which corresponds to the final setting as obtained with the Vicat needle test, the curves shift downward once again, reaching the lower bound edge (around 1 a.u.) in an asymptotic way.

When studying the R0.30 specimen, a drop in signal is found right after point of final setting (dashed area). Water seems to be migrating towards the exposed surface, which is found on the right hand side. The same is even more prominent when studying the R0.354 system. The curves shift downward and the surface dries out more quickly compared to the inner structure. Water from these regions thus moves towards the outer surface due to plastic drying. When investigating the A_e specimen, only prior to final setting there seems to be water movement. Visually this was seen as a quick release of water of the SAPs at the exposed surface but the water content as compared to the inner cementitious structure was maintained. This first water movement is responsible for the observed 2.5 mm settlement, which is still less compared to the reference settlement, being 3 mm and 4.5 mm for R0.30 and R0.354, respectively. The inner atmosphere seems to be maintained as the curves are positioned horizontally with respect to each other.

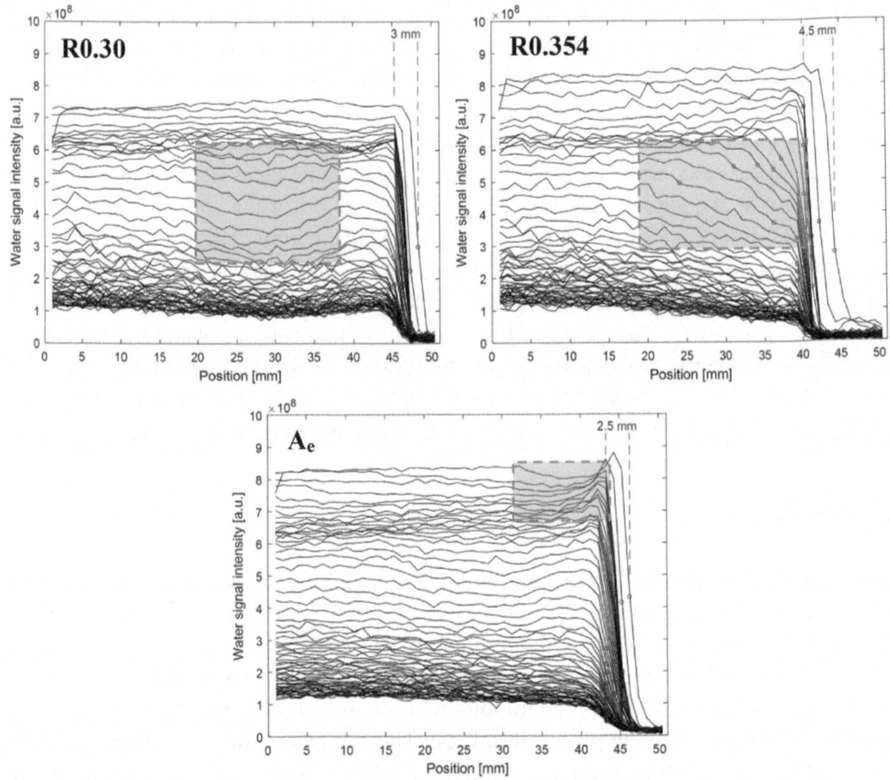

Fig. 1. Water signal intensity [arbitrary unit] as a function of the height of the specimen with the exposed surface to the right hand side with 44 min intervals from top to bottom (total time of 48 h) (Snoeck et al. 2018a).

Internally, the specimens are not showing a decrease in water signal. This result corresponds to data when being tested under sealed conditions without an air flow. When studying the internal curing by the SAP particles of specimens in sealed conditions, it is very important to use a SAP with the ideal properties. If the water is released too soon (i.e. before setting), the microstructure will be completely different and the effective water-to-cement ratio changes. If the water is released too late, the purpose of internal curing vanishes. This was therefore studied by NMR by investigating a sample positioned in the middle of the NMR setup. Figure 2 shows the comparison of the free water fraction and the entrained water fraction stored in a SAP which can mitigate autogenous shrinkage (Snoeck et al. 2015) and is able to maintain the internal relative humidity (Snoeck et al. 2017). It was found that the free water was consumed after final setting (around 11 h as determined with a Vicat needle test) and from 23 h onwards, the stored entrained water was consumed. This sheds new light on the internal curing mechanism of SAPs. NMR is thus able to provide information on the water kinetics of entrained water. This seems to be non-linear with respect to the Powers and Brownyard model, as depicted by the solid line in Fig. 2.

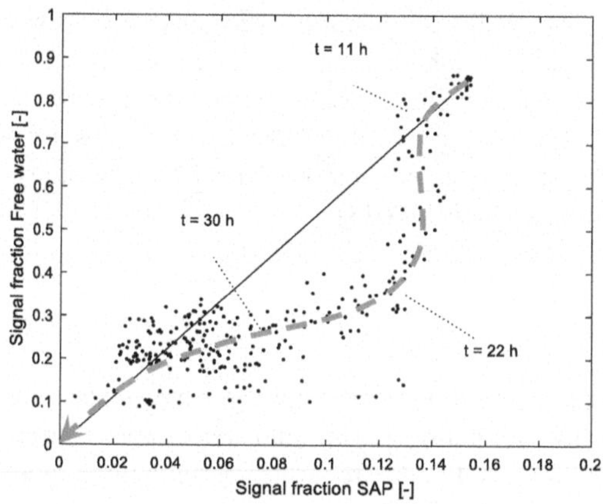

Fig. 2. The signal fraction of free water the cement paste and entrained water in SAPs [-] showing the kinetics upon internal curing (Snoeck et al. 2017).

Next to shrinkage measurements, NMR was also used to study the healing effect by SAPs. As they release absorbed water towards the cementitious matrix, they may induce further hydration and calcium carbonate crystallization (Snoeck 2018). NMR provides information on different water signals and by studying these signals, the amount of formed healing products could be estimated. As the sensitivity of the custom NMR used was not able to provide data on strongly bound water, only the free water in gel pores and larger capillaries with longer relaxation times could be studied. As we know, based on mass scale readings, how much water was added to the system, the calibrated NMR water signal provided us with the amount of water used for autogenous healing, as shown in Fig. 3.

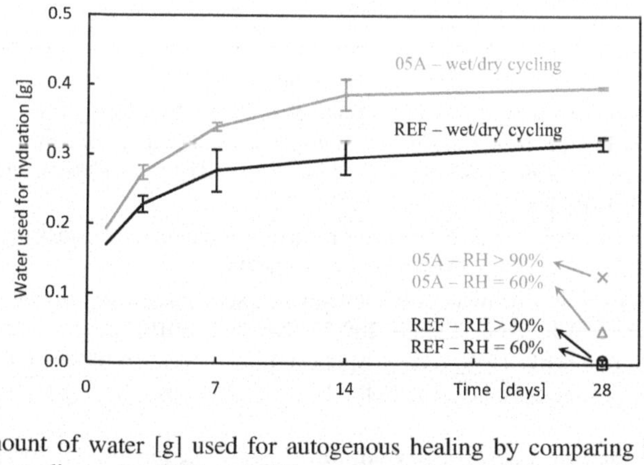

Fig. 3. The amount of water [g] used for autogenous healing by comparing the NMR signal with mass scale readings, for REF and 05A samples, after healing in wet/dry cycles, and in relative humidity (RH) of more than 90% and of 60%.

When comparing the obtained amount of consumed water for autogenous healing, the SAP samples always show a higher amount of healing. This is due to the storage of absorbed water and the subsequent release towards the cementitious matrix, leading to a stimulated autogenous healing. As the SAPs can also absorb moisture, there is a minor healing in SAP specimens when stored at relative humidity conditions, as was also reported in literature (Snoeck 2018).

4 Conclusions

NMR proved to be an effective tool to study the water release by SAPs towards the cementitious matrix for mitigating plastic and autogenous shrinkage. The entrained water signal could be compared to the free water signal, giving more information on the kinetics of water release by the SAPs.

NMR was also useful to study the amount of water used for autogenous healing of the specimens. It was confirmed that SAP specimens show more healing in different healing conditions of wet/dry cycling and storage at high RH conditions.

Acknowledgements. As a Postdoctoral Research Fellow of the Research Foundation-Flanders (FWO-Vlaanderen), D. Snoeck would like to thank the foundation for the financial support for this fellowship (12J3617 N) and for the research stay as Guest Researcher at the Technical University of Eindhoven (TU/e) (V420617 N). The authors want to thank Dr. A. Assmann and Dr. G. Herth from BASF for providing SAP A.

References

Dudziak, L., Mechtcherine, V.: Enhancing early-age resistance to cracking in high-strength cement based materials by means of internal curing using super absorbent polymers. In: Brameshuber, W. (ed.) International RILEM Conference on Material Science, RILEM Publications S.A.R.L., pp. 129–139 (2010)

Friedemann, K., Stallmach, F., Kärger, J.: NMR diffusion and relaxation studies during cement hydration—a non-destructive approach for clarification of the mechanism of internal post curing of cementitious materials. Cem. Concr. Res. **36**(5), 817–826 (2006)

Huang, H., Ye, G., Pel, L.: New insights into autogenous self-healing in cement paste based on nuclear magnetic resonance (NMR) tests. Mat. Struc. **47**, 1–15 (2016). https://doi.org/10.1617/s11527-015-0664-9

Jensen, O.M., Hansen, P.F.: Water-entrained cement-based materials I. principles and theoretical background. Cem. Concr. Res. **31**(4), 647–654 (2001)

Lee, H.X.D., Wong, H.S., Buenfeld, N.R.: Potential of superabsorbent polymer for self-sealing cracks in concrete. Adv. Appl. Ceram. **109**(5), 296–302 (2010)

Mechtcherine, V., Reinhardt, H.W. (eds.): Application of Super Absorbent Polymers (SAP) in Concrete Construction. Springer, The Netherland (2012). https://doi.org/10.1007/978-94-007-2733-5

Mechtcherine, V., Schröfl, C., Wyrzykowski, M., Gorges, M., Cusson, D., Margeson, J., et al.: Effect of superabsorbent polymers (SAP) on the freeze-thaw resistance of concrete: results of a RILEM interlaboratory test. Mat. Struc. **50**(14), 1–19 (2017). https://doi.org/10.1617/s11527-016-0868-7

Mechtcherine, V., Snoeck, D., Schröfl, C., De Belie, N., Klemm, A.J., Ichimiya, K., et al.: Testing superabsorbent polymer (SAP) sorption properties prior to implementation in concrete: results of a RILEM Round-Robin test. Mat. Struc. (2018). https://doi.org/10.1617/s11527-018-1149-4

Nestle, N., Kühn, A., Friedemann, K., Horch, C., Stallmach, F., Herth, G.: Water balance and pore structure development in cementitious materials in internal curing with modified superabsorbent polymers studied by NMR. Microporous and Mesoporous Mater. **125**(1–2), 51–57 (2009)

Schröfl, C., Snoeck, D., Mechtcherine, V.: A review of characterisation methods for superabsorbent polymer (SAP) samples to be used in cement-based construction materials - Report of the RILEM TC 260-RSC. Mat. Struct. **50**(197), 1–19 (2017). https://doi.org/10.1917/s11527-017-1060-4

Snoeck, D. (2015): Self-healing and microstructure of cementitious materials with microfibres and superabsorbent polymers. Doctor in Civil Engineering: Construction Design, Ghent University, Ghent

Snoeck, D.: Superabsorbent polymers to seal and heal cracks in cementitious materials. RILEM Tech. Lett. **3**, 32–38 (2018)

Snoeck, D., Jensen, O.M., De Belie, N.: The influence of superabsorbent polymers on the autogenous shrinkage properties of cement pastes with supplementary cementitious materials. Cem. Concr. Res. **74**, 59–67 (2015)

Snoeck, D., Pel, L., De Belie, N.: The water kinetics of superabsorbent polymers during cement hydration and internal curing visualized and studied by NMR. Sci. Rep. **7**(9514), 1–14 (2017). https://doi.org/10.1038/s41598-017-10306-0

Snoeck, D., Pel, L., De Belie, N.: Superabsorbent polymers to mitigate plastic drying shrinkage in a cement paste as studied by NMR. Cem. Concr. Compos. **93**, 54–62 (2018a)

Snoeck, D., Schröfl, C., Mechtcherine, V.: Recommendation of RILEM TC 260 RSC: testing sorption by superabsorbent polymers (SAP) prior to implementation in cement-based material. Mat. Struct. **51**(116), 1–7 (2018b)

Snoeck, D., Van Tittelboom, K., Steuperaert, S., Dubruel, P., De Belie, N.: Self-healing cementitious materials by the combination of microfibres and superabsorbent polymers. J. Intell. Mat. Syst. Struct. **25**(1), 13–24 (2014). https://doi.org/10.1177/1045389X12438623

Rheology and Admixtures

The Role of Chemical Admixtures in the Formulation of Modern Advanced Concrete

Johann Plank(✉) and Manuel Ilg

Chair for Construction Chemistry, Technische Universität München, Lichtenbergstraße 4, 85747 Garching, Germany
sekretariat@bauchemie.ch.tum.de

Abstract. Chemical admixtures constitute indispensable ingredients for the production of modern advanced concrete. In developed countries, at least 80% of the concrete produced contains one or several admixtures. They include plasticizers, superplasticizers, retarders, accelerators, stabilizers, defoamers, foamers and shrinkage reducers, to name the most important classes. With their help it is possible to optimize the properties of fresh and hardened concrete in such way as to adapt better to local climate and processing conditions and to enhance the mechanical properties and durability. Furthermore, highly sophisticated products such as ultra-high strength concrete (UHPC) or self-levelling and self-compacting concrete (SCC) became possible only with the invention of specific high performance admixtures.

This article gives an overview of major classes of chemical admixtures (e.g. PCE superplasticizers, C-S-H-PCE nanocomposites, stabilizers for SCC, shrinkage-reducing agents) and their current status of development. The main technologies will be described and their role in the formulation of modern advanced concrete will be highlighted. Finally, an outlook on potential developments in the future (e.g. improved curing agents, admixtures which enhance the ductility of concrete) will be provided.

Keywords: Admixtures · PCE superplasticizers · Shrinkage-reducing agents · C-S-H-PCE nanocomposites · Viscosity modifying agents

1 Polycarboxylate (PCE) Superplasticizers

1.1 Current PCE Technology

Polycarboxylate superplasticizers are added to the fresh concrete for the dispersion of the cement particles. They produce a highly flowable concrete which can be placed at the construction site much easier. Additionally, lower w/c ratios can be applied thus facilitating the manufacturing of building materials with high mechanical strength and long durability. PCE-based admixtures have taken an unprecedented rise since their invention in 1981 (Hirata 1981). It is estimated that in 2014, the global volume of PCE produced exceeded 3 mio. tons, based on 30% liquid concentration. Meanwhile, the term "PCE" includes a huge variety of chemically often substantially different

W. P. Boshoff et al. (Eds.): SAP 2019, RILEM Bookseries 24, pp. 143–157, 2020.
https://doi.org/10.1007/978-3-030-33342-3_16

polymers, with significant variances in performance characteristics. In the following, the main classes of PCE products on the market are described and their general chemical composition is exhibited in Fig. 1.

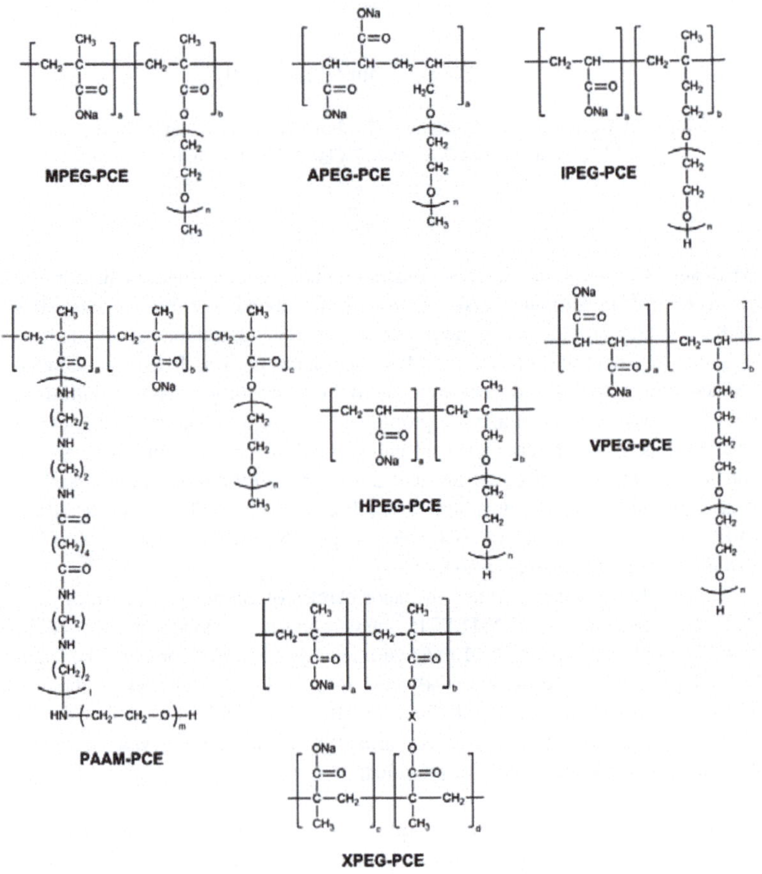

Fig. 1. Chemical structures of the different classes of PCE products currently produced by the industry.

MPEG-Type PCEs: They constitute the first type of PCE which was invented in Japan. MPEG-PCEs can be synthesized either via aqueous free radical copolymerization of methacrylic acid with an ω-methoxy poly(ethylene glycol) methacrylate ester macromonomer (this route is predominantly used by the industry) (Plank et al. 2008) or by esterification ("grafting") of short chain poly(meth)acrylic acid with ω-methoxy poly(ethylene glycol) (Guicquero et al. 1999). Note that both synthesis routes can lead to substantially different products, even when exactly the same molar ratios of monomers are used. Via esterification, a PCE polymer exhibiting a regular (statistical) repartition of side chains along the main chain is achieved while gradient polymers exhibiting a decreasing side chain density along the backbone are formed from the copolymerization process as a result of the higher reactivity of the ester macromonomer versus methacrylic acid (Pourchet et al. 2012). Performance tests have revealed that in many cases gradient polymers perform better, because their blocks of polymethacrylic acid allow higher adsorption on cement. One major disadvantage of MPEG-PCEs is their limited stability (especially when acrylate instead of methacrylate ester macromonomers are used) which derives from hydrolysis of the ester linkage between the main and the side chain. Furthermore, the diol or diester content present in the raw materials must be kept below 1% to avoid undesirable crosslinking (Paas 2015).

APEG-Type PCEs: This kind is prepared via free radical copolymerization from α-allyl-ω-methoxy or ω-hydroxy poly(ethylene glycol) ether and maleic anhydride or acrylic acid as key monomers, either in bulk or in aqueous solution (Akimoto 1992). APEG-PCEs always possess a strictly alternating monomer sequence (ABAB), because the allyl ether macromonomer does not homopolymerize as a consequence of mesomeric stabilization of the allyl radical. This stabilization causes allyl ethers to react rather slowly and can lead to low conversion rates for the macromonomer. Polymerization in bulk works well for side chain lengths of up to 34 EO units while polymerization in water typically yields copolymers possessing very short trunk chains ("star polymers") made of ~10 repeating units only which however were found to exhibit superior dispersing performance. The disadvantages of aqueous copolymerization are longer reaction times, lower conversion rates and lower concentration of the finished PCE solution.

Initially, APEG-PCEs suffered from a reputation of causing delayed plastification (i.e. the slump of concrete first increased over ~30 min to reach a maximum, and then dropped). Meanwhile, this problem has been solved, for example by incorporation of specific comonomers as spacer molecules such as styrene or allyl maleate which can modulate the conformational flexibility of the trunk chain (Plank and Lange 2012). This method provides PCE molecules with pronounced stiffness which can adsorb faster and thus avoid the effect of delayed plastification.

VPEG-Type PCEs: Such PCEs are obtained by aqueous free radical copolymerization of e.g. 4-hydroxy butyl poly(ethylene glycol) vinyl ether and maleic anhydride or acrylic acid (Albrecht 1996). Their polymerization must be conducted at temperatures <30 °C to avoid vinyl ether monomer degradation. As a result, a specific low temperature initiator such as Vazo 50 (2,2'-Azobis (2-methyl propionamidine) dihydrochloride) is required. The advantage of the vinyl over the allyl ether technology is the much higher reactivity of vinyl ethers.

HPEG-Type PCEs: Here, α-methallyl-ω-methoxy or ω-hydroxy poly(ethylene glycol) are used as macromonomers in copolymerization with e.g. acrylic acid (Hamada et al. 2001). This kind of PCE which is easy to manufacture in large industrial scale emerged a few years ago, especially in China. There, even a process has been developed where copolymerization is performed at room temperature and is applied in many factories (Wang et al. 2013). Most HPEG-PCEs can outperform the MPEG- or APEG-PCEs with respect to their dispersing ability.

IPEG-Type PCEs: This type of PCE (sometimes also referred to as TPEG-PCE) is synthesized from isoprenyl oxy poly(ethylene glycol) ether as macromonomer by copolymerization with e.g. acrylic acid (Yamamoto 2004). In recent years, this PCE has become quite popular, especially in Japan and China, because of its excellent performance which often exceeds that of any other type of PCE, and its simple preparation utilizing free radical copolymerization. A disadvantage of IPEG-PCEs is their potential to decompose into isoprene, water and glycol (Nagare 2006). To prevent this undesired process, the IPEG macromonomer and the IPEG-PCE should not be handled in bulk, but always kept in aqueous solution.

XPEG-Type PCEs: It has been established before that the ability of an individual PCE molecule to cover as much surface area on cement as possible directly correlates to its dosage (Ohta 1997). Hence, polymers which stretch out further on the surface are believed to present more effective PCEs. Following this concept, slightly crosslinked PCE molecules utilizing diesters (e.g. synthesized from PEG and methacrylic acid or maleic anhydride) were shown to provide enhanced dispersion (Tahara 1995).

PAAM-Type PCEs: These zwitterionic PCEs possess mixed side chains composed of polyamidoamine (PAAM) and PEO segments. This structural motif distinguishes them fundamentally from all other PCEs which exclusively contain PEO/PPO side chains. The PAAM-type PCE is said to fluidify cement at w/c ratios as low as 0.12 (Amaya 2000). Its disadvantage is the high cost of the PAAM side chain.

1.2 New PCE Products

Industrial and academic researchers continue to develop and introduce new and improved polymers, despite of the great diversity of already existing PCE products. Those include:

Organo-Silane (OSi) Modified PCEs: They can be prepared by incorporating either 3-trimethoxysilyl propyl methacrylate (MAPTMS) or N-maleic γ-amidopropyl triethoxy silane (MAPS) as a new comonomer into a conventional PCE, e.g. the MPEG-type (Fig. 2) (Fan et al. 2012, Witt 2012). The consideration behind this concept was to achieve a chemical bond between C-S-H and the superplasticizer, made possible through condensation of silanol (-Si-OH) groups present in both compounds. If formed, such a bond would anchor the PCE molecule irreversibly on the surface of hydrating cement and prevent its desorption e.g. by sulfate ions or anionic retarders resulting from competitive adsorption.

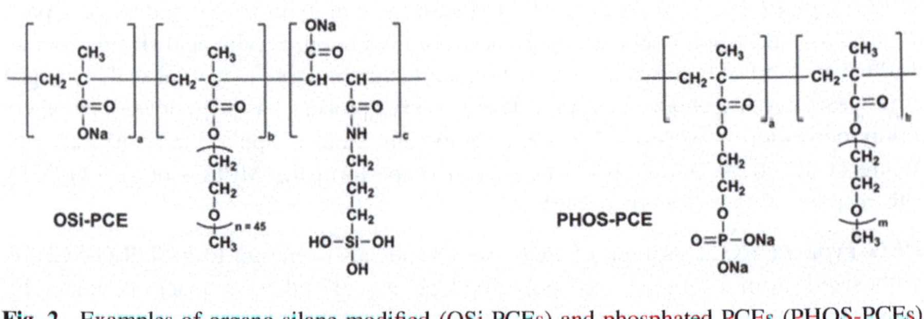

Fig. 2. Examples of organo-silane modified (OSi-PCEs) and phosphated PCEs (PHOS-PCEs).

Phosphated (PHOS) PCEs: Superplasticizers generally achieve their dispersing power through adsorption on the surface of cement, especially on ettringite (Yoshioka et al. 2002). Such adsorption is facilitated through anionic anchoring groups which typically include carboxylate or dicarboxylate groups. Some years ago it has been shown that phosphonate presents a more powerful anchoring group than carboxylate (Mosquet et al. 1997). Very recently, novel superplasticizers have been presented which incorporate phosphate as an anchoring group (Kraus 2011, Dalas et al. 2015). Phosphatation can be accomplished by esterification of e.g. hydroxyethyl methacrylate with phosphoric acid, leading to the PCE copolymer shown in Fig. 2. The phosphated PCEs are said to adsorb on cement almost instantaneously which presents a major advantage in specific concrete and dry-mix mortar applications. Furthermore, they appear to be more sulfate-tolerant, compared to conventional PCE superplasticizers, and often require lower dosages (Stecher and Plank 2019).

1.3 Tailoring PCEs to Specific Applications

Recently, substantial progress has been made in the optimization of current PCE products for difficult applications. Those include concretes of particularly low w/c ratios (<0.30) and the compatibility of PCEs with clay contaminants occurring in aggregates.

Stickiness of Concrete at Low w/c Ratio: The problem of stickiness and slow flow of concrete prepared at low w/c ratio is well-known and was solved as follows: It was found that the hydrophilic-lipophilic balance (HLB) value of a PCE molecule determines whether the concrete admixed with this polymer exhibits slow or fast flow (Lange et al. 2014). According to this study, PCE molecules should be as hydrophilic as possible and their HLB value should be >18.5. Such PCEs (preferably of IPEG- and APEG-type) produce cement pastes with particularly low plastic viscosity and exhibit fast flow without any stickiness. Such rheologically optimized concrete is easier to pump, spread and compact and presents a huge step forward in improving the workability of high-strength concretes of low w/c ratios.

Enhanced Clay Tolerance: Over the last years, applicators have observed that PCE superplasticizers – unlike polycondensates – exhibit a pronounced sensitivity to clay and silt contaminants (Jeknavorian et al. 2003, Atarashi et al. 2004). As a result, their performances are greatly reduced or the PCEs become entirely ineffective. Montmorillonite, a 2:1 smectite clay, has been found to be more harmful than other clay minerals such as kaolinites or muscovites (Lei and Plank 2014). Generally, the capacity of clays to sorb water, hydrate and swell leads to more viscous cement pastes. This effect results in a loss of workability or a higher water demand, independent of whether a superplasticizer is present or not.

Previous research has established that in cement pore solution, the surfaces of bentonite clay particles become positively charged as a result of Ca^{2+} adsorption onto the negative alumosilicate layers. Onto these surfaces, polyanionic superplasticizers such as polycondensates or polycarboxylates adsorb, thus resulting in a partial depletion of superplasticizer from the pore solution. This way, clay competes with cement for superplasticizer molecules. Moreover, PCE polymers can intercalate chemically into the interlayer space between the individual alumosilicate layers of specific clay minerals, especially montmorillonite (bentonite), resulting in an organo-mineral phase whereby their poly(ethylene glycol) side chains occupy the interlayer space, as is shown in Fig. 3. This reaction with clay is specific for PCEs and is a consequence of their PEO side chains, as was evidenced by XRD measurements (Ng 2012a). Consequently, PCEs can be used up by clay by both surface adsorption and chemical sorption whereas polycondensates such as BNS are consumed only by surface interaction (Jardine 2002, Ng and Plank 2012b). This explains why PCEs are significantly more affected by clay than polycondensates.

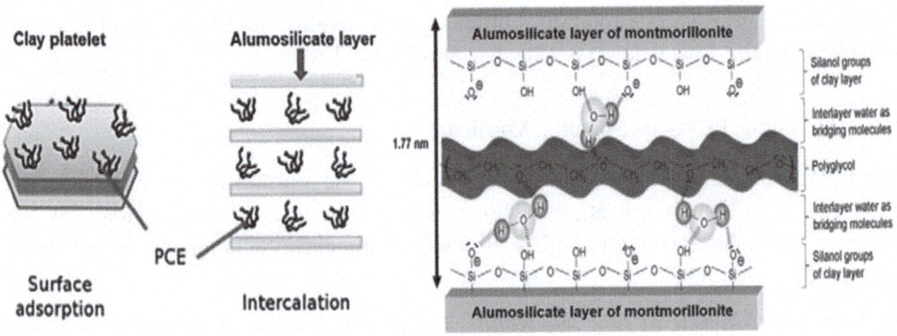

Fig. 3. Fundamental types of interaction between PCE and montmorillonite clay (left) and chemical sorption (intercalation) of a poly(ethylene glycol) side chain in between alumosilicate layers (right).

The industry has developed several strategies to mitigate the negative effects of clay on PCEs. The first concept includes the use of sacrificial agents.

Analysis of sorbed amounts of individual PCE constituents (backbone, represented by poly(methacrylic acid) and side chain, represented by poly(ethylene glycol)) revealed that the side chain sorbs in large amounts on clay ($\sim$400 mg MPEG/g clay) while the polymer trunk is consumed much less ($\sim$30 mg PMA/g clay) (Ng 2012a). This not only signifies that the PEO side chain present in PCE provides the main interaction with clay; it also offers a remedy for the problem whereby pure PEG or MPEG are utilized as sacrificial agents to occupy the interlayer spaces while the PCE molecule which exhibits a lower tendency to intercalate as a result of its anionic charge is preserved and can thus interact with the cement to achieve dispersion (Ng and Plank 2012b). As another remedy, addition of cationic polymers which inhibit the swelling of clay entirely has been proposed (Jacquet 2006). This method offers the advantages of zero water consumption because the clay will not hydrate at all. Additionally, the interlayer spacing will not be accessible for the PCEs.

Obviously, the best solution to the incompatibility problem of PCE and clay would be a novel PCE structure which does not contain PEO side chains. Recently, such polymers have been synthesized using either hydroxy alkyl esters of methacrylic acid or vinyl ethers as side chain bearing macromonomers (Lei and Plank 2012).

Utilizing XRD analysis, it was found that indeed these novel polycarboxylates do not undergo side chain intercalation with clay and adsorb in small quantities only (~25 mg polymer/g clay). Consequently, they exhibit robust performance even in the presence of clay contaminants. This behavior perfectly confirms the concept of non-PEO side chains as a remedy for the intercalation problem of conventional PCEs into clay structures.

2 Early Strength Enhancing Admixtures

A recent invention includes the application of C-S-H-PCE nanocomposites as seed crystals for the hydration of the silicate phases C_3S and C_2S (Nicoleau et al. 2011, 2013). The nanocomposites can be prepared by combining aqueous solutions of e.g. sodium silicate and calcium nitrate with a PCE solution. The resulting precipitate contains nanofoils of C-S-H with surface adsorbed and possibly intercalated PCE (Fig. 4).

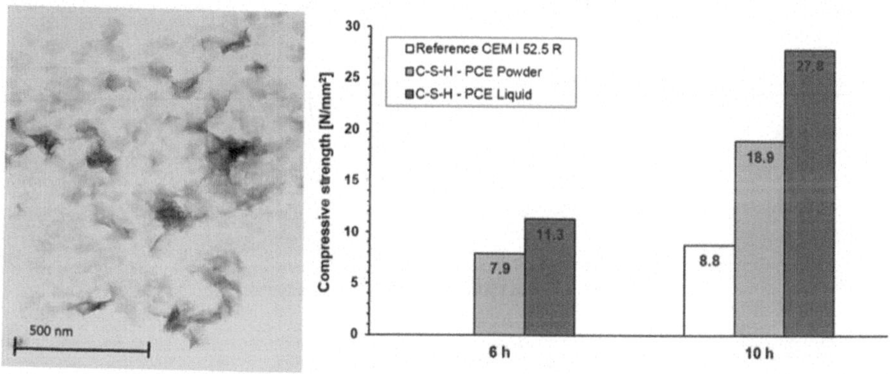

Fig. 4. TEM image of C-S-H-PCE nanocomposite foils (left) and their effectiveness as strength enhancing seeding material for a mortar prepared from CEM I 52.5R (right).

The nanofoils greatly accelerate the silicate hydration by reducing the free activation energy ΔG of the crystallization to zero. In cement hydration this barrier needs to be overcome to initiate C-S-H nucleation. The result is a much enhanced early strength development, especially after 6–12 h of hydration, without sacrificing the final strength as is the case for most common accelerators such as e.g. calcium nitrate, sodium silicate, sodium aluminate or aluminum dihydroxy formate (Fig. 4). Recently, it has been found that C-S-H-PCE nanocomposites also enhance the early strength of blended cements containing e.g. fly ash or calcined clays by accelerating the pozzolanic reaction (Kanchanason and Plank 2018).

3 Stabilizers

For highly dispersed concretes such as e.g. self-compacting concrete (SCC), polymeric stabilizers (also referred to as viscosity modifying agents, VMAs) are frequently applied to prevent disintegration and bleeding. Common stabilizers include welan gum, curdlan, hydroxypropyl cellulose, polyethylene glycol (Hibino 2000), and ATBS-based copolymers. Among the latter, two types have become quite popular in SCC mixes. The first one constitutes a terpolymer prepared via aqueous free radical copolymerization from 2-acrylamido-2-methylpropane sulfonic acid (ATBS), N-vinyl acetamide (NVA), acrylonitrile (ACN) and acrylamide (AA) while the second one comprises ATBS, N,N-dimethyl acrylamide (NNDMA) and, in some versions, tristyrylphenol poly(ethylene glycol) methacrylate ester as a third monomer. The ATBS-NNDMA copolymers can be prepared either via aqueous free radical copolymerization or through gel polymerization utilizing the Norrish-Trommsdorf effect (Futami 2003, Schinabeck 2005). The chemical structures of the ATBS-based stabilizers are displayed in Fig. 5. Both ATBS copolymers constitute linear molecules exhibiting high stiffness, owed to hydrogen bridging between ATBS and the neighboring NVA or NNDMA monomer.

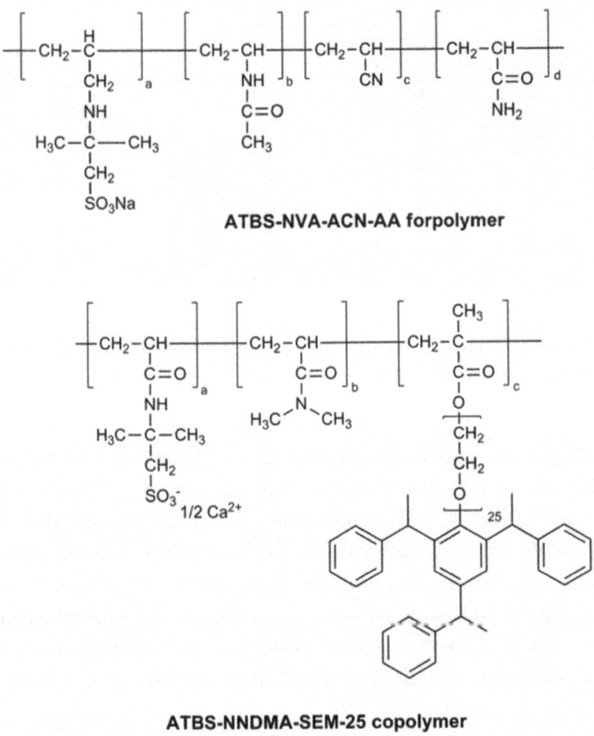

Fig. 5. Chemical structures of the ATBS-NVA-ACN-AA and ATBS-NNDMA-SEM copolymers commonly used as stabilizers in SCC.

Applicators of VMAs are well aware that these products not only can provide the desired effect, but also significantly impact on the rheology of concrete in a way that the fluidizing effect from PCE can be lost. Hence, a counterproductive (antagonistic) effect can occur which renders application of those stabilizing polymers tricky. To improve this situation, the interaction of PCE superplasticizers with ATBS/NNDMA and welan gum VMAs has been studied thoroughly.

Surprisingly, for the ATBS/NNDMA stabilizer it was found that at low dosages (0–0.1 wt%) it acts as a viscosifier in concrete while at higher additions it provides a strong dispersing effect (Fig. 6).

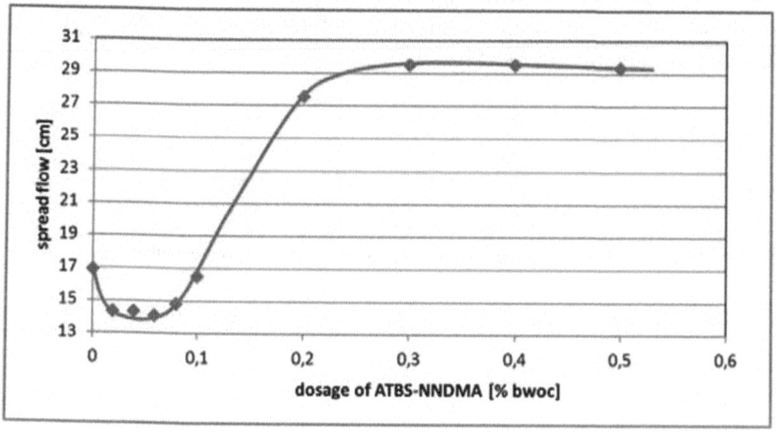

Fig. 6. Cement paste flow as a function of ATBS-NNDMA stabilizer addition (no PCE present).

Furthermore, when combined with PCE it is the stabilizing polymer which determines the flow regime, and not the PCE (Plank 2015). The reason behind this effect is that the ATBS copolymer preferably adsorbs on cement and thus prevents the PCE from adsorbing and becoming effective. The results suggest that when PCEs are combined with this ATBS copolymer, then a stabilizer dosage of >0.1% bwoc should be applied to avoid its thickening effect.

For welan gum VMA, a different scenario was found. According to these results, the stabilizing effect of welan gum biopolymer solely relies on its strong viscosifying effect on the cement pore solution which originates from its high adsorption on cement. Thus, with increased concentrations welan gum starts to destroy the fluidity generated by PCEs (Üzer and Plank 2016). Consequently, opposite to the ATBS/NNDMA stabilizer which requires a minimum dosage to avoid thickening, additions of welan gum to the PCE concrete should be kept as low as possible to avoid its negative effect on concrete rheology.

The investigations presented here suggest that admixture combinations are by no means trivial, and that understanding their mechanism of interaction with cement can help to optimize their performance.

4 Shrinkage-Reducing Admixtures

During its hydration and hardening, mortar and concrete undergo autogenous (= chemical) and dry (= physical) shrinkage (Lura et al. 2003, Tazawa et al. 1995). The latter is the consequence of water evaporation at the surface which causes a contractive force in the capillary pores and thus results in compaction. Previous studies have revealed that occurrence of physical shrinkage is dependent on the presence of pores exhibiting specific diameters, namely from ~10–50 nm (Wittmann 1982). Effective shrinkage-reducing agents (SRAs) are those which reduce the surface tension of the pore solution and which can modulate the pore size distribution in the cementitious matrix in such way that the shrinkage causing pore diameters are avoided. Diols, glycols, glycol ethers and amino terminated poly(ethylene-propylene) glycols have been identified as suitable SRAs (Fig. 7).

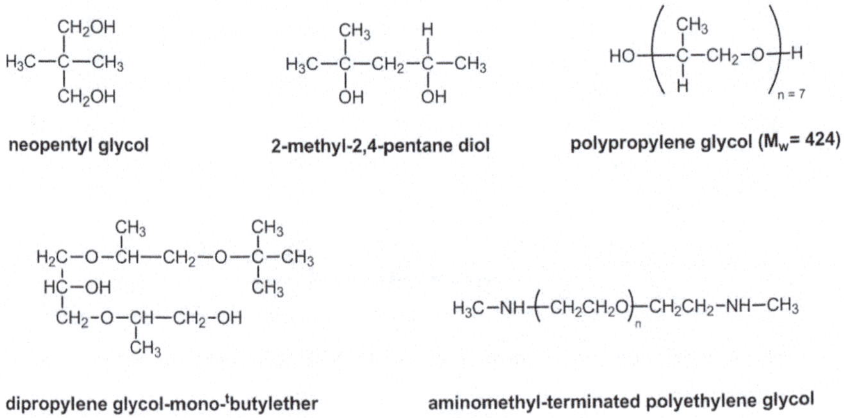

Fig. 7. Chemical structures of effective shrinkage-reducing agents (SRAs).

It is, however, well established that diols of quite similar structure and surface activity than those displayed in Fig. 7 do not provide any shrinkage-reducing effect at all, whereby the reason is still unknown. Furthermore, effective SRAs require extremely high dosages of 2–4% bwoc which are far beyond those for common functional admixtures used in concrete. Also, the reduction in shrinkage achieved from these admixtures is limited. Hence, it becomes obvious that a considerable gap with respect to the potency of SRAs and a thorough understanding of their working mechanism exists.

In recent years, two contributions on this subject were published. The first work presented that the pore-size modulating effect of SRAs is linked to their ability to form micelles of specific, large enough diameters which are the templates for pores which do not induce shrinkage (Kayello 2014). These micelles form at a stage in cement hydration when a significant amount of water has already been consumed and the SRAs are present in the pore solution at concentrations of 6–10%. Compounds which form micelles too early or too late in cement hydration cannot provide any shrinkage-reducing effect.

The second contribution utilized molecular modeling to identify potentially effective SRAs and then tested them in mortar to confirm the concept (Shlonimskaya et al. 2014). Based on a computer-aided molecular design (CAMD) approach that used the signature molecular descriptor, 2-propoxyethanol and 3-ethoxypropylamine were found to provide exceptional reduction in the surface tension of water. Their high shrinkage-reducing potential was confirmed in actual mortar tests.

Inspite of all this it obvious that our current technology of SRAs is quite limited and – compared to that existing in the field of e.g. superplasticizers or retarders – is far behind. More intense research is required to fill this gap in the future and to bring its technology to a level which allows a more effective control of physical shrinkage compared to the state of the art.

5 New Admixture Technologies – What Can We Expect in the Future?

5.1 Improved Curing Agents

Until now, a significant gap in current curing technology exists. The current situation on construction sites where large concrete slabs or decks are poured is that significant efforts have to be undertaken to reduce dry shrinkage and cracking on the surfaces. The most common practices include the spraying of water onto the concrete surface or coverage with a plastic foil to reduce water evaporation. Both methods are often not very effective, and on top they require a substantial amount of labor. Hence, the industry is challenged with developing admixtures which e.g. can be mixed into the fresh concrete and then prevent its surface desiccation, thus eliminating the need for post-curing of concrete. In light of this, superabsorbent polymers (SAPs) seem to be a promising candidate for the internal curing of concrete (Mignon et al. 2017). These cross-linked polymers which are typically synthesized from acrylic acid and/or acrylamide start to swell upon the contact with the pore solution. Consequently, a hydrogel is formed which gradually releases the absorbed water during the self-dessication of the concrete, thus mitigating the autogeneous shrinkage during hardening (Snoeck et al. 2017). Another type of curing agents are water evaporation retardants (e.g. poly lauryl methacrylate emulsions) which are applied on the surface of the plastic concrete to prevent the formation of plastic shrinkage cracks (Liu et al. 2010).

5.2 Admixtures Improving the Ductility of Concrete

Concrete presents a unique building material because of its easy preparation from abundantly available raw materials, its low cost and its enormous strength. Those excellent features have propelled the global volume of concrete poured to more than 30 billion tons per year. Inspite of these extraordinary properties, concrete suffers from one major deficiency which greatly limits its application: low ductility (= tensile or bending strength) and low fracture toughness (Fig. 8).

Compared to human bone for example, the fracture toughness of concrete is about 100 times lower. For a conventional concrete (w/c ratio ~0.5), the tensile strength reaches only $\sim 10\%$ of its compressive strength, thus rendering concrete a very brittle material. The problem becomes even worse when the w/c ratio is low. For example, in ultra-high strength concrete (UHPC, w/c = 0.25) the tensile strength develops to only 5% of the compressive strength. Consequently, such concrete is prone to crack formation through vibrational impact (on bridges e.g. from traffic, on buildings from wind forces, etc.).

In the future, the industry will be challenged with developing concepts which can reduce in-situ the brittleness of concrete. Potential solutions involve the addition of textile fibers or the generation of organo-mineral phases which are more flexible than conventional cement hydrates (e.g. meso crystals similar to those described for $CaCO_3$-PCE precipitates (Keller and Plank 2013), or Ca_2Al-polymer-LDH composites (Plank and Ng 2012)). In this respect, an interesting concept would be the in-situ formation of C-S-H-polymer nanocomposites similar to those described in Sect. 2 for C-S-H-PCE which potentially can improve the bending strength of concrete. Considering the magnitude of the task it might be useful to study concepts from nature such as they occur in mollusk shells which consist of calcite tablets with interstitial chitin (Mann 1993). Such biomimetic approaches will hopefully inspire researchers to propose solutions for this problem.

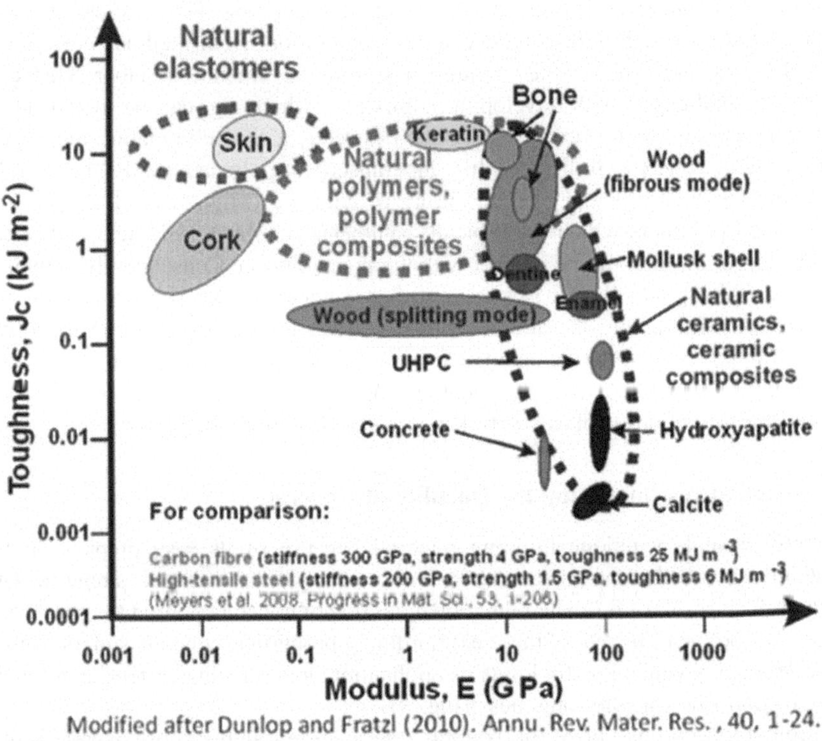

Fig. 8. Comparison of the fracture toughness of different natural or man-made materials including human bone and concrete.

6 Conclusion

Chemical admixtures have truly revolutionized modern concrete technology. They present a major driver for innovation in concrete and will continue to do so for many years to come. In the future, it would be extremely attractive to have admixtures which allow the safe application of self-compacting concrete delivered as ready-mixed concrete to the job site. Even more, to be able to control concrete consistency (fluidity) during delivery through the energy uptake of the rotating container of the concrete truck and energy-dependent PCE dosage would be most intriguing. Undoubtedly, the current admixture products will be refined further to become even more effective, and they will be tailored more specifically to distinct applications.

References

Akimoto, S., Honda, S., Yasukohchi, T.: Additives for Cement, EP 0,291,073 (1992)

Albrecht, G., Weichmann, J., Penkner, J., Kern, A.: Copolymers based on Oxyalkylene Glycol Alkenyl Ethers and Derivatives of Unsaturated Dicarboxylic Acids, EP 0,736,553 (1996)

Amaya, T., Ikeda, A., Imamura, J., Kobayashi, A., Saito, K., Danzinger, W., Tomoyose, T.: Cement Dispersant and Concrete Composition containing the Dispersant, WO 0,039,045 (2000)

Atarashi, D., Sakai, E., Obinata, R., Daimon, M.: Interactions between superplasticizers and clay minerals, Cement Sci. Concr. Technol. **58**, 387–392 (2004)

Dalas, F., Nonat, A., Pourchet, S., Mosquet, M., Rinaldi, D., Sabio, S.: Tailoring the anionic function and the side chains of comb-like superplasticizers to improve their adsorption. Cem. Concr. Res. **67**, 21–30 (2015)

Fan, W., Stoffelbach, F., Rieger, J., Regnaud, L., Vichot, A., Bresson, B., Lequeux, N.: A new class of organosilane-modified polycarboxylate superplasticizers with low sulfate sensitivity. Cem. Concr. Res. **42**, 166–172 (2012)

Futami, T., Yamaguchi, T., Tagoshi, H.: Use of a Polymer as a High-Flow Concrete Additive and Concrete Material Containing the Additive, EP 0,757,998 (2003)

Guicquero, J.P., Maitrasse, P., Mosquet, M.A., Sers, A.: A Water Soluble or Water Dispersible Dispersing Agent, FR 2,776,285 (1999)

Hamada, D., Yamato, F., Mizunuma, T., Ichikawa, H.: DE 10,048,139 A1 (2001)

Hibino, M.: Effect of viscosity enhancing agent on self-compactibility of fresh concrete. In: Sixth International Conference on Superplasticizers and other Chemical Admixtures in Concrete (CANMET/ACI), Nice, SP-195, pp. 305-320 (2000)

Hirata, T.: Cement dispersants, JP 842,022 (S59-018338) (1981)

Jacquet, A., Villard, E., Watt, O.: Method for inserting impurities, WO 2006,032,785 (2006)

Jardine, L., Koyata, H., Folliard, K., Ou, C.C., Jachimowicz, F., Chun, B., Jeknavorian, A.A., Hill, C.L.: Admixture and method for optimizing addition of EO/PO superplasticizer to concrete containing smectite clay-containing aggregates, U.S. 6,352,952 (2002)

Jeknavorian, A.A., Jardine, L., Ou, C.C., Koyata, H., Folliard, K.J. (2003) Interaction of superplasticizers with clay-bearing aggregates, In: Malhotra, V.M. (ed.) 7th CANMET/ ACI International Conference on Superplasticizers and Other Chemical Admixtures in Concrete, Berlin/Germany, ACI, SP-217, pp. 1293–1316

Kanchanason, V., Plank, J.: Effectiveness of a calcium silicate hydrate – polycarboxylate ether (C-S-H-PCE) nanocomposite on early strength development of fly ash cement. Constr. Build. Mater. **169**, 20–27 (2018)

Kayello, H.M., Naresh, K.R., Tadisina, R., Shlonimskaya, N., Biernacki, J.J., Visco, D.P.: An application of computer-aided molecular design (CAMD) using the signature molecular descriptor – Part 1. identification of surface tension reducing agents and the search for shrinkage reducing admixtures. J. Am. Ceram. Soc. **97**(2), 365–377 (2014)

Keller, H., Plank, J.: Mineralisation of $CaCO_3$ in the presence of polycarboxylate comb polymers. Cem. Concr. Res. **54**, 1–11 (2013)

Kraus, A., Dierschke, F., Becker, F., Schuhbeck, T., Grassl, H., Groess, K.: Method for producing phosphate polycondensation products and the use thereof, US patent 2011/0281975 A1 (2011)

Lange, A., Hirata, T., Plank, J.: Influence of the HLB value of polycarboxylate superplasticizers on the flow behavior of mortar and concrete. Cem. Concr. Res. **60**, 45–50 (2014)

Lei, L., Plank, J.: A concept for a polycarboxylate superplasticizer possessing enhanced clay tolerance. Cem. Concr. Res. **42**, 1299–1306 (2012)

Lei, L., Plank, J.: A study on the impact of different clay minerals on the dispersing force of conventional and modified vinyl ether based polycarboxylate superplasticizers. Cem. Concr. Res. **60**, 1–10 (2014)

Liu, J.P., Li, L., Miao, C.W., Tian, Q., Ran, Q.P., Wang, Y.J.: Characterization of the monolayers prepared from emulsions and its effect on retardation of water evaporation on the plastic concrete surface. Colloids Surf. A: Pysicochem. Eng. ASP. **366**(1–3), 208–212 (2010)

Lura, P., Jensen, O.M., van Breugel, K.: Autogenous shrinkage in high-performance cement paste: an evaluation. Cem. Concr. Res. **33**, 223–232 (2003)

Mann, S.: Molecular tectonics in biomineralization and biomimetic materials chemistry. Nature **365**, 499–505 (1993)

Mignon, A., Snoeck, D., Dubruel, P., Van Vlierberghe, S., De Belie, N.: Crack mitigation in concrete: superabsorbent polymers as key to success? Materials **10**(3), 237 (2017)

Mosquet, M., Chevalier, Y., Brunel, S., Guicquero, J.-P.: Polyethylene di-phosphonates as efficient dispersing polymers for aqueous suspensions. J. Appl. Pol. Sci. **65**, 2545–2555 (1997)

Nagare, K.: Storage and/or Transportation Method of Polyalkylene Glycol Monomers, US 7,030,282 B2 (2006)

Ng, S., Plank, J.: Study on the interaction of Na-montmorillonite clays with polycarboxylate based superplasticizers. In: Malhotra, V.M. (ed.) 10th CANMET/ACI Conference on Superplasticizers and Other Chemical Admixtures in Concrete (Proceeding Papers), ACI, Prague, pp. 407–421 (2012a)

Ng, S., Plank, J.: Interaction mechanisms between Na montmorillonite clay and MPEG-based polycarboxylate superplasticizers. Cem. Concr. Res. **42**, 847–854 (2012b)

Nicoleau, L., Albrecht, G., Lorenz, K., Jetzlsperger, E., Fridrich, D., Wohlhaupter, T., Dorfner, R., Leitner, H., Vierle, M., Schmitt, D., Braeu, M., Hesse, C., Montero, Pancera, S., Zuern, S., Kutschera, M.: Plasticizer-Containing Hardening Accelerator Composition, US 2011,0269,875 A1 (2011)

Nicoleau, L., Gädt, T., Chitu, L., Maier, G., Paris, O.: Oriented aggregation of calcium silicate hydrate platelets by the use of comb-like copolymers. Soft Matter **9**, 4864–4874 (2013)

Ohta, A., Sugiyama, T., Tanaka, Y.: Fluidizing mechanism and application of polycarboxylate-based superplasticizers, In: Malhotra, V.M. (ed.) 5th CANMET/ACI Conference on Superplasticizers and Other Chemical Admixtures in Concrete (Proceedings volume), Rome, ACI, SP-173, pp. 359–378 (1997)

Paas, J., Müller, M.W., Plank, J.: Influence of diester content in macromonomers on performance of MPEG-Based PCEs. In: Malhotra, V.M., Gupta, P.R., Holland, T.C. (eds.) 11th CANMET/ACI Conference on Superplasticizers and Other Chemical Admixtures in Concrete (Proceedings), ACI SP-302, Ottawa (Canada), pp. 199–210 (2015)

Plank, J., Pöllmann, K., Zouaoui, N., Andres, P.R., Schaefer, C.: Synthesis and performance of methacrylic ester based polycarboxylate superplasticizers possessing hydroxy terminated poly (ethylene glycol) side chains. Cem. Concr. Res. **38**, 1210–1216 (2008)

Plank, J., Lange, A.: Concrete Admixtures, EP 12,002,354.4 (2012)

Plank, J., Ng, S., Foraita, S.: Intercalation of Microbial Biopolymers Welan gum and EPS I into Double Layered Hydroxides, Zeitschrift für Naturforschung B 67b, 479-487 (2012)

Plank, J., Meyer, L.: New insights into physicochemical interactions occurring between polycarboxylate superplasticizers and a stabilizer in self-compacting concrete. J. Sustain. Cem.-Based Mat. **4**, 164–175 (2015)

Pourchet, S., Liautaud, S., Rinaldi, D., Pochard, I.: Effect of the repartition of the PEG side chains on the adsorption and dispersion behaviors of PCP in presence of sulfate. Cem. Concr. Res. **42**, 431–439 (2012)

Schinabeck, M., Friedrich, S., Holland, U., Pfeuffer, T., Eberwein, M., Schuhbeck, T.: Water-soluble copolymers containing sulfo groups, method for the production and use thereof, EP 1,763,546 (2005)

Shlonimskaya, N., Biernacki, J.J., Kayello, H.M., Visco, D.P.: An application of computer- aided molecular design (camd) using the signature molecular descriptor – part 2: evaluating newly identified surface tension-reducing substances for potential use as shrinkage-reducing admixtures. J. Am. Ceram. Soc. **97**(2), 378–385 (2014)

Snoeck, D., Pel, L., De Belie, N.: The water kinetics of superabsorbent polymers during cement hydration and internal curing visualized and studied by NMR. Sci. Rep. **7**, 9514 (2017)

Stecher, J., Plank, J.: Novel concrete superplasticizers based on phosphate esters. Cem. Concr. Res. **119**, 36–43 (2019)

Tahara, H., Ito, H., Mori, Y., Mizushima, M.: Cement Additive, Method for Producing the same, and Cement Composition, US 5,476,885 (1995)

Tazawa, E., Miyazawa, S., Kasai, T.: Chemical shrinkage and autogenous shrinkage of hydrating cement paste. Cem. Concr. Res. **25**, 288–292 (1995)

Üzer, E., Plank, J.: Impact of welan gum stabilizer on the dispersing performance of polycarboxylate superplasticizers. Cem. Concr. Res. **82**, 100–106 (2016)

Wang, Z.M., Xu, Y., Wu, H., Liu, X., Zheng, F.Y., Li, H.Q., Cui, S.P., Lan, M.Z., Wang, Y.L.: A Room Temperature Synthesis Method for Polycarboxylate Superplasticizer, CN 101974135 B (2013)

Witt, J., Plank, J.: A novel type of PCE possessing Silyl functionalities. In: Malhotra, V.M. (ed.) 10th CANMET/ACI Conference on Superplasticizers and Other Chemical Admixtures in Concrete (Proceedings), ACI, Prague, SP-288.04, pp. 57–70 (2012)

Wittmann, F.H.: Creep and Shrinkage in Concrete Structures, pp. 129–161. John Wiley & Sons Ltd, Hoboken (1982)

Yamamoto, M., Uno, T., Onda, Y., Tanaka, H., Yamashita, A., Hirata, T., Hirano, N.: Copolymer for Cement Admixtures and its Production Process and Use, US 6,727,315 (2004)

Yoshioka, K., Tazawa, E., Kawai, K., Enohata, T.: Adsorption characteristics of superplasticizers on cement component minerals. Cem. Concr. Res. **32**, 1507–1513 (2002)

Bio-Based Rheology Modifiers for High Performance Concrete – Possible Modes of Actions and Case Study for Cassava Starch in West Africa

Wolfram Schmidt[1(✉)], Kolawole Adisa Olonade[2], Rose Njeri Mbugua[3], Francis Julissa Lenz[1], and Inès Tchetgnia Ngassam[1]

[1] Department Safety of Structures, Bundesanstalt für Materialforschung und -prüfung, 12205 Berlin, Germany
wolfram.schmidt@bam.de

[2] Department of Civil and Environmental Engineering, University of Lagos, Akoka, Lagos 100213, Nigeria
olonade143lah@gmail.com

[3] Civil Engineering Department, Walter Sisulu University, Ibika Campus, Butterworth 4960, Eastern Cape, South Africa
mbuguarose12@gmail.com

Abstract. Polymers that help tailoring rheological properties during the casting process have become inevitable constituents for all kinds of high-performance concrete technologies. Due to lacking industries, these typically crude-oil based admixtures are not readily available in many parts of the world, which limits the implementation of more sustainable high-performance construction technologies in these regions. Alternative polymers, which often demand for less processing, can be derived from local plant-based resources.

The paper provides experimental data of flow tests of cement pastes with polysaccharides from Triumfetta pendrata A. Rich, acacia gum and cassava without and in the presence of polycarboxylate ether superplasticizer. The flow tests are amended by observations of the zeta potentials and the hydrodynamic diameters in the presence of and without calcium ions in the dispersion medium. The results show that in the presence of and without calcium ions all polysaccharides provide negative zeta potentials, yet, they affect flowability and thixotropy in different ways. Cassava starch, acacia gum, and the gum of Triumfetta pendrata A. Rich qualified well for robustness improvement, strong stiffening, and additive manufacturing, respectively. The reason for the different effects can be found in their average sizes and size distribution.

Due to the promising results, a flow chart for local value chains is derived on the example of yet unused cassava wastes, which can be converted in parallel into a chemical admixture, energy, and supplementary cementitious material.

Keywords: Admixtures · Polysaccharides · Rheology · Thixotropy · Concrete

W. P. Boshoff et al. (Eds.): SAP 2019, RILEM Bookseries 24, pp. 158–166, 2020.
https://doi.org/10.1007/978-3-030-33342-3_17

1 Introduction

Concrete technology has significantly changed over the course of the last decades. Major driving force for innovation has been the development of chemical admixtures that help modifying rheological properties of the fresh state concrete, and along with the new potentials to tailor the rheology of new concrete materials, processing and casting technologies came up [1]. Apart from the technological advancements that were made possible, chemical admixtures are also key for more sustainable concrete technologies. Today, concrete technologies are accountable for approximately 10% of the global carbon emissions with an increasing trend due to the increasing global demand [2]. Chemical admixtures help to use cement in concrete more efficiently by reducing the required water and they help coping with the challenges arising from the use of supplementary cementitious materials that are applied to reduce the amount of Portland cement clinker in the binder.

However, many regions in the world, particularly in the Southern hemisphere lack supply of chemical admixtures, that often have to be imported from Europe, North America or East Asia. As a result, they are available only in limited variability and at high costs [3]. This scenario limits the applicability of sustainable and high performance applications in regions of the Southern hemisphere, particularly on the African continent, which is an obstacle for Africa's enormous innovation potential [4].

Nevertheless, innovative and efficient admixture technologies are possible based on locally available plant-based materials. It has already been shown that high performance concrete can be developed [5], and complex rheological properties can be achieved [6–8] with bio-based polysaccharides. In addition, new value chains can be created [9], which will be discussed later in this paper in more detail.

Despite their tremendous potential as future admixtures, polysaccharides are complex and occur in high variety, and for most of them the mode of interactions is hardly understood today in the cementitious system. Based on rheological experiments and analyses of the polymers in different dispersion phases, the effects of three polysaccharides without and in the presence of PCE are discussed.

2 Experimental Setup

2.1 Materials

The cement that was used for the present investigations was an ordinary Portland cement (CEM I 42.5 R) with a density of 3100 kg/m^3.

The superplasticizer used in the study (PCE) was a powder type purely polycarboxylate based agent.

The gum of the Triumfetta pendrata A. Rich (Nk) was derived from Cameroonian barks. The gum was dissolved in water without agitation. The liquid was dried until mass constancy and the resulting solid residuum was ground.

The acacia gum (GA) was derived from a South African acacia tree of the Acacia Heynes species. The treatment in advance to the test was identical to the procedure described for Nk.

The cassava starch (CS) was derived from cassava tubers from Thailand. The tubers were peeled, and the starch was boiled at 60° for 15 min under constant stirring. The resulting pulp was dried until mass constancy was obtained and then ground.

2.2 Flow Diameter

The flow diameter provides information on the rheology of a plastic sample. The value is linked to the yield stress. The test was conducted according to DIN EN 1015-3:2007. A cone with an upper and lower diameter of 70 mm and 100 mm, respectively and a height of 60 mm was filled with material, and then the cone was lifted. The average of two measured perpendicular diameters was defined as flow diameter without strokes. Then the cone was lifted and dropped 15 times and the flow diameter was measured again. For each cement paste, the w/c was varied between 0.2 and 0.5. The dosage of each polysaccharide was 0.2% by weight of cement (bwoc). Pastes were tested without PCE and in the presence of 0.066% (bwoc) PCE.

2.3 Zeta Potentials and Hydrodynamic Diameters

The zeta potential provides information on the charges upon adsorption of ions around particles in the shear plane. The hydrodynamic radius provides information on small particles, and polymers under the assumption that the polymer provides similar diffusion properties as a particle of the same size. For zeta potential and hydrodynamic diameter measurements dynamic light scatter techniques were used in a Malvern Zetasizer Nano zs. Since zeta potentials are depending on the pH as well as the ionic content in the surrounding dispersion phase (and in cementitious systems particularly the calcium ions), and hydrodynamic radii can also be affected by the surrounding ions, both values were determined in deionized water and in fully saturated limewater.

3 Results and Discussion of the Flow Tests

The results of the flow behaviour without strokes for pastes without PCE can be seen in Fig. 1a. Below w/c = 0.35 no flow can be observed, and then with increasing w/c, the diameters increase for all systems. However, no significant effect can be observed for the admixtures apart from the GA that strongly reduces the diameter. After lifting and dropping the samples 15 times (Fig. 1b), the diameters significantly increase for all systems. Only at w/c = 0.25 and below, no significant flow can be observed. The Nk reduces the flow spread below w/c = 0.4, but above, the flow is similar or even wider than without the admixture. CS reduces the flow at all w/c, while the diameter reduction maintains identical regardless of the w/c. GA shows the most prominent effect. Even at high w/c the diameter remains much smaller than without the admixture.

In the presence of PCE and without strokes (Fig. 1c) the reference curve does not show significant changes compared to the paste without PCE (Fig. 1a) with the only difference that beyond w/c = 0.35 the diameters are wider than without PCE. However, the performance of the polysaccharides varies greatly. All polymers cause reduced flowability. The curves for pastes with Nk and CS in the presence of PCE are more or less identical to the respective curves without PCE. GA is the most efficient stabilizing agent, though, in the presence of PCE the effect is less pronounced than in the absence of PCE. After the strokes, the diameters increase for all systems with and without polysaccharides (Fig. 1d). The diameter reduction is strongest for GA but the difference between GA and CS is less pronounced compared to the other observations. The effect of Nk on the flow reduction can only be observed for w/c = 0.4 or lower. Beyond this w/c the diameters with Nk are similar to those without Nk.

Based on the observation of the flow, the polymers show different effects on the rheology, thus, qualify for different applications. The GA seems to be a strong stabilising agent, which, however, in applications with PCE seem to lose efficiency more than other polysaccharides, once high shear forces are applied. In addition, it is known that GA at certain w/c and polymer dosages can convert from stabilising agents to plasticizing agents [6, 10].

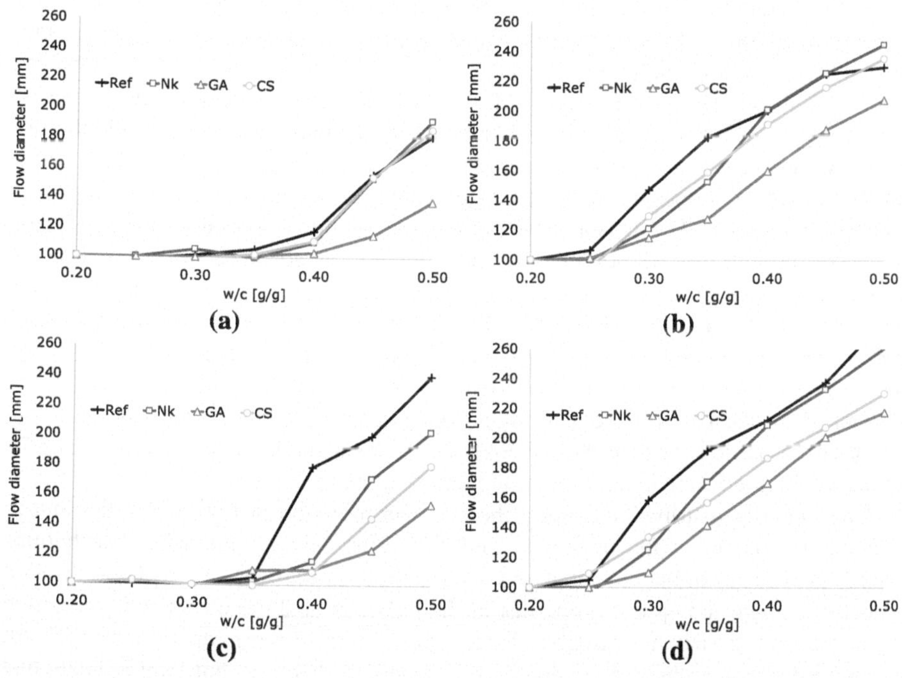

Fig. 1. Flow diameters of cement pastes at different w/c for samples without PCE and without strokes (a), without PCE and with strokes (b), with PCE and without strokes (c), and with PCE and strokes (d)

The CS is particularly strong in systems with PCE. The effect of starch based admixtures in the presence and without PCE has been discussed before, and may result from their availability as non-adsorbed deformable particles in the aqueous phase of the cement paste [7]. They qualify well for robustness improvement of very flowable systems with high plasticizer dosages. In addition CS can also cause a certain plastifying effect [6].

The most interesting behaviour can be observed with the addition of Nk in the presence of PCE. While Nk has a strongly flow-limiting effect in the absence of high shear forces (Fig. 1c) it does not show any strong effect at higher w/c after application of strokes (Fig. 1d). This means, the pastes with this polymer are particularly stiff at low shear rates (e.g. at rest), while they are soft, once higher shear forces are applied (e.g. during processing). Hence, this material, qualifies well for all processes, where the material has to be soft for the application process, but has to be stable once applied, e.g. sprayed concrete or 3D-printed concrete.

Although the polymer parameters are not well investigated, zeta potentials and hydrodynamic diameters can provide some indications for the polymers' modes of actions with and without PCE. The zeta potentials and intensities in water and limewater are presented in Figs. 2 and 3.

The zeta potentials depend strongly on pH values and calcium ion content in the dispersing medium. For Nk and GA, the zeta potentials are negative in water and calcium hydroxide. The less negative values are results of adsorption of calcium ions on anionic groups in the polymer structure. This also applies for the PCE, however, there is a much stronger tendency of the PCE to attract calcium ions, which is assumedly based on the availability of carboxy groups in the backbone, while the two gums Nk and GA with their more complex polymeric structure sterically shield their anionic groups. CS does not provide a significant zeta potential value in water, but in calcium hydroxide. The reason is deprotonation of the OH-groups in the glucose molecules at high pH. All polymers are thus likely to be adsorbed in the highly ionic pore solution. The adsorption of stabilising agents is a known phenomenon, though the adsorption rates are often low [11]. In the presence of PCE, the adsorption of polysaccharides becomes negligible [7, 11] despite the low zeta potentials, most likely due to their better diffusion mobility compared to the complex polysaccharides.

Hence, it can be assumed that all polysaccharides presented here can be found adsorbed on particles and in the aqueous phase. Their mode of action, thus, may be explained by their individual sizes in the aqueous phase.

The intensity distributions versus the d_{hyd} can be found in Fig. 3 in water and in calcium hydroxide, in the beginning and 17 h after mixing. Intensity distributions cannot provide information about the volumetric distribution, but without detailed knowledge about the optical properties of the polymers and the dispersion medium, a volume distribution is not possible. Hence, peak sizes in the intensity curve do not provide information about the volume of polymers of that size, but they indicate that polymers of that size are present. In general, larger polymers have stronger influence on the intensity (to the power of 6), thus, in a bi-modal distribution, a higher intensity of the coarser fraction, does not at all mean that there are more polymers of the coarser fraction.

The PCE's diameters are between 100 and 200 nm, and there is no big change in water and calcium hydroxide or over the course of time. This is not a common behaviour for PCE, which typically form larger complexes with calcium ions [12]. The behaviour can be explained by the specific modification of the used polymers, which contain one area with many carboxy groups, and an additional long trunk without anionic groups.

The Nk has two fractions, one below 100 nm and one with a peak around 400–500 nm. In the beginning, there is no big influence of the calcium ions, but it can be seen, that at later stages the two fractions significantly shift to the larger fractions, and that polymer agglomerates in the order of magnitude of 4–5 μm have been formed.

The GA polymers have a very broad bi-modal distribution of sizes in calcium hydroxide, which broadens over the course of time. The CS polymer is very stable in calcium hydroxide, while in water the polymer distribution broadens over the course of time, due to hydrolysis. Starch polymers consists of an amylose and amylopectin fraction, the latter being significantly larger than the amylose. The smaller diameters in the presence of calcium as well as the stability with time indicate that calcium ions stabilise the tree-like structure of the amylopectin molecule efficiently.

Their rheology effect can be hypothetically explained by the polymers' individual sizes and size distributions. The strong effect in the presence and without PCE of the GA is assumedly based on their wide range size distribution in the liquid phase. Though smaller polymers will assumedly be adsorbed on surfaces, which adds up to their effect in absence of PCE, as well as partly adsorbed larger polymers may even bridge particles, the size distribution of the non-adsorbed polymers will hinder larger particles to move among each other, which explains the strong effect. The CS is smaller and more uniform; thus, its non-adsorbed polymer fraction creates less polymer contacts, and thus less effect on stabilisation. The Nk is significantly larger than the other polymers and it seems to form huge complexes in the presence of calcium ions. This explains their strong thixotropic effect. The polymers form bridges between the particles, which are very strong in the absence of high shear forces, but once high shear forces are applied, the structure breaks down quickly and allows for good flowability.

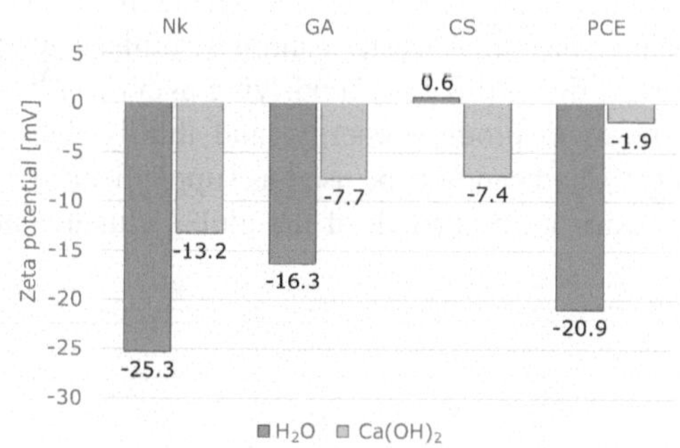

Fig. 2. Zeta potentials of the polymers in deionised water and in limewater

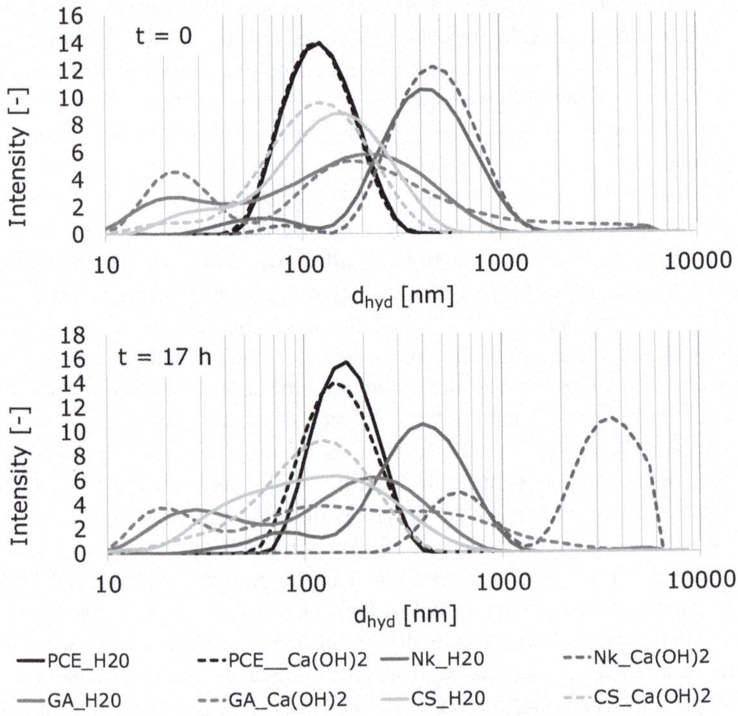

Fig. 3. Intensity distribution curves of the polymers in water and limewater as dispersion medium at early stage and after 17 h

4 Case Study: Cocrete Value Chains with Cassava Waste

The present study shows that bio-based rheology modifying agents can be very efficient, and once their effect is fully understood, can well be used for high-performance applications, thus contribute to positive economic developments. Cassava points out to be particularly interesting, since their wastes cannot be utilised adequately today. From their today unused wastes, first a rheology modifying agent can be obtained, secondly, the residuals can be burnt to produce energy, and thirdly the residual ashes show pozzolanic properties [13], and can thus be used as supplementary cementitious binder. This way, new value chains to the benefit of the global climate can be created [9], as shown in Fig. 4.

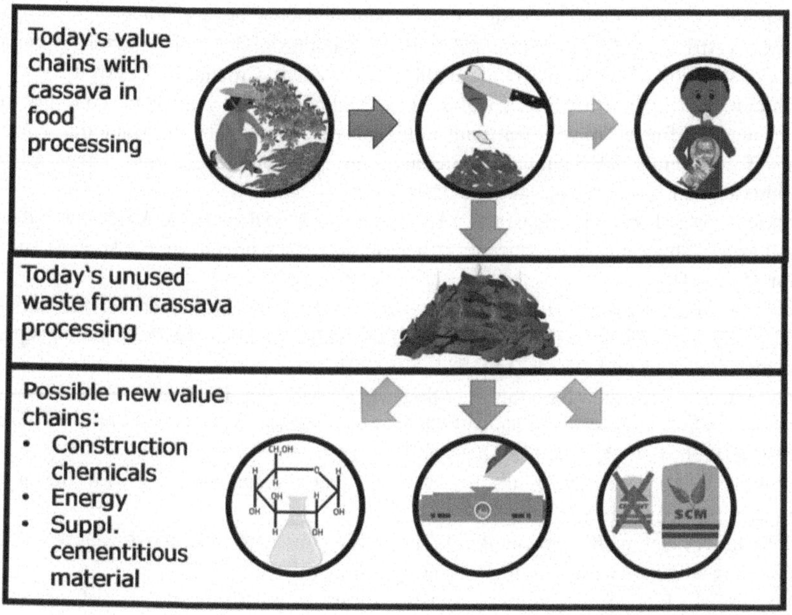

Fig. 4. Market potentials for cassava wastes in high performance concrete

5 Conclusions

Bio-based polymers can be strong rheology modifying agents, and once their effect is fully understood, can effectively help tailoring rheological concrete properties. Their individual effect on flow and thixotropic properties can be explained by their charges, interactions with ions in the liquid phases as well as their sizes and size distribution.

For countries of the global South, where the supply with chemical admixtures is not good, bio-based rheology modifiers show promising market potentials.

Acknowledgements. The study was supported by the German African Innovation Incentive Award provided by the German Federal Ministry of Education and Research.

References

1. Schmidt, W., Sonebi, M., Brouwers, H.J.H., Kühne, H.-C., Meng, B.: Rheology modifying admixtures: the key to innovation in concrete technology - a general overview and implications for Africa. Chem. Mater. Res. **5**, 115–120 (2013)
2. WBCSD: Getting the Numbers Right, Project Emissions Report 2014 (2016)
3. Schmidt, W., Msinjili, N.S., Kühne, H.-C.: Materials and technology solutions to tackle the challenges in daily concrete construction for housing and infrastructure in sub-Saharan Africa. Afr. J. Sci. Technol. Innov. Dev. **11**, 401–415 (2018)

4. Schmidt, W.: Potentials for sustainable cement and concrete technologies – comparison between Africa and Europe. Presented at the Proceedings of the 1st International Conference on Construction Materials for a Sustainable Future, Zadar, Croatia, 19–21 April 2017 (2017)
5. Schmidt, W., Msinjili, N.S., Uzoegbo, H.C., Makunza, J.K.: Admixture concepts for the sub-saharan african environment with indigenous raw materials. In: Malhotra, V.M. (ed.) SP-302 Eleventh International Conference on Superplasticizers and Other Chemical Admixtures in Concrete, pp. 491–505(2015)
6. Schmidt, W., Tchetgnia Ngassam, I.L., Mbugua, R., Olonade, K.A.: Natural rheology modifying admixtures for concrete (Rheologische Messungen an Baustoffen). tredition GmbH (2017)
7. Schmidt, W., Brouwers, H.J.H., Kühne, H.-C., Meng, B.: Interactions of polysaccharide stabilising agents with early cement hydration without and in the presence of superplasticizers. Constr. Build. Mater. **139**, 584–593 (2017)
8. Schmidt, W., Tchetgnia Ngassam, I., Olonade, K., Mbugua, R., Kühne, H.-C.: Plant based chemical admixtures – potentials and effects on the performance of cementitious materials. RILEM Tech. Lett. **3**, 124–128 (2019)
9. Schmidt, W., Barucker-Sturzenbecher, M.J.: Bio-based concrete, p. 7:51. Ed. Berlin (2019) https://vimeo.com/310549146
10. Mbugua, R., Salim, R., Ndambuki, J.: Effect of gum Arabic karroo as a water-reducing admixture in concrete. Materials **9**(2), 80 (2016). (in English)
11. Palacios, M., Flatt, R., Puertas, F., Sanchez-Herencia, A.: Compatibility between polycarboxylate and viscosity-modifying admixtures in cement pastes. In: 10th International Conference on Superplasticizers and Other Admixtures in Concrete, Prague, Czech Republic, vol. SP-288, pp. 29–40. ACI (2012)
12. Schmidt, W., Weimann, C., Chaves Weba, L.: Influences of hydration effects on the flow phenomena of concrete with admixtures. In: Advances in Cement and Concrete Technology in Africa, Dar es Salaam, Tanzania, pp. 79–88. BAM (2016)
13. Salau, M.A., Olonade, K.A.: Pozzolanic potentials of cassava peel ash. J. Eng. Res. **16**(1), 10–21 (2011)

Influence of Supplementary Cementitious Materials and Superplasticisers on the Rheological Properties of Concrete

Johandre, M. H. Bessinger, Luqmaan Parker, and Riaan Combrinck(✉)

Unit for Construction Materials, Civil Engineering Department, Stellenbosch University, Private Bag X1, Matieland 7602, South Africa
rcom@sun.ac.za

Abstract. The influence of different supplementary cementitious materials (SCM) and superplasticisers on the rheological properties of concrete was investigated to identify potential compatibility issues. Superplasticisers and SCM often have unexpected interaction with certain cementitious compounds, resulting in concrete that is difficult to place in the fresh state due to poor rheological properties. Various mixes were designed containing different superplasticisers and/or SCM in different quantities. Slump, slump flow and concrete rheometer tests were conducted to determine the yield stress, plastic viscosity and thixotropic behaviour of the concrete. Obtained results showed that the specific Sulphonate Naphthalene Formaldehyde (SNF) and Polycarboxylic Ethers (PCE) superplasticisers used, reduced the yield stress, thixotropic behaviour and plastic viscosity of concrete. Modified Acrylic Polymer (ACR) superplasticiser showed a similar effect except for the plastic viscosity which increased at higher dosages. The addition of fly ash and slag to concrete containing superplasticiser had little effect on the rheology and showed similar results as mixes only containing superplasticiser. The use of superplasticiser in conjunction with silica fumes caused a decrease in yield stress and thixotropic behaviour while plastic viscosity increased. The use of superplasticiser in conjunction with higher than normal dosages of gypsum also caused a decrease in yield stress and thixotropic behaviour but had negligible effect on plastic viscosity. It was also found that the use of PCE superplasticiser in conjunction with gypsum, used to control the set of concrete, can cause potential slump loss issues.

1 Introduction

Cement and water reacts in a process called hydration to form a hardened paste which binds the stone and sand together to form concrete. Various other materials and substances, also known as mineral and chemical admixtures, are often added to the basic concrete constituents to improve the fresh and hardened properties of the concrete (Domone and Illston 2010). Mineral admixtures or supplementary cementitious materials (SCM) are generally in powder form such as fly ash, silica fumes and slag, while chemical admixtures are generally in liquid form such as plasticisers and superplasticisers. Superplasticisers are used to maintain workability at lower w/c ratios

W. P. Boshoff et al. (Eds.): SAP 2019, RILEM Bookseries 24, pp. 167–175, 2020.
https://doi.org/10.1007/978-3-030-33342-3_18

or maintain a desired w/c ratio while increasing fluidity i.e. workability. Superplasticisers interact with cementitious materials in three ways. The first is the adsorption of superplasticiser by the hydrating cement particle, during the formation of ettringite, resulting in the decrease in active dosage available for particle dispersion. Secondly, negatively charged superplasticiser molecules bind with positively charged cement particles, forming an electrical double layer. The electrical double layer results in repulsive forces, improving fluidity. The third mechanism in which superplasticisers increase concrete fluidity and workability is through non-absorbed particles. These particles lubricate cement particles, reducing friction via the arrangement of polymer chains parallel to the direction of flow. Furthermore, these particles also reduce the water surface tension within the mix allowing for greater dispersion of cement particles (Chakkamalayath et al. 2011).

Silica fumes, fly ash and slag are used to replace the pure Portland cement content to produce a more environmentally friendly and affordable concrete. Set retarders such as gypsum are added to allow a dormant period and prevent flash setting which allows time for placement. Flash setting occurs due to unregulated tricalcium aluminate (C_3A) hydration. Normal setting occurs when C_3A hydrates in the presence of gypsum to from ettringite, the quantity and dissolution rate of set retarder playing a key role in early rheological properties of cementitious materials. Rheology is defined as the study of the flow and deformation of matter under an applied shear stress. Rheology thus investigates the shear stress, shear strain relationship which is used to describe the fluidity of a material i.e. the fresh state properties of concrete (Quanji 2010).

Various admixtures and SCM are used to manipulate the rheology of concrete. Superplasticisers are used to increase fluidity and prevent rapid slump loss which can cause placement issues during construction (Chandra and Björnström 2002). However, cases have been reported where admixtures have adverse effects on concrete characteristics. The cause of this is the interaction of different concrete constituents, including superplasticiser and SCM, with each other. When compatibility issues of such nature do arise, it is generally termed as concrete incompatibilities (Burgos-Montes et al. 2012). Understanding the interaction between concrete constituents of a mix is important to predict the performance thereof. However, understanding the influence of mix constituents is challenging due to the fast development of admixtures. The continuous investigation of concrete-admixture compatibility is therefore necessary to avoid problems such as slump loss, set retardation, strength loss etc.

This study investigates the rheological properties of concrete with and without superplasticiser and SCM to identify compatibility issues between the concrete constituents and superplasticisers.

2 Materials

The reference mix was designed to be robust and therefor allow several alterations without showing signs of segregation. The mix has a good fines distribution, w/c ratio of 0.5 and a slump of 105 mm (>75 mm) which is necessary for the mix to be used in the ICAR rheometer (Koehler and Fowler 2004). The proportions and constituents of the reference mix is shown in Table 1.

Table 1. Reference mix design

Material	Mass (kg/m^3)	Relative density	Volume (l)
Water	220	1	220
Binder: OPC CEM I 52.5N	440	3.14	140
Sand: Fine natural quarry (Locally known as Malmesbury sand)	874	2.6	336
Stone: 6 mm Greywacke stone	850	2.8	304

The influence of three superplasticisers and four SCM as well as superplasticiser in conjunction with the SCM, on the rheology of the reference mix, was investigated to identify potential compatibility issues. Table 2 indicates the dosages stipulated in the Chryso (2007) and Mapei (2017) manufacturer's catalogue for the respective superplasticiser along with its relative density.

Table 2. Superplasticisers used with recommended dosages

Superplasticiser	Dosage (liter per 100 kg cement)	Relative density
Sulphonate naphthalene formaldehydes (SNF: Chryso Fluid)	0.6–2.0	1.150 ± 0.010
Polycarboxylic ethers (PCE: Chryso Fluid Premia 310)	0.3–3.0	1.050 ± 0.010 at 20°
Modified acrylic polymer (ACR: Mapei Dynamon SP1)	0.6–1.2	1.08 ± 0.02 at 20°

SCM used in this study include Class S fly ash (FA) supplied by Ulula Ash (Pty) Ltd, Pretoria Portland Cement (PPC) supplied ground granulated corex slag (SL), Microfume silica fumes (SF) supplied by SiliconSmelters and PPC supplied gypsum (GYP). Although a gypsum particle size of 45 μm was initially aimed for, a final particle size of 850 μm was achieved after grinding with a pulveriser. Although this was not ideal, it was sufficient to give an indication of the influence of gypsum in a mix.

3 Test Methodology

For mixes containing FA, SL and SF, a percentage of the total binder content was replaced by the mentioned SCM keeping the total binder content constant, whereas GYP was added to the reference mix, therefore increasing the binder content. Superplasticisers were added to the mix at the minimum and 1.5 times the minimum dosage suggested by suppliers. SCM were added to the reference mix as a percentage of binder content. Three different quantities of each SCM was added to the reference mix; FA was added at 10%, 20% and 30%, SL at 15%, 30% and 45%, SF at 5%, 10% and 15% and GYP at 2%, 4% and 6%.

Before mixing, all materials were stored in a climate controlled room at 23 °C and relative humidity of 60%. The concrete mixes were mixed and produced in accordance with SANS 5861-1 (2006), while the slump and slump flow tests that followed were performed in accordance with SANS 5862-1 (SANS and: SANS 5862-1 2006). The slump and slump flow test were necessary to determine whether mixes could be tested using the ICAR rheometer. The ICAR rheometer was designed by the International Centre for Aggregate Research at the University of Texas in Austin, USA. This testing device has the ability to retrieve fundamental rheological characteristics that other more conventional tests such as slump tests and slump flow tests cannot.

After the slump tests were performed the ICAR rheometers container was filled with 19 litre of concrete and the torque motor positioned in the centre with the vane fully submerged. The stress growth test was performed first by using an input speed of 0.025 rev/s to determine the static yield stress. The flow curve test was performed next to obtain the Bingham parameters i.e. dynamic yield stress and plastic viscosity. Due the thixotropic behaviour of cementitious materials it was ensured that the timing control was kept constant for all mixes. The mixing process took approximately 10 min whereas the slump/slump flow and rheometer tests took 35 min in total. Only one slump/slump flow as well as stress growth and flow curve test was conducted per mix for a total of 33 mixes.

4 Results and Discussion

4.1 Influence of Superplasticisers

The influence of SNF, PCE and ACR superplasticisers on the rheological properties of concrete were investigated. All three superplasticisers exhibited similar results i.e. a decrease in peak torque and yield stress with an increase in slump. The measured rheological parameters obtained for mixes containing only superplasticisers are presented in Table 3. It should be noted that obtained results are specific to the superplasticisers used, this is particularly true for PCE superplasticisers that consist of a wide range of chemically diverse superplasticisers that vary in behaviour.

Table 3. Influence of superplasticisers, at different dosages, on the rheology of the reference mix.

Mix	Static yield stress_SYS (Pa)	Dynamic yield stress_DYS (Pa)	Thixotropy (DYS - SYS)	Plastic viscosity (Pa.s)	Peak torque (Nm)	Slump or slump flow if more than 300 (mm)
REF	2440	1146	1294	19.1	10.5	105
SNF_{min}	1469	1039	431	4.3	6.3	190
$SNF_{min} \times 1.5$	1066	831	234	4.2	4.6	210
PCE_{min}	1246	827	419	4.0	5.4	220
$PCE_{min} \times 1.5$	534	415	119	2.7	2.3	260
ARC_{min}	251	174	78	9.2	1.1	510
$ARC_{min} \times 1.5$	140	47	93	23.5	0.6	770

The results obtained for the SNF and PCE superplasticiser showed similar trends. With the addition of superplasticiser the measured peak torque required to shear the mix decreased, this effect being more evident the higher the superplasticiser dosage. The decrease in torque required is due to the decrease in static yield stress of the mix; this decrease being larger with increasing superplasticiser dosage. A decrease in static yield stress was also evident when considering the increase in slump. The addition of both SNF and PCE resulted in a decrease in plastic viscosity, with viscosity decreasing with increasing superplasticiser dosage. The thixotropic behaviour showed a decreasing trend with higher dosages of superplasticiser. The observed decrease in thixotropic behaviour along with the decreased viscosity suggest that the mix is more prone to segregation with the addition of superplasticiser, although no segregation was observed. ARC superplasticiser showed similar behaviour to SNF and PCE regarding the static yield stress, thixotropic behaviour and slump, however an increase in plastic viscosity was obtained when the dosage was increased. This would indicate that ACR is less prone to cause segregation at higher dosages.

4.2 Influence of SCM

The effect of several SCM on the rheological properties of the reference mix was also studied. The measured rheological parameters obtained for mixes containing different SCM are presented in Table 4.

Table 4. Influence of SCM on the rheology of the reference mix.

Mix	Static yield stress_SYS (Pa)	Dynamic yield stress_DYS (Pa)	Thixotropy (DYS - SYS)	Plastic viscosity (Pa.s)	Peak torque (Nm)	Slump or slump flow if more than 300 (mm)
REF	2440	1146	1294	19.1	10.5	105
10%FA	2443	1436	1007	4.9	10.5	110
20%FA	1753	741	1012	29.2	7.5	95
30%FA	1307	906	401	8.3	5.6	85
15%SL	2596	1259	1337	11.1	11.1	120
30%SL	2578	125	1327	11.9	11.1	110
45%SL	3163	1366	1797	11.8	13.6	95
5%SF	2877	114	2763	382.8	12.4	80
2%GYP	2458	1181	1277	12.5	10.5	100
4%GYP	2839	1214	1625	14.2	12.2	105
6%GYP	2871	1323	1548	12.5	12.3	105

The torque required to shear the mix decreased with increasing percentage of fly ash used to replace cement resulting in a decrease in static yield stress of the mix being measured. Thixotropic behaviour also decreased with increased percentage of FA used. Unexpectedly the slump decreased with increasing percentage fly ash used. This could be due to the source of FA having a more angular shape than conventional FA. However, it is not certain why the static yield stress decreased, which indicates an increase in ease of flow under own weight, would cause a decrease in slump. A clear pattern regarding the influence of fly ash on the plastic viscosity of a mix was not observed.

The addition of slag caused an increase in static yield stress, this effect increased with an increase in amount of slag used. Although a weak correlation was found between measured static yield stress and slump values, the decrease in slump with increasing use of slag substantiates the increase in static yield stress. It is believed that this is due to the angular shape of slag particles. Thixotropic behaviour of the mix remained constant with increasing use of slag until 45% slag replacement, where an increase in thixotropic behaviour was observed due to the slag properties dominating the mix characteristics. The plastic viscosity was reduced with the addition of slag but did not vary with greater percentage additions.

The addition of silica fumes had similar effects to slag on the characteristics of the reference mix. An increase in static yield stress, causing a decrease in slump, was found with the addition of silica fumes. Only 5% SF addition was tested with the ICAR rheometer due to the slump being too low when higher percentages of silica fumes were added. This is due to the particle fineness resulting in mix stiffening. The addition of silica fumes showed a significant increase in plastic viscosity coupled with a greater thixotropic behaviour.

The addition of gypsum also caused a small increase in static yield stress, the effect increasing with increasing percentage of gypsum added. However, the slump remained nearly constant. The addition of gypsum indicated a reduced plastic viscosity and increased thixotropic behaviour.

4.3 Interaction Between SCM and Superplasticisers

This section specifically focuses on results obtained for the reference concrete mix with added SCM in combination with superplasticisers. The measured rheological parameters obtained for mixes containing different SCM in combination with superplasticisers are presented in Table 5.

The results indicate that, the addition of the SNF and PCE superplasticiser in combination with fly ash and slag, respectively, causes a decrease in yield stress, plastic viscosity and thixotropic effects. The addition of the ACR superplasticiser indicates a decrease in yield stress and thixotropic effects however, plastic viscosity remained similar to that of the reference mix. Therefor fly ash and slag does not present the same compatibility with all superplasticisers. The decrease in yield stress could be due to the lower aluminate cement content available for superplasticisers to react with (Chakkamalayath et al. 2011).

All superplasticisers exhibited similar effects on the rheology of the reference mix in combination with silica fumes and gypsum respectively. Superplasticisers in combination with silica fumes and gypsum, respectively, caused a decrease in yield stress and thixotropic behaviour, with plastic viscosity remaining approximately constant. It should be noted that ACR did not cause an increase in plastic viscosity in combination with silica fumes and gypsum as was the case when combined with fly ash and slag, respectively. All the superplasticisers used in combination with silica fumes and gypsum, respectively, therefore show similar behaviour. However, PCE and ARC, in

Table 5. Influence of SCM in combination with superplasticisers on the rheology of the reference mix.

Mix	Static yield stress_SYS (Pa)	Dynamic yield stress_ DYS (Pa)	Thixotropy (DYS - SYS)	Plastic viscosity (Pa. s)	Peak torque (Nm)	Slump or slump flow if more than 300 (mm)
REF	2440	1146	1294	19.1	10.5	105
20%FA+SNF	1374	984	390	8.5	5.9	180
20%FA+PCE	956	831	125	4.5	4.1	195
20%FA+ARC	212	116	96	18.7	0.9	610
2%GYP+SNF	1952	1108	845	11.0	8.4	180
2%GYP+PCE	341	186	155	15.1	1.5	210
2%GYP+ARC	325	216	109	12.7	1.4	550
30%SL+SNF	884	670	214	6.7	3.8	225
30%SL+PCE	816	613	203	4.9	3.5	220
30%SL+ARC	213	97	116	21.8	0.9	690
10%SF+SNF	1639	1178	461	4.5	7.0	160
10%SF+PCE	741	609	132	4.3	3.2	200
10%SF+ARC	654	623	31	2.6	2.8	245

combination with silica fumes and gypsum, showed a more pronounced effect on the rheological properties of the mix.

4.4 Effect of PCE on Stress Growth in the Presence of Gypsum

Two concrete mixes namely, PCE_{min} and GYP6PCE (refenence mix with PCE and 6% Gypsum), were continuously sheared in the rheometer over a 45 min period to identify any compatibility issues between the PCE superplasticiser and gypsum. Both gypsum and PCE bind with the aluminates in cement and rapid slump loss could occur if the formation of ettringite is prevented (Plank et al. 2010). Figure 1 presents the variation in measured torque over time.

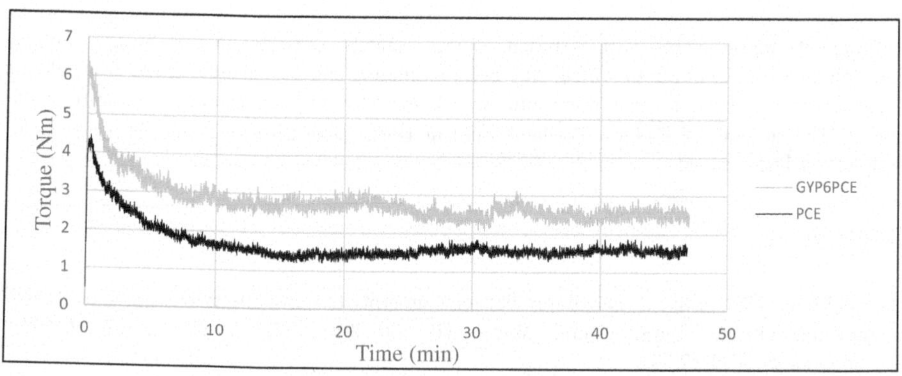

Fig. 1. Influence of superplasticiser on the measured torque in the presence of gypsum

From Fig. 1 it is evident that the presence of gypsum has an influence on the measured torque decay over time. Applied torque to GYP6PCE decreases gradually for roughly 20 min after which it increases and decreases inconsistently. The torque applied to PCE decreases for approximately 25 min after which it increases slightly. Slump loss was also measured before and after the strength growth test was conducted. A slump loss of 100 mm and 60 mm was observed for PCE_{min} and GYP6PCE respectively. It is believed that this is an indication of increased intercalation of superplasticiser into the interlayer of the metastable C_3A hydration products when the sulphate concentration in a cementitious material is low, which would reduce the active superplasticiser available in the mix to induce fluidity. However, a larger data set is needed to draw a distinct conclusion.

5 Conclusion

The objective of this study was to identify compatibility issues between SCM, superplasticisers and SCM in combination with superplasticisers, with concrete by considering the influence on rheology. From the tests conducted the following conclusions can be drawn:

SNF and PCE superplasticisers reduce the yield stress, thixotropic behaviour and plastic viscosity of a mix. ACR on the other hand causes an increase in plastic viscosity at high dosages while still causing a decrease in yield stress and thixotropic behaviour, thus making the mix less prone to segregation. When these superplasticisers were used in conjunction with fly ash and slag, the results did not deviate much from those obtained by using solely superplasticiser.

The use of superplasticiser in conjunction with silica fumes and gypsum, respectively, showed similar results with regards to yield stress and thixotropic behaviour, causing a decrease in the aforementioned. However, when superplasticiser and gypsum are used in conjunction, negligible change in plastic viscosity occurred whereas silica fumes resulted in an increase. This could be caused by the tendency of both the superplasticiser and gypsum to react with the available aluminates in the cement.

It was also found that the used PCE, in conjunction with gypsum as set retarder, could cause potential slump loss issues, however more test would have to be conducted to verify.

Acknowledgements. This work is based on the research supported in part by the National Research Foundation of South Africa. Any opinion, finding and conclusions or recommendation expressed in this material is that of the author(s) and the NRF does not accept any liability on this regard. The support of Pretoria Portland Cement (PPC) and Chryso South Africa are also gratefully acknowledged.

References

Burgos-Montes, O., et al.: Compatibility between superplasticiser admixtures and cements with mineral additions. Constr. Build. Mater. **31**, 300–309 (2012). https://doi.org/10.1016/j.conbuildmat.2011.12.092

Chakkamalayath, J., et al.: Cement-superplasticiser compatibility - Issues and challenges. Indian Concr. J. (2011). https://www.researchgate.net/publication/286714849_Cement-superplasticiser_compatibility_-_Issues_and_challenges. Accessed 3 Apr 2019

Chandra, S., Björnström, J.: Influence of superplasticiser type and dosage on the slump loss of Portland cement mortars - Part II. Cem. Concr. Res. **32**(10), 1613–1619 (2002). https://doi.org/10.1016/S0008-8846(02)00838-4

Chryso, S.A.: General Catalogue 1ste Edition: 2007. Chryso South Africa, Boksburg (2007)

Domone, P., Illston, J. (eds.): Construction Materials, 4th edn. Spon Press, New York (2010)

Koehler, E.P., Fowler, D.W.: Development of a Portable Rheometer for Fresh Portland Cement Concrete (ICAR Report 105-3). International Center for Aggregates Research, Austin, TX (2004)

Mapei, S.A.: General Catalogue 1ste Edition: 2017. Mapei South Africa, Roodekop Germiston (2017)

Plank, J., et al.: Fundamental mechanisms for polycarboxylate intercalation into C_3A hydrate phases and the role of sulfate present in cement. Cem. Concr. Res. **40**(1), 45–57 (2010). https://doi.org/10.1016/j.cemconres.2009.08.013

Quanji, Z.: Thixotropic behavior of cement-based materials: effect of clay and cement types (2010). https://lib.dr.iastate.edu/etd/11724. Accessed 15 Feb 2019

SANS 5861-1, S.: SANS 5861-1:2006 : 2006 SOUTH AFRICAN NATIONAL STANDARD Concrete tests Part 1 : Mixing fresh concrete in the laboratory, pp. 2–6 (2006)

SANS 5862-1, S.: SANS 5862-1:2006: 2006 SOUTH AFRICAN NATIONAL STANDARD Concrete tests - Consistence of freshly mixed concrete - Slump test, pp. 1–5 (2006)

Acacia Karroo as Potential Admixture for Hot African Weather

Rose Mbugua[1(✉)], Salim Wanjala[2], and Julius Ndambuki[2]

[1] Department of Civil Engineering, Walter Sisulu University, Butterworth, South Africa
rmbugua@wsu.ac.za

[2] Department of Civil Engineering, Tshwane University of Technology, Pretoria, South Africa
{SalimRW, NdambukiJM}@tut.ac.za

Abstract. Africa needs a new approach for the use of admixtures that provide specific modification properties to concrete in addition to being compatible with African climatic conditions. For example, when concrete is mixed at elevated temperatures, there is quick loss of workability due to high evaporation of mixing water and the tendency is to add more water to the mix. Thus, there is need to develop admixtures that are natural (do not have to undergo derivation processes), readily available, cheap and environmentally-friendly. Gum Arabic (GA) is a sticky natural fluid which oozes from the Acacia tree when an insertion is made and contains natural resin which has arabin. GA comes from two species of Acacia tree, i.e. Acacia Senegal and Acacia Seyal. Gum Acacia Karroo (GAK) which is readily available in these hot areas was used as an admixture for mortar and concrete. GAK comes from Acacia Karroo Haynes which grows mainly in the Southern countries of Africa (Zimbabwe, Mozambique, Zambia and Angola) while Gum Arabic from Acacia Senegal or Seyal comes from countries in Northern Africa (Sudan, Chad, Nigeria). Preliminary results showed that mixes containing GAK have improved compressive strengths and chloride penetration resistance compared to the mixes without GAK, when temperature was increased from 23 °C to 40 °C at the age of 56 days. This suggests that GAK can be used at high temperatures as an admixture to improve these properties of concrete.

Keywords: Gum acacia karoo · Admixture · Chloride penetration

1 Introduction

Admixtures are now a common ingredient in concrete due to the fact that the quantities involved are quite small but their effect is astonishing. Different types of admixtures from different sources e.g. from crude petroleum oil and bye-products of industries are used in the developed world. Research in this area is quite advanced in the developed world. However, in Africa there is scarce evidence in literature of research in this area despite the fact that admixtures currently being used in Africa are mainly imported as finished products. This has serious impact to the environment due to carbon footprint

W. P. Boshoff et al. (Eds.): SAP 2019, RILEM Bookseries 24, pp. 176–183, 2020.
https://doi.org/10.1007/978-3-030-33342-3_19

during transportation. In addition, few studies have been done to proof suitability of these admixtures with African weather.

This call for rigorous research in Africa to come up with local admixtures to enhance performance of concrete mixed and placed in African weather. One such admixture is Gum from Acacia species. Other potential admixtures are Nkui and cassava peel (Schmidt et al. 2017). Acacia which grows in Southern countries of Africa (Zimbabwe, Mozambique, Zambia and Angola) is known as Acacia Karroo while the species that grows in the upper part of Africa (East, North and West) is known as Acacia Seyal. Acacia Karoo can tolerate temperatures as high as 40 °C and as low as minus 12 °C and drought season up to 9 months (Cabi.org 2018). It survives in different types of soils including saline soils and can be found at different altitudes. The reddish wood is used as fuel, bark is used for leather turning, roots are used as medicine, while thorns are used for stitching clothes. One hectare of Acacia can produce 25 to 30 tons of gum. Gum Arabic (GA) is a sticky natural fluid which oozes from Acacia tree when an insertion is made and contains natural resin which has arabin and is exported for use in beverages and cosmetics. The gum from Karoo Haynes which was investigated in this study, contains toxins thereby making it not suitable for export. The gum was investigated as a bio-admixture in cement systems at 23 °C and 40 °C.

2 Objectives

The most common performance measurement used by engineers is compressive strength. Admixtures have an effect on compressive strength of concrete. For example, according to Ramachandran (1995) concrete with retarding admixtures showed higher compressive strength at 28 days. This was due to greater degree of hydration at latter ages which resulted in formation of denser products of hydration leading to higher compressive strength. High curing temperatures result in accelerated hydration and early strength development due to more hydration products being formed. Compressive strength values are therefore greatly influenced by the curing temperature.

Another factor that affect compressive strength is the water cement (w/c) ratio. According to Ait-Aider et al. (2007) increase in w/c ratio in harsh hot weather conditions does not significantly influence the strength of concrete. This was due to the fact that high w/c ratio was compared to adequacy in curing where excess water was used for hydration at early age of curing.

To meet design requirements, concrete mixtures must be designed to provide a wide range of durability and mechanical requirements of structures. The most common mechanical performance measurement used by engineers is compressive strength. On the other hand, durable concrete should resist potentially destructive conditions to which it might be exposed to during service. One important tests used to measure durability of concrete is compressive strength and chloride ingress. Chloride penetration plays a significant role in loss of durability of concrete.

The purpose of the present study was to investigate experimentally the effect of GAK on compressive strength of concrete as well as chloride conductivity considering the effects of varying temperatures at different ages. For this purpose, two types of concrete, having w/c ratio of 0.61 and 0.5, were investigated and cured at two curing temperatures, 23 °C and 40 °C. The performance of concrete samples treated with different dosages of GAK and cured at the two temperature levels were evaluated while concrete without GAK was used as control to gauge performance.

3 Durability of Concrete

Compressive strength of concrete is an indirect indication of degree of hydration. Durability of concrete can be measured either by compressive strength or ingress of gases and liquids into concrete. These properties depend on many factors such as the curing temperature of concrete, humidity and environmental conditions. Researchers for example Kjellsen et al. (1990) and Wild et al. (1995) confirmed loss of attainable strength at higher curing temperatures. At low temperatures, slow hydration resulted in formation and uniform distribution of hydration products in the interstitial space leading to higher compressive strength. High curing temperatures result in accelerated hydration and early strength development due to more hydration products being formed. Compressive strength values are therefore greatly affected by the curing temperature. Researchers using admixtures in concrete at high temperature mostly used Self Consolidating Concrete (SCC). Ghafoori and Diawara (2010) studied the influence of super-plasticizer in SCC. Their study reported loss of workability as well as decrease in flow or plastic viscosity for SCC at high temperatures. GAK was found to improve plasticity and compares well with other chemical admixtures (Schmidt et al. 2018).

4 Expimental Framework

Ordinary Portland Cement (OPC) CEM I 52.5 N mixed with 25% unclassified fly ash conforming to SANS 1491:2005 was used for all concrete mixes. The relative density of cement is 3.12 and blain surface area is 324 m^2/kg. X-ray Fluorescence (XRF) was used to determine the composition of cement and the fly ash as shown in Table 1.

Crusher sand with Fineness Modulus of 3.2 and 19 mm coarse aggregates conforming to SANS 1083:2006 were used, both donated by Rosslyn Quarry in Pretoria, South Africa. The coarse aggregate was crushed from Norite rock. Gum Acacia Karroo (GAK) was harvested from Pretoria Botanical Garden. The tears were hand-picked from Acacia Karroo tree where they had formed on the bark as shown in Fig. 1a. The gum was then left to dry and harden. It was cleaned by removing pieces of bark and any foreign matter (Fig. 1b) and dissolved in part of gauge water.

Table 1. Cement and fly ash composition

Constituents	Cement	Fly ash (percentage)
SiO_2	20.17	41.7
Al_2O_3	3.96	22.3
Fe_2O_3	2.35	2.4
Mn_2O_3	0.753	–
TiO_2	0.29	–
CaO	60.57	18.2
MgO	2.71	1.8
P_2O_5	0.09	–
Cl	96.8 ppm	–
SiO_3	2.39	–
Na_2O	0.148	0.2
K_2O	0.37	0.2
CaO_2	1.5	–
SO_3	1.5	–

(a) **(b)**

Fig. 1. Gum tears (a) Ozing from the bark; (b) cleaned tears of GAK

Compressive strength tests were carried out according to SANS5863:2006. The Cement and Concrete Institute method of mix design was used to achieve a target strength of 35 MPa after 28 days (Owens 2009). All constituent materials were batched by weight. Samples with 0, 0.3, 0.7 and 0.8% (weight of cement) dosage of GAK with water/bidder ratio (w/b) of 0.61 were prepared. In addition, samples with w/b ratio of 0.5 at 0.7 and 0.8% dosage level were prepared. Reduction in w/c ratio was possible due to the water reducing effect of Gum at this level of dosage (Mbugua et al. 2016). Cubes with dimensions 100 × 100 × 100 mm were cured at 23 °C and 40 °C. Curing of samples at 40 °C was done in a climate chamber with 99% humidity. All samples were tested for compressive strength at 28 and 56 days of curing. Average compressive strength for three specimens was recorded.

Chloride Conductivity Index (CCI) was determined using South African Durability Index tests (Alexander 2008). Concrete cubes were prepared, cured and tested at the age of 56 days. The test was carried out after 56 days due to the fact that there was still high hydration process witnessed even after 28 days due to the slow hydration process in presence of GAK. Coring of cubes was done perpendicular to casting direction no more than 3 days after the age of testing. The same dosages of GAK used for compressive strength were used for this test. The chloride conductivity was calculated using Eq. 1.

$$CCI = \frac{it}{VA} \tag{1}$$

Where, *CCI* - conductivity of specimen (mS/cm), t - average thickness of specimen (cm), i - electric current (mA), V - voltage difference (V), A - cross-sectional area of the specimen (cm^2). Chloride conductivity index was determined as the average of the chloride conductivity of at least three valid test.

5 Results and Discussion

Results for compressive strength at different temperatures are shown in Fig. 2 at 28 days of curing. High compressive strength values were observed at curing temperature of 40 °C for all mixes at all dosage levels compared to 23 °C. The highest compressive strength value observed was at 0.7% dosage when the w/b ratio was 0.5. It was also noted that compressive strength of all samples at 40 °C were higher than the design value of 35 MPa at both 28 and 56 days of curing.

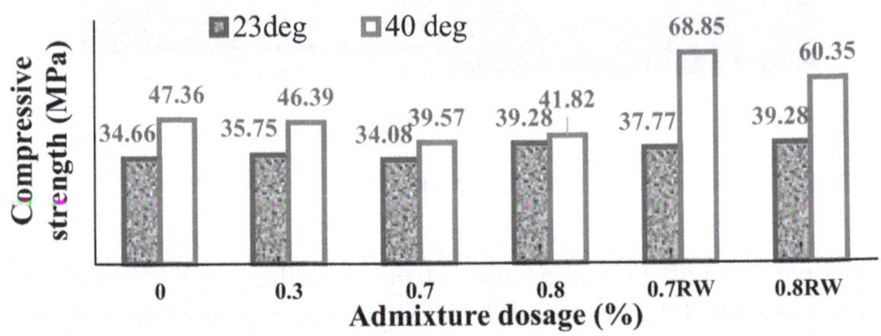

Fig. 2. Compressive strength of concrete at 28 days of curing

Comparing the compressive strength for mixes with w/b ratio of 0.61 after 56 days of curing (Fig. 3), the control increased strength by 3.6% when temperature was increased from 23 °C to 40 °C while at 0.7% and 0.8% dosage level an increase of 11.7 and 14.4% was observed. This shows that GAK influenced the compressive strength at high ambient temperature since all other conditions were kept constant and only the GAK dosage was changed.

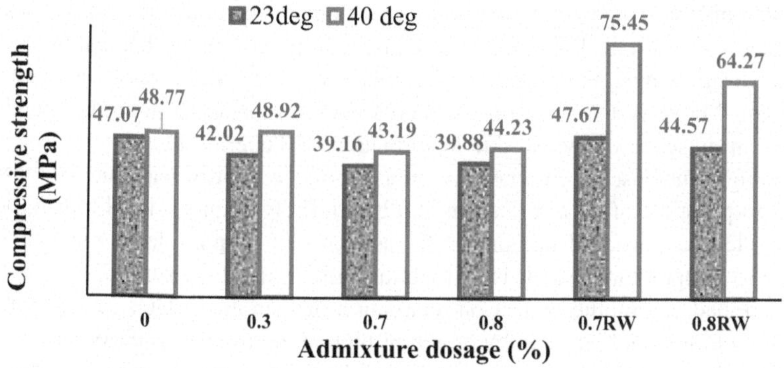

Fig. 3. Compressive strength at 56 days of curing

CCI values for samples with reduced w/b ratio (0.7RW and 0.8RW) (Fig. 4) were much lower due to impermeable micro-structure as confirmed by the compressive strength results. Samples cured at 40 °C with different GAK dosages showed higher chloride ingress. This can be attributed to the fact that at these dosages there was higher air entrainment as indicated from other studies where carbonation also increased (Mbugua 2017).

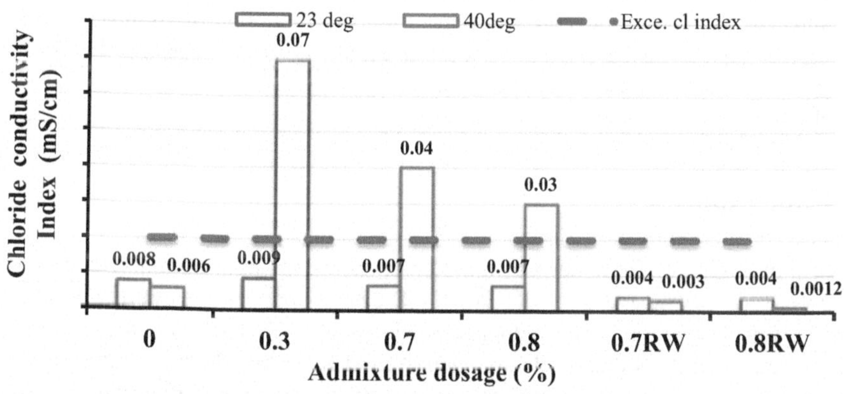

Fig. 4. Chloride conductivity index at 56 days of curing

It can be concluded that at low w/b ratio and at 40 °C there was accelerated hydration. However, in presence of GAK the hydration process was not accelerated in the same fashion as for concrete without GAK. This was evident by the lower values of compressive strength observed at all curing ages of mixes with w/b ratio of 0.61 which were lower than the control. The low values of compressive strength observed as the dosage increased from 0.7 to 0.8% can be associated with increase in retardation of hydration process with increase in dosage. This phenomenon was also observed in a study conducted by Ait-Aider et al. (2007) which showed higher compressive strength in concrete with lower w/c ratio and plasticizer cured under different conditions.

High ambient temperature increase water demand of concrete and increases the temperature of concrete. Temperature has a large influence on mechanical properties of concrete at early age. Furthermore, the presence of fly ash cannot be overlooked. Nasser and Marzouk (1979) reported increase in compressive strength at high temperatures in presence of fly ash due to formation of Tobermorite.

The slight increase in compressive strength of the control was due to pozzolanic reaction in presence of fly ash Ortiz et al. (2005). Early strength development is influenced by temperature and amount of fly ash present as pozzolanic reaction is also sensitive to temperature than it is to hydration of Portland cement. GAK and fly ash could have had a combined effect on retardation of hydration. After curing started the rate of hydration was high thereby attaining high strength even at early age for mixes with low w/b ratio. However, since all mixes had the same amount of fly ash incorporated in them any change in compressive strength could only be explained by GAK dosage and w/b ratio. High increase in compressive strength can be associated by slow formation of hydration products due to the retardation properties (Mbugua et al. 2015) of GAK coupled with increase in temperature and water reduction. Therefore, it seems that GAK can be used where high strength concrete is required even at elevated temperatures. However, the increase in compression strength is significantly more compared to the influence of other admixtures and therefore requires further investigation.

Different mechanisms of transport are responsible for chloride ingress. Capillary absorption of chloride-containing liquids or permeation of salt solution are some of the transport mechanisms associated with chloride ingress and can occur simultaneously or in sequence. Dominant transport will depend on the saturation of capillary pores but both mechanism can act simultaneously or in sequence. Using GAK as admixture has shown that chloride ingress can be reduced by slowly forming hydration products which reduce permeability.

6 Conclusions

Increase in dosage of GAK decreased the compressive strength of concrete at 28 days and 56 days of curing and at 23 °C and 40 °C curing temperature when w/b ratio was at 0.61. However, due to water reducing capability of GAK there was increase in both compressive strength and chloride conductivity index at both 28 and 56 days of curing for samples with reduced w/b ratio and cured at 40 °C. These preliminary results shows that GAK can be used at elevated temperatures to attain not only high performance but also more durable concrete. GAK is a future admixture for Africa and other hot places in the world.

References

Ait-Aider, H., Hannachi, N., Mouret, M.: Importance of W/C ratio on compressive strength of concrete in hot climate conditions. Build. Environ. **42**, 2461–2465 (2007)

Alexander, M.G., Ballim, Y., Stanish, K.: A framework for use of durability indexes in performance-based design and specifications for reinforced concrete structures. Mater. Struct. **41**(5), 921–936 (2008)

Cabi.Org. 2018. Invasive Species Compendium: Acacia Karroo (sweet thorn). 2019 CAB International. https://www.cabi.org/isc/datasheet/2289. Accessed 24 Feb 2019

Ghafoori, N., Diawara, H.: Influence of temperature on fresh performance of self-consolidating concrete. Constr. Buil. Mater. **24**, 946–955 (2010)

Kjellsen, K.O., Detwiler, R.J., Gjørv, O.E.: Pore structure of plain cement pastes hydrated at different temperatures. Cem. Concr. Res. **20**, 927–933 (1990)

Mbugua, R., Salim, R., Ndambuki, J.: Utilisation of gum acacia karoo as set-retarding water-reducing admixture in cement mortar and concrete at optimum dosage. Knowledge Exchange for Young Scientists (KEYS), p. 97 2015

Mbugua, R.: Development and modelling of Acacia exudate as a retarding admixture in cement hydraation at different temperatures. Tshwane Univerity of Technology, Doctorate (2017)

Mbugua, R., Salim, R., Ndambuki, J.: Effect of gum Arabic karroo as a water-reducing admixture in concrete. Materials **9**(2), 80 (2016)

Nasser, K., Marzouk, H.: Properties of mass concrete containing fly ash at high temperatures. J. Proc. 537–550 (1979)

Ortiz, J., Aguado, A., Agullo, L., Garcia, T.: Influence of environmental temperatures on the concrete compressive strength: simulation of hot and cold weather conditions. Cem. Concr. Res. **35**, 1970–1979 (2005)

Owens, G. (ed.) Fulton's Concrete Technology. Cement & Concrete Institute (2009)

Ramachandran, V.: Concrete Admixtures Handbook: Properties, Science, and Technology (1995)

Schmidt, W., Ngassam, I.L.T., Olonade, K.A., Mbugua, R.N., Kühne, H.C.: Plant based chemical admixtures–potentials and effects on the performance of cementitious materials. RILEM Tech. Lett. **3**, 124–128 (2018)

Schmidt, W., Ngassam, I.T., Mbugua, R., Olonade, K.A., Für Materialforschung Und-Prüfung, B., Schmidt, D.-I.W.: Natural Rheology Modifying Admixtures For Concrete. Rheologische Messungen An Baustoffen, pp. 75–87 (2017)

Wild, S., Sabir, B., Khatib, J.: Factors influencing strength development of concrete containing silica fume. Cem. Concr. Res. **25**, 1567–1580 (1995)

Used Oil as an Admixture to Improve the Rheological Properties of Concrete

Gerrit M. Moelich, Rick van Huffel, and Riaan Combrinck(✉)

Unit for Construction Materials, Civil Engineering Department, Stellenbosch University, Private Bag X1, Matieland 7602, South Africa
{gmm, rcom}@sun.ac.za

Abstract. In the past, industrial waste and by-products have successfully been used to improve the properties of concrete. Used engine oil is a waste product which is burdensome to discard of and, due to frequent replacement, is produced in high quantities in the construction industry. The utilisation of used engine oil in concrete has shown potential as an admixture by reducing slump and increasing air-content. The main disadvantage is a reduction in long term compressive strength. This study investigates used engine oil (UEO) and used hydraulic oil (UHO) as admixtures to concrete, focusing on its effect on the rheological properties. Slump, air-content, compressive strength and rheometer tests are conducted for concrete containing different dosages of UEO and UHO. Adding low dosages of UEO and UHO have no noteworthy effect on the compressive strength, although increasing air-content and altering the rheological properties significantly. UEO and, to a lesser extent UHO, reduced the energy required to initiate flow (static yield stress) as well as decreased the plastic viscosity. Adding UEO has a similar effect on the static yield stress and plastic viscosity as increasing water content or substituting cement with a proportion of fly-ash. In conclusion, UEO shows potential as an air-entrainer or water-reducing admixture.

Keywords: Concrete · Admixtures · Used oil · Air-entrainer · Water-reducer

1 Introduction

Industrial waste and by-products, such as fly ash, silica fume and blast furnace slag, have successfully been used as additives to improve the properties of concrete. Reusing waste products can reduce disposal costs, the environmental impact and if applied correctly, can improve construction materials. Based on these reasons, research is increasing on the addition of waste products to concrete. A waste product that has attracted attention in the recent past is used mineral oil, commonly used as a lubricant in mechanical equipment. Mineral oil needs to be replaced regularly due to a loss in efficiency and is, for this reason, produced in high quantities. It is estimated that up to 55% of used oil ends up in the natural environment where it contaminates natural resources and poses a threat to public health (El-Fadel and Khoury 2001).

This study investigates the influence of two types of used mineral oils (UMO), namely used engine oil (UEO) and used hydraulic oil (UHO), as admixtures to concrete.

W. P. Boshoff et al. (Eds.): SAP 2019, RILEM Bookseries 24, pp. 184–191, 2020.
https://doi.org/10.1007/978-3-030-33342-3_20

Only a few authors have investigated this topic. The existing research on UMO can be categorised in four different applications:

(1) UMO as an admixture which improves the properties of concrete (Hamad et al. 2003). An increase in the slump and the air-content occurs with an increase in UEO dosage from 0% to 0.3%. For the same mixture, an air-entrainer increased the air-content fourfold where UEO doubled the air-content. Adding UEO decreases the porosity and initial setting time while increasing the consistency of the concrete (Shafiq et al. 2006). A similar reduction in slump occurs when adding UEO and water-reducing admixtures to concrete (Abdelaziz 2009). Shafiq et al. (2011) showed that a similar reduction in slump can be achieved by adding UEO as what can be achieved by a replacement of 40% or 50% of cement with fly ash. The main attribute of concrete is its compressive strength. A reduction in compressive strength occurred due to the addition of 0.15% UEO in the study by Shafiq et al. (2011) but not in the study by Hamad et al. (2003). The addition of UEO reduces the flexural and tensile strength and modulus of elasticity (Hamad et al. 2003). However, Chin et al. (2012) showed that UEO increased the flexural strength of reinforced beams. Hussein (2015) demonstrated that the addition of UEO resulted in a reduction in compressive strength, even at a lower W/C ratio.

(2) The use of oil-contaminated aggregates in concrete where other alternatives are not available. Using fine aggregates with higher levels of crude oil contamination (>4.5 per cement weight) can significantly reduce the compressive strength of concrete (Ajagbe et al. 2012).

(3) The effect of oil on the surface of mature concrete. It has been reported that the leakage of engine oil on the concrete surface can increase its resistance to freezing and thawing (Hamad et al. 2003). Pukhov (2001) investigated and stated that concrete deterioration can be caused by UEO on the concrete surface, but the extent of deterioration is dependent on its permeability. Diab (2011) showed that UEO can prevent strength gain when applied to the surface of mature concrete.

(4) UEO as a mould releasing agent is an application that has not been extensively investigated. From the author's experience as a site engineer, several concrete contractors prefer the use of UEO as opposed to commercially available releasing agents.

For this study, to understand the effect of UEO and UHO on the properties of concrete high dosages are added to a reference mixture. The workability and flowability can then be evaluated with a rheometer in terms of static and dynamic yield stress as well as plastic viscosity (Domone and Illston 2010).

2 Research Significance

The effect of adding low to relatively high dosages of UEO and UHO on the slump, air-content, compressive strength and rheological properties are investigated. The test results can potentially provide insight into the further application of UMO as an admixture to improve the properties of concrete. Adding these UMOs to concrete can be a practical and less environmentally harmful method to discard of them.

3 Experimental Program

3.1 Test Setup

Slump, air-content, compressive strength and rheometer tests were conducted according to the appropriate standards: SANS 5862-1 (2006), SANS 6252 (2006) and SANS 5863 (2006). The results of the three slump tests, conducted 5 min after mixing, were averaged. For measuring air-content, the results of the three tests were averaged. For the compressive strength tests, 10 cubes were cast for each mixture, testing three cubes on the 1st and 7th day and four cubes on the 28th day.

An ICAR rheometer, as shown in Fig. 1, was used to determine the flow and stress growth curves of the mixtures. In a flow curve test, the torque is measured for the impeller rotating at different speeds. The speed is plotted against the resulting torque and a straight line is fitted. The intersection of the straight line with the vertical axis is related to the dynamic yield stress where the slope is related to the plastic viscosity. In a stress growth test, the stress exerted on the material is gradually increased until flow is initiated. This point is defined as the static yield stress (Koehler and Fowler 2004). These tests were conducted 3 min after mixing. Due to time constraints, the rheology test was only conducted once for each mixture.

3.2 Mixtures and Materials

Used hydraulic oil (UHO) and used engine oil (UEO) were added to the reference mixture at dosages of 0.3% (Hamad et al. 2003), 0.9% (Hussein 2015) and 2.7% per cement weight during the mixing phase. The reference mixtures can be seen in Table 1. A different, more flowable reference mixture was used for the rheology test, as illustrated in Table 1.

Table 1. Reference mixtures

Material	Reference (kg/m^3)	Rheometer reference (kg/m^3)
Water	205	230
Cement (CEM II 52.5 N)	342	329
Coarse aggregate (19 mm)	1231	766*
Fine aggregate (Natural pit sand)	646	1026

* Coarse aggregate (6 mm)

Fig. 1. Rheometer

4 Results

4.1 Slump Test

The slump test can be used to indicate the flowability or consistency of the plastic concrete. The results from the slump test can be seen in Fig. 2. At a dosage of 0.3%, the UEO results in a 40% increase in slump where the UHO results in a 20% increase. At higher dosages, the UEO resulted in a 70% increase in slump. The slump value is proportional to an increase in dosage of UEO.

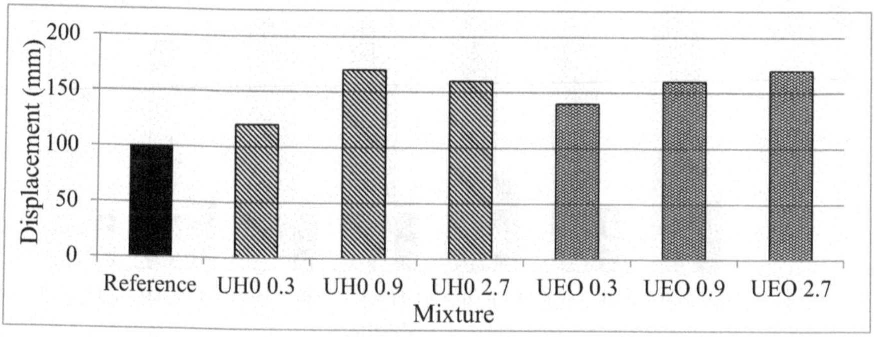

Fig. 2. Slump test result

4.2 Air-Content

The air-content is an indication of the amount of voids in the fresh concrete. An increase in voids can be favourable in applications where resistance to freeze-thaw, a lower density concrete and better insolation are required. The results from the air-content test can be seen in Fig. 3. At a dosage of 0.3%, the UEO results in a 120% increase in air-content where the UHO results in a 100% increase when compared to the reference mixture. At a dosage of 2.7%, the UEO increase the air-content by 300%. The increase in air-content is consistent with an increase in dosage of UEO. This is not the case for UHO where the addition of 0.9% UHO resulted in a lower air-content than adding 0.3% UHO.

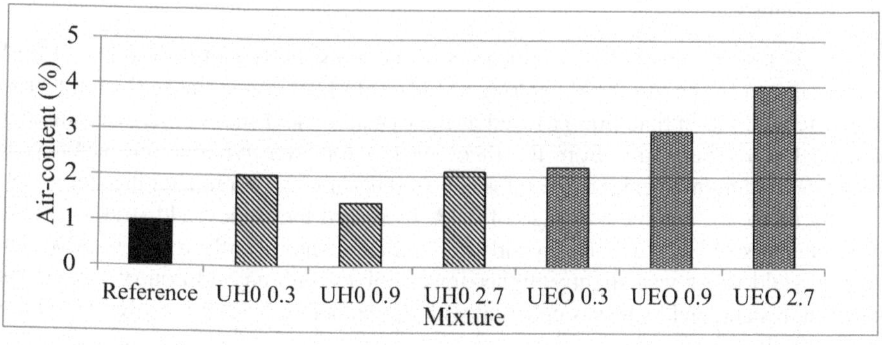

Fig. 3. Air-content

4.3 Compressive Strength

The compressive strength on 1, 7 and 28 days of both the UHO and UEO mixtures can be seen in Fig. 4. For the addition of UHO at dosages of 0.3% and 0.9%, the 7- and 28-day strengths are similar to the reference. For the higher dosage of 2.7%, the UHO resulted in a reduction in strength of 17% at 28 days when compared to the reference.

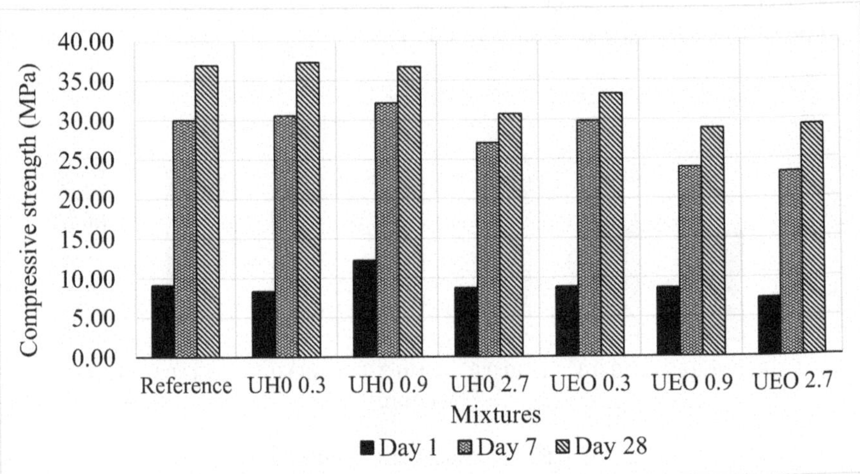

Fig. 4. Compressive strength results

For the addition of UEO at a dosage of 0.3%, the 7- and 28-day strength reduced up to 10% compared to the reference. For dosages of 0.9% and 2.7% of UEO, the strength reduced by up to 22%.

Literature shows that the relative strength of cement gels is proportional to the degree to which the gels fill the empty voids (Powers 1958). Therefore, a decrease in porosity (empty voids) results in an exponential increase in compressive strength (Chen et al. 2013). In general, the results of this study show that a decrease in compressive strength occurred with an increase in air-content and is thus as expected.

4.4 Rheology

A rheology test can give a better indication of the workability and flowability of fresh concrete than a slump test. The rheology test is used to determine the static yield stress (energy required to initiate flow) as well as the dynamic yield stress (energy required to maintain flow). The results from the rheology test for both dynamic and static yield stress can be seen in Fig. 5. UHO at dosages of 0.9% and 2.7% reduced the static yield stress by up to 38%, while a dosage of 0.3% increased the static yield stress by 15%. For all dosages of UEO the static yield stress reduced significantly by up to 50%. The dynamic yield stress was slightly higher for all mixes with an addition of UMO. The reduction in static yield stress is coherent with the reduction in slump caused by adding UEO and UHO.

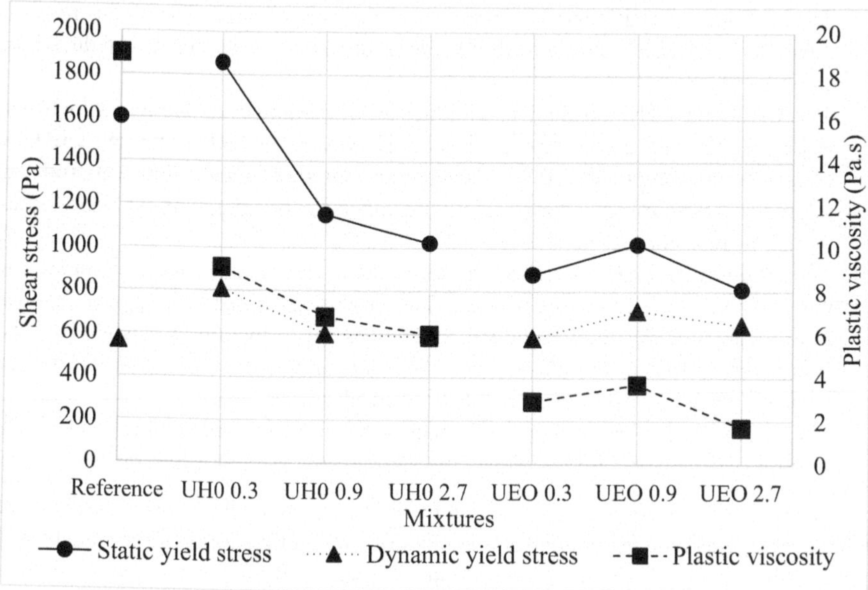

Fig. 5. Rheological properties of the mixtures

The plastic viscosity is the resistance to flow once the static yield stress is exceeded. The results of the plastic viscosity can be seen in Fig. 5. The addition of UHO reduced the plastic viscosity by between 52% and 69%. The addition of UEO, even at a dosage of 0.3%, reduced the plastic viscosity by between 85% and 91%. The reduction in plastic viscosity could increase the mix tendency to segregate, although no segregation was observed for all the mixes tested in this study.

It should also be noted that, for this study, the plastic viscosity decreases when the air content increases. This may be due to the voids reducing the contact area between particles and, therefore, the friction; causing a reduction in plastic viscosity (Domone and Illston 2010). Alternatively, the reduction of plastic viscosity can be due to the UMO directly reducing the friction between particles.

The reduction in static yield stress indicates a lower resistance under low shear stresses. Furthermore, a reduction in plastic viscosity indicates less cohesion. The combination results in a more flowable fresh concrete.

5 Conclusion

Experiments to determine the influence of used hydraulic oil (UHO) and used engine oil (UEO) on the properties of concrete were successfully executed. The rheology, slump, air content and compressive strength of the concrete with different dosages of UHO and UEO were evaluated and compared to a reference mixture. The following conclusions can be drawn:

(1) The addition of UHO and UEO increased the slump value. The slump result is coherent with the results from the rheology test since these oils also reduced the static yield stress.
(2) The UEO and UHO resulted in a reduction in compressive strength which was more notable with an increase in dosage. However, for both UHO and UEO the compressive strength at a 0.3% dosage was not significantly different from the reference. These oils can consequently be added to concrete, without a substantial reduction in compressive strength.
(3) The addition of UEO increased the air-content significantly when compared to both UHO and the reference mixture. The increase in porosity may have resulted in the reduction in compressive strength and/or the reduction in plastic viscosity. It is believed that in the plastic state the voids and/or the oil reduce the friction between particles resulting in a reduction in plastic viscosity.
(4) The addition of UHO reduced the energy required to initiate flow and plastic viscosity. However, UEO achieves the same result even at a low dosage.

In conclusion, the addition of UEO, and to a lesser extent UHO, has a similar effect on the rheology as increasing the water content or substituting some of the cement with fly-ash, when keeping all other proportions the same. Both the energy required to initiate flow and plastic viscosity reduced. Although additional research is required, the addition of UEO, and to a lesser extent UHO, shows potential as a water-reducing and air-entraining admixture in concrete, especially at low dosages where the compressive strength reduction is minimal.

Acknowledgements. This work is based on the research supported in part by the National Research Foundation of South Africa. Any opinion, finding and conclusions or recommendations expressed in this material is that of the author(s) and the NRF does not accept any liability in this regard. The financial support of Pretoria Portland Cement (PPC) is also gratefully acknowledged.

References

Abdelaziz, G.E.: Utilization of used-engine oil in concrete as a chemical admixture. HBRC J. Housing Build. Natl. Res. Centre, Egypt **5**(3) (2009)

Ajagbe, W.O., Omokehinde, O.S., Alade, G.A., Agbede, O.A.: Effect of crude oil impacted sand on compressive strength of concrete. Constr. Build. Mater. **26**(1), 9–12 (2012)

Chen, X., Wu, S., Zhou, J.: Influence of porosity on compressive and tensile strength of cement mortar. Constr. Build. Mater. **40**, 869–874 (2013)

Chin, S.C., Shafiq, N., Nuruddin, F.: Effects of used engine oil in reinforced concrete beams: the structural behaviour. Int. J. Civ. Geol. Eng. **6**, 83–90 (2012)

Diab, H.: Effect of mineral oil on reinforced concrete structures, part I: Deterioration of compressive strength. J. Eng. Sci. Assiut Univ. **39**(6), 1321–1333 (2011)

Domone, P.L.J., Illston, J.M. (ed.): Construction materials: their nature and behaviour. Milton Park, Abingdon, Oxon, Spon Press, pp. 120–126 (2010)

El-Fadel, M., Khoury, R.: Strategies for vehicle waste-oil management: a case study. Resour. Conserv. Recycl. **33**(2), 75–91 (2001)

Hamad, B.S., Rteil, A.A., El-Fadel, M.: Effect of used engine oil on properties of fresh and hardened concrete. Constr. Build. Mater. **17**(5), 311–318 (2003)

Hussein, M.N.: Properties of concrete containing new and used engines oil. Int. J. Sci. Res. (IJSR) **4**(12), 268–272 (2015). ISSN (Online)

Koehler, E.P., Fowler, D.W.: Development of a portable rheometer for fresh portland cement concrete (2004)

Powers, T.C.: Structure and physical properties of hardened Portland cement paste. J. Am. Ceram. Soc. **41**(1), 1–6 (1958)

Pukhov, I.E.: Effect of mineral oil on the reinforced-concrete floors of the Uglich and Rybinsk hydroelectric power plants. Power Technol. Eng. (formerly Hydrotechnical Construction) **35**(7), 370–372 (2001)

Shafiq, N., Nuruddin, M.F., Kamaruddin, I.: Effectiveness of used engine oil on improvement of properties of fresh and hardened concrete. In: Proceedings of the 6th Asia-Pacific Structural Engineering and Construction Conference (ASPEC 2006), pp. 159–166 (2006)

Shafiq, N., Nuruddin, M.F., Beddu, S.: Properties of concrete containing used engine oil. Int. J. Sustain. Constr. Eng. Technol. **2**(1), 72–82 (2011)

Effect of Different Molecular Weights and Chemical Composition of Superplasticizers on the Structural Build-up of Cement Paste Using Dynamic Oscillatory Rheology

Nonkululeko W. Radebe[1], Christopher O. Klein[1], Lei Lei[2], and Manfred Wilhelm[1(✉)]

[1] Institut für Technische Chemie und Polymerchemie, Karlsruher Institut für Technologie (KIT), 76131 Karlsruhe, Germany
manfred.wilhelm@kit.edu

[2] Lehrstuhl für Bauchemie, Technische Universität München (TUM), Lichtenbergstr., 4, 85747 Garching, Germany
lei.lei@bauchemie.ch.tum.de

Abstract. Two polycarboxylate ether (PCE) superplasticizers with different molar masses and functionalities were studied in relation to their strain and time-dependant rheological properties. The effect of dosage on cement paste fluidity and structural-build up was of particular interest. Through size exclusion chromatography (SEC) the relative molar mass and polydispersity index (PDI) was determined. Superplasticizer A (SP-A) was found to be three times the molar mass of superplasticizer B (SP-B). The dosage limitations were then determined through a spread flow test. The SP-B, with a lower molar mass, and a backbone functionality of a methacrylate - ester, began to flow at higher dosages and with relatively small changes in dosage, a large impact on the fluidity was notable. Dynamic oscillatory rheology was used to determine structural build-up, as a non-destructive method. The structural build-up of cement paste is a time-dependent phenomenon therefore a time sweep was done. A constant strain and angular frequency, within the linear viscoelastic regime (LVE), was used. The LVE was determined through a series of oscillatory strain sweeps for cement pastes with and without SP-A and B. Thereafter the rate of rigidification (G_{ridge}) and percolation time (t_{perc}) as a function of hydration time was investigated.

Keywords: Superplasticizers · Oscillatory dynamic rheology · Cement paste · Size exclusion chromatography (SEC) · Storage modulus (G')

1 Introduction

In the year 2016, the global concrete admixture market size in terms of volume, was estimated at 22,5 megatons. The expected monetary reach is expected at US$ 33.6 billion by 2025 (Wood 2017). Water reducing agents such as plasticizers and super-plasticizers are anticipated to dominate the global concrete admixture market over the forecast period. Water reducing agents are used to reduce the water content in the

W. P. Boshoff et al. (Eds.): SAP 2019, RILEM Bookseries 24, pp. 192–199, 2020.
https://doi.org/10.1007/978-3-030-33342-3_21

concrete by 5% to 12%. Superplasticizers (SP) are second-generation of plasticizers and are gaining momentum for their distinct ability to reduce water content whilst maintaining workable. This is especially important for applications in pumping, casting in regions with high temperatures and setting properties. Plasticizers are a product of petro-chemical products and have seen an increase in their use in the past few years. Plasticizers consist of a backbone of polyethylene, grafted chains of polyethylene oxide (PEO), and carboxylic groups as adsorbing functional groups. Polycarboxylate ether (PCE) superplasticizers undergo a dispersion mechanism to prevent agglomeration when added to cement paste or concrete. This is due to the molecular structure and shape which is a steric hindrance effect (Lei and Plank 2014). This effect is mainly due to the presence of long neutral side graft chains, more so than the anionic groups. The neutral side chains are responsible for the adsorption of the polymers on the surface of cement particles (Shi 2009). Together with adsorptive groups, grafting density, and molecular conformation, the molecular weight of PCEs has an important role on the properties of cement paste and concrete. Many studies have been conducted on the effect of molecular weights on the performance of PCEs (Peng et al. 2012; Ran et al. 2015; Winnefeld et al. 2007). In this contribution, two of the same type of SPs are studied by determining at how a difference in functionality as well as molar mass can affect the oscillatory rheology. For the molecular analysis of polymer materials there are three main characteristics to consider: molecular size, functionality and polymer topology. Molecular size is quantified via the molecular weight distribution (MWD) as determined e.g. by size exclusion chromatography (SEC). It is then important to determine the limitations (i.e. bleeding) of the SPs when added to cement through dosage testing.

Rheological properties of cement paste are generally of interest when looking at workability though yield stress measurements. As yield stress measurements destroy the structure that we would like to monitor, we instead work within the linear viscoelastic domain (LVE) using small amplitude oscillatory shear (SAOS) (Yuan et al. 2017). Remaining at or below the critical strain allows for a non-destructive monitoring of the transient behaviour of a material as a function of time. By looking at build up indices such as percolation time (t_{perc}) and rate of rigidification (Mostafa and Yahia 2016, 2017), it is possible to have an overall view of the cement pastes begins to rapidly build up structure. The percolation time describes the resting period needed for colloidal particles to reach their favourable and equilibrium positions (Mostafa and Yahia 2016, 2017). These properties are all affected by different properties such as temperature, water to cement ratio, and admixtures. Here we compare two SPs with differing molar mass and functionality for their effect on fluidity, and rate of rigidification. The aim is to increase the percolation time without compromising the structural build-up of the cement paste.

2 Materials and Methods

2.1 Materials

Portland cement type CEM I 42,5 R (HeidelbergCement Group, Germany) was used in together with distilled water. A mass of 3.5 g Portland cement was used for each mixture. The plain cement paste was mixed by hand during 2 min and then placed into the plate - plate geometry. All results were reproducible within one decade. Two PCEs were used, namely SP-A and SP-B. The differences are listed in the results section. SP-A and SP-B have vinyl ether-ethylene glycol (VPEG) and methacrylate ester-ethylene glycol (MPEG) based functionalities, respectively. The two samples SP-A and SP-B were provided by BASF (Germany).

2.2 Size Exclusion Chromatography (SEC)

Molar masses (M_w and M_n) of the PCE samples and the conversion of the macro-monomers were determined via SEC, which is also known as gel permeation chromatography (GPC). The polydispersity index (PDI) which is derived from M_w/M_n indicates how broad the molecular weight distribution is, i.e. the larger the PDI value, the broader the molecular weight distribution of the sample is. For this study, a Waters Alliance 2695 separation instrument (Waters, Eschborn, Germany) equipped with a three-angle static light scattering detector ("mini Dawn" from Wyatt Technology Corp., Santa Barbara, CA) was employed to give absolute molar mass. Separation of the polymer fractions was achieved by using three Ultrahydrogel (120, 250, and 500) columns (Waters, Eschborn, Germany). The eluent was composed of 0.1 M $NaNO_3$ and 0.1 g/L NaN_3 adjusted to pH = 12. The value of dn/dc applied to calculate the molar masses of the PCEs was 0.135 mL/g (value for polyethylene oxide).

2.3 Cement Dispersion Testing

For determination of paste flow, a modified "mini slump "test according to DIN EN 1015 was utilized and carried out as follows: First, the water-to-cement (w/c) ratio of the paste without polymer was set to 0.3. At this w/c ratio, the spread flows of cement pastes holding various dosages of the PCE samples were determined. Generally, the polymer was added to the mixing water placed in a porcelain cup. The amount of water contained in the polymer solutions was subtracted from the amount of mixing water. Next, 300 g of cement was added to the mixing water and stirred manually for 1 min with a spoon, then rested for 1 min without stirring and were again stirred for 2 min. After the stirring, the cement paste was immediately poured into a Vicat cone (height 40 mm, top diameter 70 mm, bottom diameter 80 mm) placed on a glass plate and the cone was vertically removed. The resulting spread of the paste was measured twice, the second measurement being in a 90° angle to the first and averaged to give the final spread value.

2.4 Dynamic Oscillatory Rheology

A strain-controlled rheometer (ARES G2, TA-Instruments) was used to perform oscillatory rheological testing of the cement pastes with and without SP-As a function of time (t). A plate – plate stainless steel geometry (d = 25 mm) was used. The plate gap was kept in the range of 1.5 mm. A strain amplitude (γ_0) test was conducted using a frequency of 1 Hz which is conventional for cement paste testing (Bellotto 2013; Conte and Chaouche 2016). The linear (SAOS) region was tested by performing an oscillatory stress sweep from $\gamma_o = 4 \times 10^{-5}$ to $\gamma_o = 0.5$. Each measurements took 7 min. The results are discussed in the results and discussion section. The temperature was kept constant at $20\,°C \pm 0.1\,°C$ using a Peltier system. Thereafter, a series of time sweep measurements were conducted by measuring the storage modulus, G', as a function of time.

3 Results and Discussion

3.1 Characterization of PCE Polymers

The molecular structure of the PCE polymers are presented in Table 1. The broad polydispersity (PDI > 2) values imply the wide molecular weight distribution of polymerization products with different polymerization degrees. These are relative values based on the dn/dc of the side chains (PEO).

Table 1. Molar masses (M_w, M_n), PDI and conversion rates for the PCE polymers

Polymer sample	M_n (g/mol)	M_w (g/mol)	PDI (M_w/M_n)	Conversion rate of monomers
SP-A (Precast VPEG-PCE)	47,220	98,680	2.1	81.7%
SP-B (Ready-mix MPEG-PCE)	11,770	28,090	2.4	84.4%

3.2 Cement Dispersion

Adding either SP-A or SP-B had significantly different influences on the cement paste flow. SP-B requires more than two times the dosage for a fifth of the flow as highlighted in Fig. 1. SP-A has a higher molar mass, which could indicate shorter side chains and longer backbones. With this assumption, the lower dosages needed for flow could be as a result of high surface coverage of the backbone. Fewer polymers are needed to cover a larger surface and therefore flows at lower dosages. In contrast, SP-B has a three times lower molar mass, which could indicate that three times the polymers are needed to cover the same surface. Figure 1 shows that for a flow of 33 cm, SP-A a dosage of needs 0.5% bwoc and SP-B needs 1.5% bwoc.

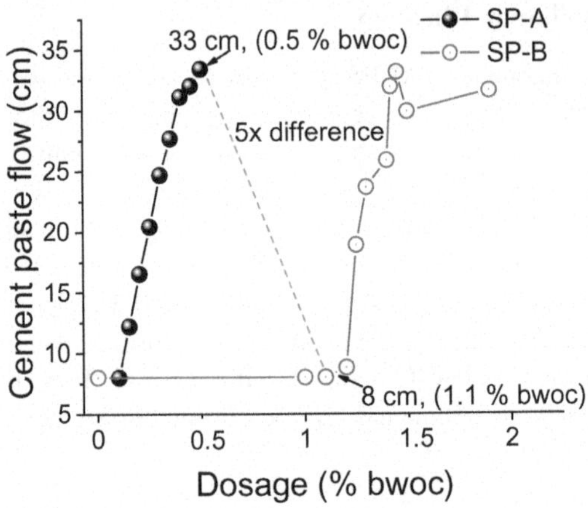

Fig. 1. Spread flow of cement pastes (w/c = 0.3) containing different dosages of PCE polymers

3.3 Dynamic Strain Sweep

Table 2 shows the mix proportions of cement water and PCE used for the rheological measurements. These were chosen based in the cement dispersion results. Two different samples with the same solid content were prepared for a time sweep. Here the strain was kept at the critical linear regime strain and the frequency was 1 Hz. The solid border, ending at $\gamma_0 = 0.05\%$,, marks the end of the linear viscoelastic region (SAOS) in Fig. 2. The dotted border is the regime in which large amplitude oscillatory strain (LAOS) measurements (See Fig. 2) are possible. For this study, we measure in the SAOS regime, as it is the non-destructive region. The molar mass notably affects the storage (G') and loss modulus (G") linear viscoelastic regime region.

Table 2. Mix proportion of the cement pastes

Sample name	Water to cement ratio (w/c)	Solid volume fraction (S_f)	Added SP-A (% bwoc)	Added SP-B (% bwoc)
CEM1	0.28	0.52	–	–
CEM2	0.34	0.48	–	–
SPCEM1	0.34	0.48	0.28	–
SPCEM2	0.28	0.52	0.28	–
SPCEM3	0.28	0.52	0.57	–
SPCEM4	0.28	0.52	–	1.42
SPCEM5	0.28	0.52	–	2.00
SPCEM6	0.28	0.52	–	2.86

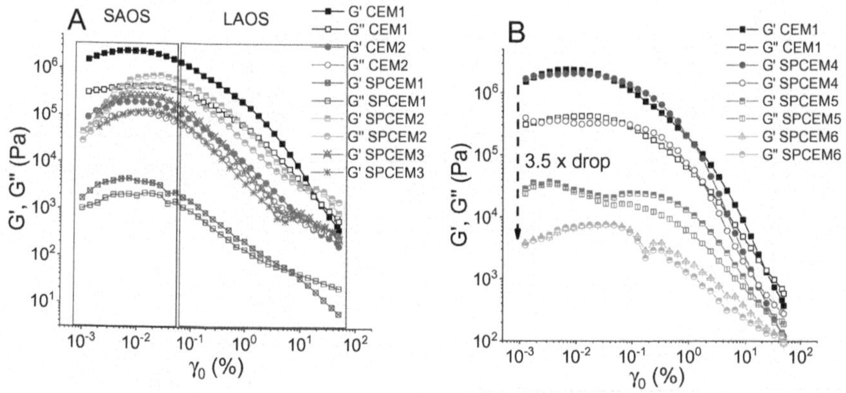

Fig. 2. Storage modulus (G') and loss modulus (G") (T = 20 °C, $\omega/2\pi = 1$), shown as a function of applied strain. (a) Influence of SP-A with vinyl ether-ethylene glycol side chains (b) influence of SP-B with methacrylate ester-ethylene glycol side chains

This region is the same for all cement pastes, with and without SP-And highlighted in Fig. 2a. SPCEM2 and SPCEM3 show a relatively small deviation from CEM1 in Fig. 2a. This is possibly because they all have a high solid volume fraction ($S_f = 0.52$), so the dispersing effect is minimized by having less water to flow in. However, SPCEM1 and CEM2 show a difference of two decades in G' and G". Even with the same S_f of 0.48, the addition of a low dosage of SP-A plays a significant role in the G' and G" of these two samples. This could be because more water means the plasticizers can flow easier and therefore the dispersing effect is higher.

3.4 Structural build up- dynamic time sweep

In Fig. 3a a delay in the increase of G' is notable. The percolation time is 400 s for the SPCEM2 (0.28% bwoc) and 730 s for SPCEM4 (0.57% bwoc). This almost two-fold increase in percolation time shows a similar behaviour as that is seen in Fig. 1 for the dispersion tests. Even though SP-B is used at a higher dosages, it does not remain in a fluid like state for it to be notable, as seen in Fig. 3b.

These percolation times for the samples containing SP (Fig. 3) are all followed by a rapid increase in G' which may be an indication of sedimentation and phase separation. Figure 4 shows the kind of behaviour that would be excepted without the influence of sedimentation (Mostafa and Yahia 2016). In Fig. 4, the rate of rigidification decreases with a decrease in solid fraction as also seen Fig. 3 the CEM1 and CEM2. However, the rate of rigidification is not expected to increase faster than the neat cement paste of the same volume fraction.

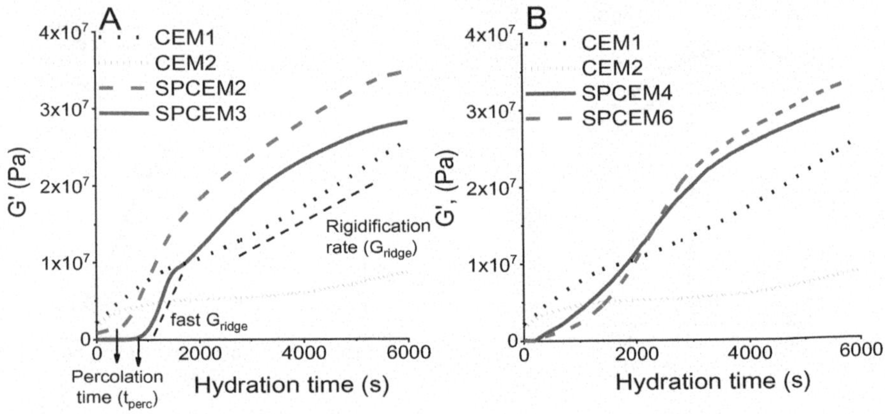

Fig. 3. Rate of rigidification of cement pastes, monitored by the storage modulus (G') of cement pastes with and without SP-A and B as a function of time (T = 20 °C, $\omega/2\pi = 1$ Hz and$\gamma_0 = 0.05\%$). (a) Cement pastes with and without SP-A (b) cement pastes with and without SP-B

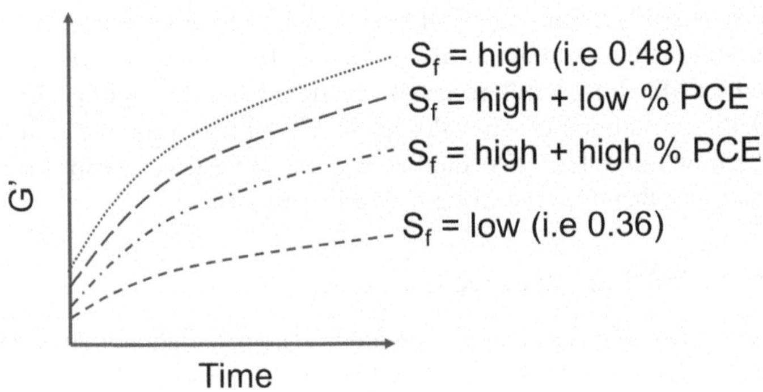

Fig. 4. A pictorial representation of the G' (Pa) behaviour that is theoretically expected without the effect of sedimentation.

4 Conclusions

In this study is clear that the chemical composition and molar mass of the superplasticizers plays a significant role in fluidity and subsequently in structural build up. The LVE regime for the cement pastes is limited to less than 0.05%, which is very low. This indicates that the particles are held together by short range particle interactions. In principle, the longer the percolation time, the longer the time to mix, pump and place the cement. However, the influence of dosage on sedimentation should is also of interests and should be investigated further. Furthermore, a more comprehensive study how the water to cement ratio plays a role in the amount and type of PCE that gives a sensible percolation time and does not negatively impact the rate of structural growth.

Acknowlegements. The authors thank the Deutsche Forschungsgemeinschaft (DFG) SPP 2005 “Opus Fluidum Futurum“ for funding. Project number 386871659

References

Bellotto, M.: Cement paste prior to setting: a rheological approach. Cem. Concr. Res. **52**, 161–168 (2013)

Conte, T., Chaouche, M.: Rheological behavior of cement pastes under large amplitude oscillatory shear. Cem. Concr. Res. **89**, 332–344 (2016)

Lei, L., Plank, J.: Synthesis and properties of a vinyl ether-based polycarboxylate superplasticizer for concrete possessing clay tolerance. Ind. Eng. Chem. Res. **53**(3), 1048–1055 (2014)

Mostafa, A.M., Yahia, A.: New approach to assess build-up of cement-based suspensions. Cem. Concr. Res. **85**, 174–182 (2016)

Mostafa, A.M., Yahia, A.: Physico-chemical kinetics of structural build-up of neat cement-based suspensions. Cem. Concr. Res. **97**, 11–27 (2017)

Peng, X., Yi, C., Qiu, X., Deng, Y.: Effect of molecular weight of polycarboxylate-type superplasticizer on the rheological properties of cement pastes. Polym. Polym. Compos. **20**(8), 725–736 (2012)

Ran, Q., Liu, J., Yang, Y., Shu, X., Zhang, J., Mao, Y.: Effect of molecular weight of polycarboxylate superplasticizer on its dispersion, adsorption, and hydration of a cementitious system. J. Mater. Civ. Eng. **28**(5), 04015184 (2015)

Shi, C.: Recent developments of pc supeplasticizers. In: Second International Symposium on Design, Performance and Use of Self-Consolidating Concrete SCC 2009-China, June 5–7 2009, Beijing, China pp. 16–25 (2009)

Winnefeld, F., Becker, S., Pakusch, J., Götz, T.: Effects of the molecular architecture of comb-shaped superplasticizers on their performance in cementitious systems. Cem. Concr. Compos. **29**(4), 251–262 (2007)

Wood, L.: Concrete Admixture Market Analysis by Type (Water Reducing, Waterproofing, Accelerating, Air-Entraining, Retarding), by Application (Residential, Non-Residential & Infrastructure) and Segment Forecasts, 2014–2025 (2017)

Yuan, Q., Xin, L., Khayat, K.H., Feys, D., Shi, C.: Small amplitude oscillatory shear technique to evaluate structural build-up of cement paste. Mater. Struct. **50**(2), 112 (2017)

Compatibility Between Cement and Superplasticiser in Combination with Fines, Gypsum and Fly Ash

Lorna Stone, Rian Pretorius, and Riaan Combrinck(✉)

Unit for Construction Materials, Civil Engineering Department, Stellenbosch University, Private Bag X1, Matieland 7602, South Africa
rcom@sun.ac.za

Abstract. The interaction between superplasticisers and cement in concrete is complex and can result in unpredictable and unwanted concrete behaviour. It is known that the positively charged tricalcium aluminate (C_3A) component of the cement does not only react with the sulphate (gypsum) present in ordinary Portland cement, but also absorbs the superplasticiser. However, the exact interaction is still not fully understood. This study aims to identify compatibility issues between superplasticisers when exposed to gypsum, fine sand dust and fly ash using the Marsh cone test. The results showed that especially gypsum, which is used in the production process of cement, influence the flow time of the concrete and affects the interaction between the C_3A component and the superplasticiser. The more gypsum added, the more superplasticiser is needed to have the same effect. This indicates that the ratio between C_3A, gypsum and superplasticiser can result in compatibility issues. Particle size and shape, as in the case of the fly ash, was found to play a role in the effectiveness of the superplasticiser. When fly ash is used to increase the flowability of a concrete mix, the superplasticiser does not result in the same significant improvement in flowability than without fly ash.

Keywords: Gypsum · Superplasticiser · Compatibility · Marsh cone · Flowability

1 Introduction

Concrete is one of the most used building materials and is composed of sand, stone, cement and water. In modern concrete, chemical admixtures are frequently used to improve and alter the performance of the concrete. Superplasticisers are some of the most commonly used admixtures, incorporated to give high strength concrete mixes with low water:binder ratios a higher flowability or to create highly flowable concrete mixes such as self-compacting concrete (Biggs et al. 2010). The effectiveness of superplasticiser on a concrete mix does not only depend on the type of admixture but also the chemical composition of the cement, particularly the proportions of C_3A, C_3S, and gypsum, as well as the particle size and shape of the materials in the mix (Aitcin et al. 1994).

W. P. Boshoff et al. (Eds.): SAP 2019, RILEM Bookseries 24, pp. 200–208, 2020.
https://doi.org/10.1007/978-3-030-33342-3_22

The superplasticisers work in one of two mechanisms namely steric hindrance and electrostatic dispersion, with the latter being the more common mechanism of dispersion (Jayasree et al. 2011). Although the mechanism of action of superplasticisers are reasonably well established, there is still a lack of knowledge as to why these chemicals occasionally do not interact as intended. This is due to the many variables in the cement/superplasticiser reaction. The amount and type of binder used in a concrete mix plays an important role in the effectivity of the admixture as a significant amount of superplasticiser is incorporated into the gypsum/C_3A reaction, reducing the amount available to increase the flowability of a mix. This reaction however varies in significance for different types of superplasticiser (Biggs et al. 2010).

The particle size of the cement is also an important aspect in the interaction between superplasticisers and cement. The final grinding operation of cement influence the surface charge on the cement particles, while a smaller particle diameter also has a higher packing density, making the paste more viscous. Not only does finer particles increase the water demand of concrete, the finer cement particles also cause a better adsorption of superplasticiser (Flatt and Bowen 2003). Variations in the characteristics of the cement, along with the type and dosage of the superplasticiser used can lead to compatibility problems, compromising the effect of the superplasticiser.

The different mechanisms of various superplasticisers, the chemical composition of the cement as well as the fineness of the materials in the concrete mix play an important role in the compatibility between the components and should all be investigated when determining what causes these problems. These compatibility issues refer to the adverse effect certain combinations of cement and superplasticiser have on the performance of the mix. Some of these commonly include flash setting or delayed setting times, rapid slump loss, improper strength gain, durability etc. (Shrivastava and Kumar 2016). An improved understanding of the interaction between cement and superplasticiser is a key factor in the process of identifying and preventing these incompatibilities, saving both the construction industry as well as the cement and admixture manufacturers significant costs while simultaneously improving the quality of concrete mixes.

2 Experimental Framework

The Marsh cone test, in accordance to the ASTM C939-10 (2010) standard code, was used to determine the time needed for a pre-determined volume of fluid cement mortar to flow through a standardised flow cone. This test is used to understand the flow behaviour of superplasticised cement paste and is applicable to cement pastes with a w/c ratio beyond 0.5 (Jayasree and Gettu 2008).

Five mixes, as shown in Table 1, was used in this study. Comparing the dust filler, gypsum filler and gypsum binder results give insight into whether particle shape and size or chemical composition of the particles influences the slump. For all tests in this study, the same mixing procedure was followed: in a 10 litre cake type mixer, the dry components were mixed for exactly 30 s at a constant speed. Next, 80% of the water was added and mixed for 2 min where after the last 20% of fluid, containing the water and superplasticiser, were added to the mix and mixed for a further 2 min. The washed Marsh cone was then filled with the mortar and the test was conducted 5 min after mixing started.

Table 1. Mix constituents and proportions

	Reference	Fines filler	Gypsum filler	Gypsum binder	Fly ash binder
	kg/m^3	kg/m^3	kg/m^3	kg/m^3	kg/m^3
CEM II 52.5 N	802	802	802	753	740
Sand	963	946	946	962	945
Dust	–	17	–	–	–
Gypsum	–	–	17	48	–
Fly ash	–	–	–	–	47
Water	401	401	401	400	394

The materials used are described in the following sections, including Scanning Electron Microscope (SEM) photos thereof. These SEM photos gives information regarding the shape and size of the different material particles. Smaller and more angular particles require more water than larger particles as it has a bigger surface area to cover (Yally and Sam 2018).

Ordinary Portland cement (OPC) supplied by Pretoria Portland Cement (PPC), a CEM II/A-L 52.5 N cement with between 6 and 20% limestone, was used in all tests. As seen in Figs. 1 and 2, the cement particles are both very fine as well as angular and is therefore expected to have a high water demand. Natural quarry sand, locally known as Malmesbury sand, was used. The sand has a much bigger particle size than that of the cement as seen in Figs. 3 and 4. The particles are also smooth which suggests a lower water content will be needed to achieve the same flowability. The sand mentioned was also milled in a ball-mill for 20 min to create dust for this study.

In Figs. 5 and 6, the milled sand is seen to have a much smaller particle size compared to that of the unmilled sand, however it is still much bigger than the other materials mentioned. The particle shape is also notably more angular due to the milling process. Natural gypsum, used in the production process of cement and sourced at Saint Gobain near Vanrhynsdorp in the Western Cape, was provided by PPC. The particles of the ground gypsum used is shown in Figs. 7 and 8. The particles are similar in size to that of cement and also have an angular shape. DuraPozz fly ash was used in this study. Although the fly ash has a very small particle size, as seen in Figs. 9 and 10, the particles have a very smooth finish which would lower the expected water requirement of the concrete mix.

In this study, both polycarboxylate ether (PCE) and sulphonated naphthalene formaldehyde (SNF) based superplasticisers were used. Commercially, the PCE used in this study is known as CHRYSO Fluid Premia 310 and the SNF as CHRYSO Fluid L. For all tests the addition of the superplasticiser was used to replace the water to keep the total amount of fluid in the mix constant.

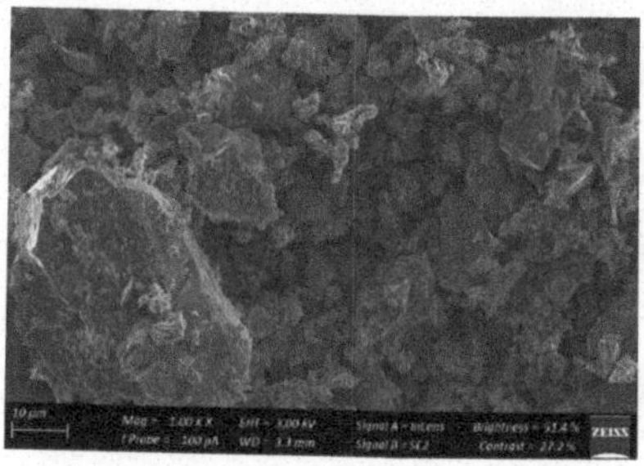

Fig. 1. CEM II (x 1 K).

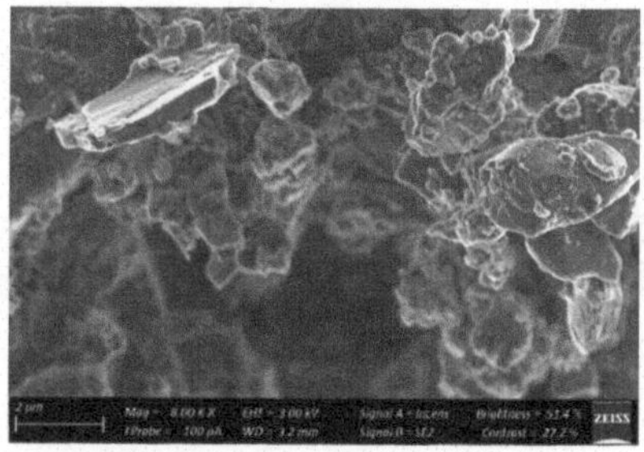

Fig. 2. CEM II (x 8 K).

Fig. 3. Sand (x 200).

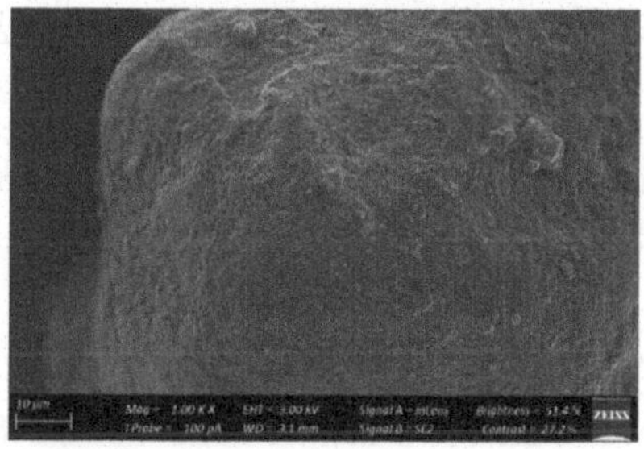

Fig. 4. Sand (x 1 K).

Fig. 5. Dust (x 200).

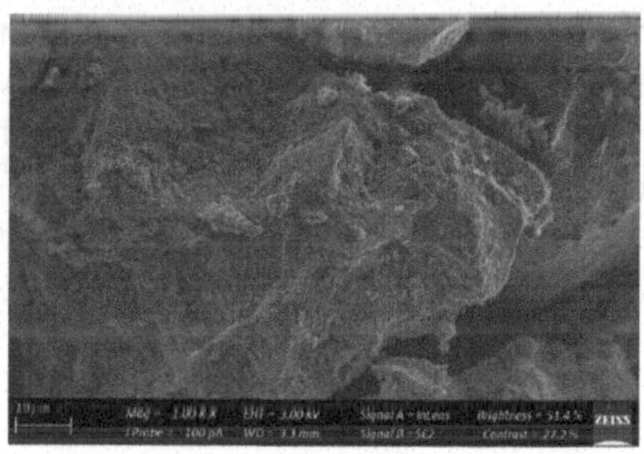

Fig. 6. Dust (x 1 K).

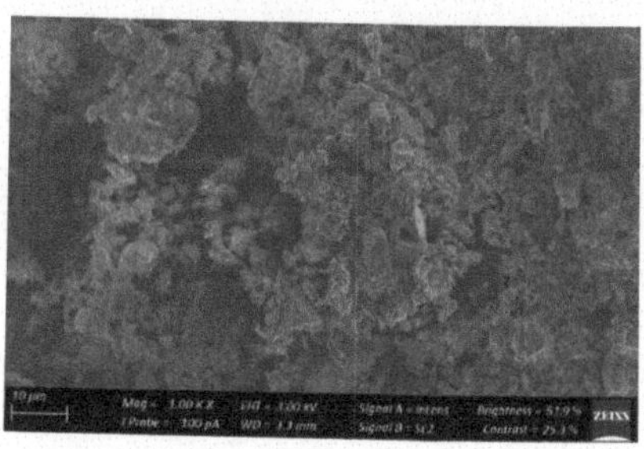

Fig. 7. Gypsum (x 1 K).

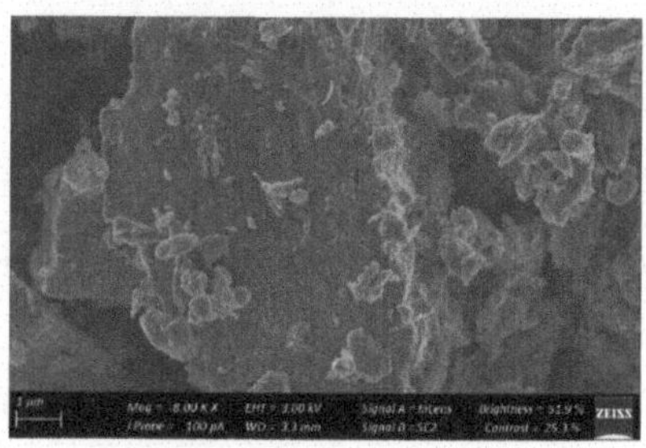

Fig. 8. Gypsum (x 8 K).

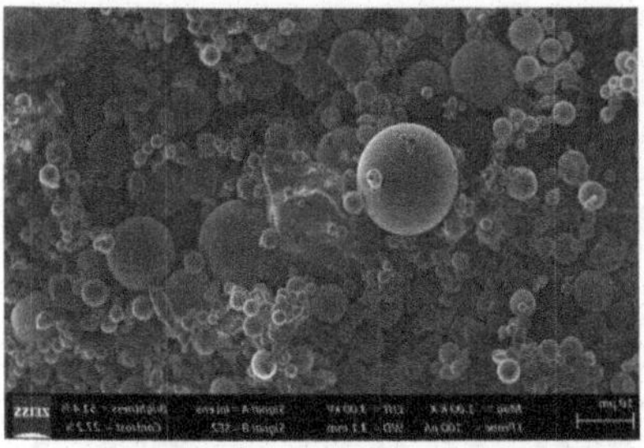

Fig. 9. Fly ash (x 1 K).

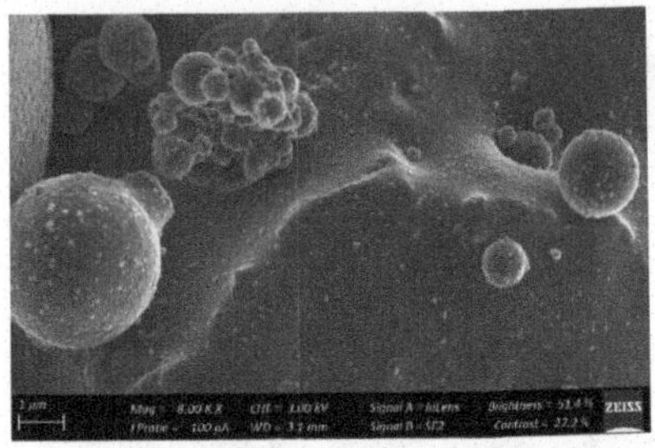

Fig. 10. Fly ash (x 8 K).

3 Experimental Results and Discussion

The Marsh cone results of the reference mix as influenced by superplasticisers individually and in combination with gypsum, fly ash and filler are discussed in the following sections.

3.1 Influence of Superplasticiser

From Fig. 11, it can be seen that the PCE showed a more pronounced effect at the lowest dosage used (0.2%) compared to the SNF, but that the SNF continued to show an improvement up to higher dosages of 0.5% after which it showed no further improvement, while the PCE only resulted in small improvements after the 0.2% dosage up to a dosage of 0.7% where it stabilised. The results also show that although the PCE has the greatest effect initially at lower dosages, the SNF ultimately has the greatest effect at higher dosages. In addition, the results show that overdosing does not result in any improvement and that the most significant improvements are made at the lower dosages. It is believed that the stabilisation of the flow times at higher dosages indicate that the C_3A component of the cement gets saturated by the superplasticisers.

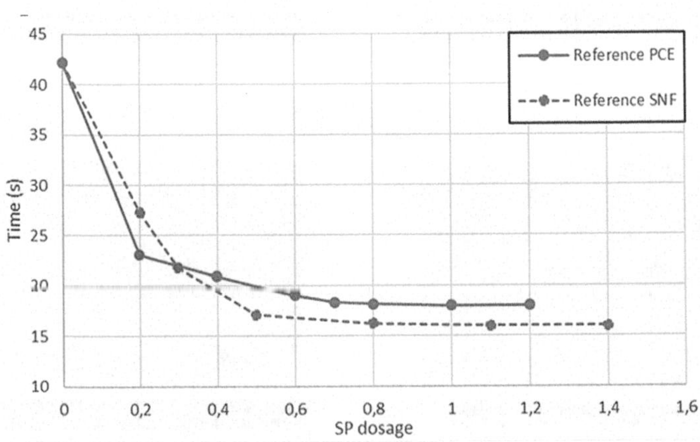

Fig. 11. Marsh cone test results for the Reference mix using PCE and SNF.

When only considering the C_3A component's reaction with the superplasticiser, the superplasticiser would not be able to be adsorbed efficiently (Bassioni 2010), thus making the superplasticiser act as normal water in the mix.

3.2 Influence of Gypsum and Superplasticiser

Figure 12 shows that the replacement of binder by 6% gypsum significantly increased the flow time of the reference mix from 42 to 67 s. For the reference mix, both superplasticisers showed the most significant effect at the lowest dosage of 0.2%, while at the same dosage for the gypsum mix the addition of both superplasticisers showed nearly no improvement. A much higher dosage of around 0.8% was needed for both superplasticisers to get a similar flow time when gypsum was added to the mix. For the PCE mix with gypsum, the flowability decreased after the 0.8% dosage while this was not the case for the SNF.

It is believed that due to the surplus gypsum, the C_3A gets more enclosed by the ettringite making it more difficult for the superplasticiser to react with the C_3A. The positive charge of the C_3A is believed to be reduced due to reaction with the negatively charged gypsum causing a lower attraction force to the superplasticiser (Bassioni 2010). A higher dosage superplasticiser is thus needed to react with the C_3A. The absorption of the superplasticiser still occurs as normal for both the PCE and the SNF but for the PCE a negative effect occurred at higher dosages, as seen in Fig. 12, causing a loss in flowability. From this it is noted that when more gypsum is added, a higher dosage superplasticiser is needed, indicating that there is a definite interaction between the gypsum and the superplasticiser. Furthermore, the electrostatic repulsion mechanism of the SNF behaves more effectively than the steric hinderance of the PCE superplasticiser in the presence of a surplus gypsum at high dosages of superplasticiser, as seen from the results.

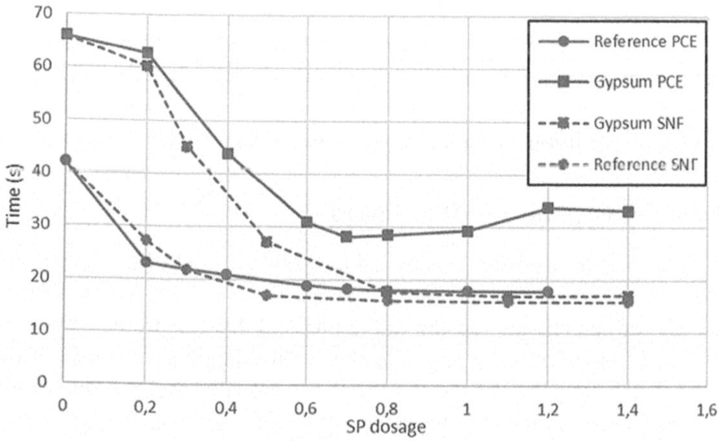

Fig. 12. Marsh cone test results for the Gypsum Binder mix.

3.3 Influence of Fly Ash and Superplasticiser

Figure 13 shows that the addition of fly ash decreased the flow time of the reference mix from 42 to 25 s. The increase in flowability of the fly ash mix is caused by the spherical particle shape of the fly ash which increases particle mobility and reduces the water entrapped by the particles. However, with the addition of fly ash, the superplasticiser did not have the same notable effect on the flowability compared to the reference mix results. For both mixes, both superplasticisers showed the most significant effect at the 0.2% dosage, although the superplasticiser were much less effective in the fly ash mix, as seen in Fig. 13. No significant difference could be noted between the PCE and SNF superplasticiser in the fly ash mix.

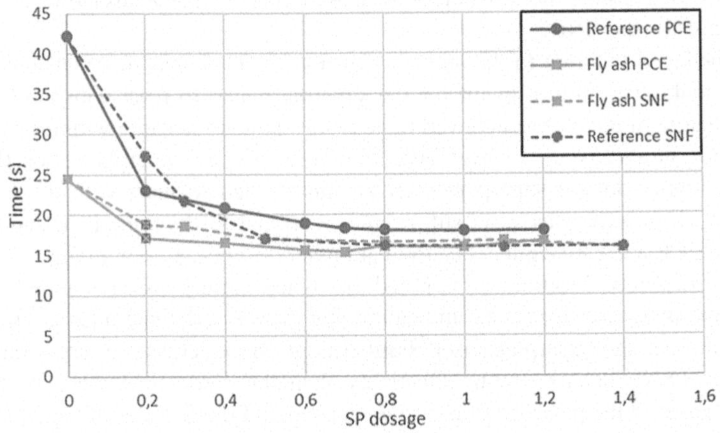

Fig. 13. Marsh cone test results for the Fly Ash Binder mix.

The results suggest that the superplasticiser is still absorbed by the C_3A and that this effects the flowability of the mix but to a lesser degree when fly ash is present. For fly ash mixes the particle shape and size seem to govern or dominate the increase in flowability and the addition of the superplasticiser does not have a significant effect.

3.4 Influence of Gypsum and Dust Fillers

As seen in Fig. 14, after gypsum was added as a replacement of a portion of sand, the flow time of the reference mix increased from 42 to 64 s. This is due to the reaction between the higher percentage gypsum in the mix and the C_3A in the cement (Pourchet et al. 2009). With the addition of superplasticiser an immediate increase in flowability was observed at the lowest dosages used (0.2%), showing that the quantity of added gypsum was not enough to have the same significant influence as in Sect. 3.2. Figure 15 show that the addition of dust or fines only slightly increased the flow time of the reference mix from 42 to 47 s. This is due to the higher water demand of the finer, more angular sand particles (Yally and Sam 2018). Furthermore, slightly higher dosages of both superplasticisers were needed to result in similar flowability than the reference mix, indicating that the particle shape and size of inert sand particles can also influence the effectiveness of superplasticiser. However, from the gypsum and dust filler results it is evident that the chemical composition of a particle has a more dominate role regarding the effectiveness of the superplasticiser than the particle shape. This does however need further investigation and quantification.

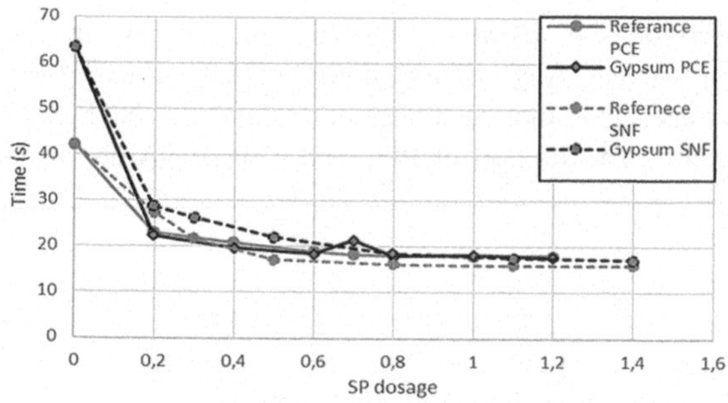

Fig. 14. Marsh cone test results for the Gypsum Filler mix.

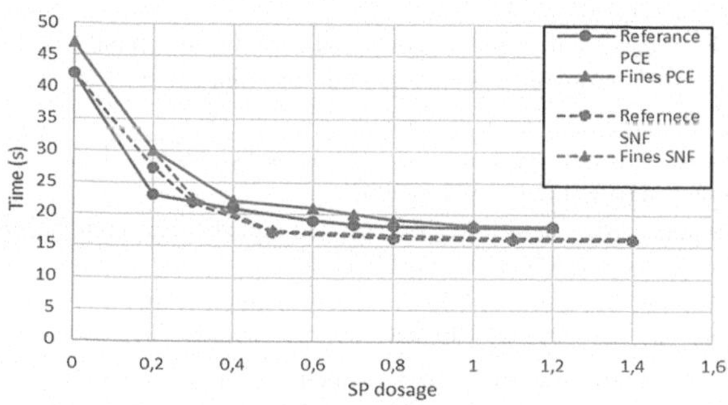

Fig. 15. Marsh cone test results for the Dust Filler mix.

4 Conclusion

This study investigated compatibility issues between cement and superplasticiser in combination with fines, gypsum and fly ash. The following conclusions can be drawn:

High dosages of gypsum significantly influence the effectivity of the superplasticiser and a larger dosage superplasticiser is needed to gain similar results. By adding high dosages of superplasticiser, the risk of overdosing arises with all the negative effects associated with it. The fly ash however, causes the superplasticiser to be less effective. The fly ash significantly increases the flowability of the concrete thus leaving little room for the superplasticiser to improve the flowability. Particle size and shape showed to be important in the flowability and effectiveness of the superplasticiser.

By using a finer sand, the flowability decreases but gets restored by adding higher dosages of superplasticiser. Finally, the amount of gypsum in a mix increases the sensitivity of the mix to superplasticiser compatibilities and a careful investigation must be done before introducing a new cement or superplasticiser to a mix.

5 Acknowledgements

This work is based on the research supported in part by the National Research Foundation of South Africa. Any opinion, finding and conclusions or recommendation expressed in this material is that of the author(s) and the NRF does not accept any liability on this regard. The support of PPC and Chryso SA are also acknowledged.

References

Aitcin, P.-C., Jolicoeur, C., MacGregor, J.G.: Superplasticizers: how they work and why they occasionally don't. Concr. Int. **16**(5), 45–52 (1994)

ASTM:C939-10: Standard Test Method for Flow of Grout for Preplaced-Aggregate Concrete (Flow Cone Method). ASTM Int. **04**(c), 9–11 (2010)

Biggs, B., McCoil, I., Moon, B.: Construction materials. In: Domone, P., Illston, J. (eds.), 4th edn. Spoon Press, Hong Kong (2010)

Flatt, R.J., Bowen, P.: Electrostatic repulsion between particles in cement suspensions: domain of validity of linearized Poisson-Boltzmann equation for nonideal electrolytes. Cem. Concr. Res. **33**(6), 781–791 (2003)

Jayasree, C., Gettu, R.: Experimental study of the flow behaviour of superplasticized cement paste. Mater. Struct. **41**(9), 1581–1593 (2008)

Jayasree, C., Santhanam, M., Gettu, R.: Cement-superplasticiser compatibility - Issues and challenges. Indian Concr. J. **85**(7), 48–60 (2011)

Shrivastava, A.K., Kumar, M.: Compatibility issues of cement with water reducing admixture in concrete. Perspect. Sci. **8**, 290–292 (2016)

Yally, P.P., Sam, A.: Effect of sand fines and water/cement ratio on concrete properties. Civ. Eng. Res. J. **4**(3) (2018)

Bassioni, G.: The influence of cement composition on superplasticizers' efficiency. Int. J. Eng. (IJE) **3**(6), 577–587 (2010)

Pourchet, S., Regnaud, L., Nonat, A., Perez, J.: Early C3A hydration in the presence of different kinds of calcium sulfate. Cement Concr. Res. **39**, 989–996 (2009)

Applications and Innovations

Application of Nano-silica Particles to Improve the Mechanical Properties of High Performance Concrete Containing Superabsorbent Polymers

Thyala A. Cunha[1], Lívia B. Agostinho[1], and Eugênia F. Silva[2(✉)]

[1] Post-Graduate Program in Civil Construction – PECC/UnB, Brasília, Brazil
thyalacunha@gmail.com, liviaborbagostinho@gmail.com
[2] Civil Engineering Department, Universidade de Brasília, Distrito Federal, Brazil
eugenia@unb.br

Abstract. Autogenous shrinkage is a phenomenon that affects the High Performance Concrete (HPC), mainly due to its refined microstructure, the high cement content, the low w/c ratio and the presence of mineral additions. This is closely associated to the hydration process, not depending on any external interference. The internal curing with Superabsorbent Polymer (SAP) is reported as the most effective mitigating strategy, whereas that it provides the necessary water for hydration, preventing the appearance of tensile stresses that may generate autogenous shrinkage and consequently the cracking of structural elements. The use of SAP increases the porosity of the concrete, mainly because the unconnected voids leaved inside the material and the additional water added for SAP absorption, which would reduce mechanical strength. In this work was used the Nano-silica (NS) particles to compensate this effect. Nine micro concretes with different amounts of SAP and NS were produced. For this work, the mechanical strength was evaluated and the autogenous shrinkage was determined from Time Zero (T_0) until 28 days. The results indicate that SAP was efficient for the mitigation of autogenous shrinkage (reduction of 84% for the 0.3% content of SAP), while the addition of NS increases the mechanical properties (there was an improvement in the compressive strength of about 10% for the 2% content of NS). The concrete containing SAP and NS were very promising, since the beneficial action of SAP in mitigation of autogenous shrinkage was not impaired by the presence of NS, while the NS maintained the mechanical strength values of the mixtures containing both SAP and NS additions approximated of the values of the reference.

Keywords: High performance concrete · Internal curing · Autogenous shrinkage · Superabsorbent polymer · SAP · Nano-silica

W. P. Boshoff et al. (Eds.): SAP 2019, RILEM Bookseries 24, pp. 211–221, 2020.
https://doi.org/10.1007/978-3-030-33342-3_23

1 Introduction

When the compressive strength of concrete reached some greater values than usual, it was legitimate to call this as a high strength concrete (HSC). Producing concrete with appropriate workability using low water/cement (w/c) ratio, highly reactive mineral additions and high cement content is a complex task. The reason is that with the increase of compressive strength, the w/c ratio is no longer the only controlling the properties, essential parameter to usual concrete due to the porosity of the hydrated paste [1]. In HSC, the refinement of the pores due to the greater presence of hydrates densified the microstructure and raise up water demand, increasing the self-desiccation and resulting in greater levels of autogenous shrinkage and cracking, impairing the quality and performance of the structure [2]. By the mechanism of capillary depression, the autogenous shrinkage happens when the liquid phase is pulled by creating meniscus at the liquid-gas interface, which induces a negative pressure in the walls [3].

1.1 The Superabsorbent Polymer (SAP)

Since 1990s, mitigation strategies of autogenous shrinkage have been studied, highlighting the use of absorbents materials (internal curing agents), acting providing water to the capillaries during the hydration, avoiding the appearance of tensile stresses. Jensen and Hansen [4] proposed one of these. It consists of the addition of superabsorbent polymer (SAP) particles. SAPs absorb huge quantities of water and subsequently releases it, providing necessary water to the surrounding matrix and mitigating or even eliminating the self-desiccation. SAP are a group of synthetic polymeric materials that absorb significant amount of fluid and retain that, and it can present absorption up to 5000 times its own weight [5]. During the mixture, it creates macro inclusions containing free water that was consumed during the hydration as the relative humidity decreases. In this way, SAP contribute with additional hydration. However, the use of SAP increases the concrete porosity, leaves unconnected voids in the hardened material, which reduces mechanical properties [6–9]. The question to consider is to what extent such loss is admissible.

1.2 The Use of Nano Materials

It is possible to attribute the advantages of the mineral additions to the increasing of properties, reducing clinker consumption, the environmental impact and CO_2 emissions from cement furnaces [10]. Nano-silica (NS) is a highly reactive pozzolanic addition that improves the mechanical properties of cementitious materials, reduces water penetration and influences the durability. It provides the filler effect to improve the microstructure of concrete and mortar, as an activator for the pozzolanic reaction and as nucleation sites, contributing to hydration [9]. This behavior leads to an improvement in the microstructure due to the refinement of the pores, which rise up the mechanical properties. The effects can be explained by the increase the viscosity, helping the suspension of particles, improving the resistance to the segregation; fill the empty spaces between particles (filler effect); act as crystallization center, accelerate the hydration; formation of small and uniform group of C-S-H; improve the transition zone, strengthening the link between aggregate and paste [11].

2 Materials and Mixtures Proportions

The experimental program is composed of nine concretes with 0.3 w/c ratio: one concrete of reference (REF); two of it with SAP only, in content of 0.2 and 0.3%, respectively (SAP 02 and SAP03); two mixtures containing only NS in proportion of 1 and 2% (NS1 and NS2); and four mixtures containing both SAP and NS, called hybrid mixtures. Table 1 shows the mixtures. All of it contain fixed contents of cement, silica fume and sand, only variation the SAP and NS amounts. The superplasticizer (SP) content of each one was determined experimentally in order to obtain 190 ± 10 mm on the spread table test recommended by the TC 225-SAP [12]. The cement used in this work was a CP-V ARI in Brazilian classification (equivalent to CEM – I). A non-densified silica fume in powder also was used in all mixtures. Tables 2a and 2b shows the chemical and physical properties of Portland cement and silica fume applied in this work.

Table 1. Mixtures Proportions.

Mixture	Cement Kg/m^3	Silica fume Kg/m^3	Sand Kg/m^3	SAP %	NS (%)	SP (%)	Water Kg/m^3	Water in NS	Water for SAP
REF	700	70	1340	–	–	2.6	210	–	–
SAP02	700	70	1340	0.2	–	3.0	210	–	21.0
SAP03	700	70	1340	0.3	–	3.2	210	–	31.5
NS1	700	70	1340	–	1.0	3.0	210	16.4	–
NS2	700	70	1340	–	2.0	3.5	210	32.7	–
SAP02-NS1	700	70	1340	0.2	1.0	3.2	210	16.4	21.0
SAP02-NS2	700	70	1340	0.2	2.0	3.5	210	32.7	21.0
SAP03-NS1	700	70	1340	0.3	1.0	3.5	210	16.4	31.5
SAP03-NS2	700	70	1340	0.3	2.0	3.9	210	32.7	31.5

Table 2a. Chemical composition of cement and silica.

Material	SO_3	MgO	SiO_2	Fe_2O_3	Al_2O_3	CaO	Free CaO	$CaSO_4$	Na_2O
Cement	3.28	4.36	24.4	3.02	7.09	53.40	2.16	2.16	0.29
Silica fume	–	0.49	93.5	0.16	0.15	00.37	–	–	0.26

Table 2b. Physical properties of cement.

Test method		Unit	Result	NBR 5733	Reference
Density		g/cm^3	3.03	–	NBR NM 23:2001
Specific surface		cm^2/g	5723	≥ 3000	NBR NM 23:2001
Setting time	Initial	h:m	03:00	$\geq 01:00$	NBR NM 65:2003
	Final	h:m	04:10	$\leq 10:00$	NBR NM 65:2003
Compressive strength	1 day	MPa	22.3	≥ 14.0	NBR 7215:1996
	3 days	MPa	30.7	≥ 24.0	
	7 days	MPa	39.5	≥ 34.0	

The Professor Ole Mejlhede Jensen provided the SAP used in this work. It is produced by the inverse suspension polymerization process and it specially developed for use in high alkaline environment, as the concrete [5]. It is dry white powder, with spherical particles. Figure 1 show the size distribution of SAP by laser granulometry and an image of the dry particles by optical microscope. Table 3 presents other characteristics.

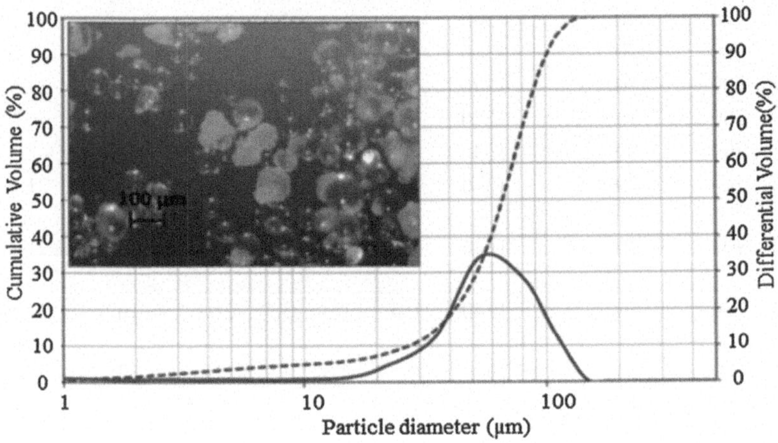

Fig. 1. Particle size distribution of the SAP studied and image of its particles in optical microscope [10].

The NS was an aqueous and translucent solution with stabilized colloidal silica with 30% solid contents. Because of this, was necessary to discount the percentage of water in NS of the total amount of water added in the mixtures. NS amorphous particles have a negative surface charge, are discontinuous, slightly rough, spherical in shape and narrow particle size distribution, specially designed for use in concrete. Table 4 shows the NS used by the Transmission Electronic Microscopy (TEM) method and other properties.

Table 3. Properties of the studied SAP.

Property	Method	Results
Specific mass	Pycnometer	1.456 g/ml
Absorption in aqueous environment	Graduated cylinder Optical microscope	95.80 g/g 80.30 g/g
Absorption in cementitious environment	Slump-flow	15.00 g/g
Spherical particles composition	Scanning Electron Microscope (SEM)	C, O, Na and S
Particle size according the laser granulometry test	D10 D50 D90	27.0 66.0 102.0

Table 4. (a) TEM Micrograph of NS [10]; (b) Properties of NS.

(a)

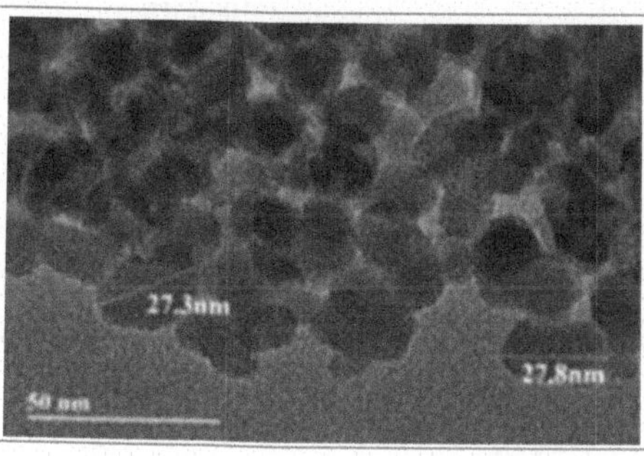

(b) Colloidal Nano-silica properties	
Nano-silica content (%)	30.0
Density (g/cm^3)	1.20
Na_2O content (%)	0.55
Particle size (nm)	3.0 to 40.0
Surface area (m^2/g)	80.0

3 Test Methods

3.1 Mechanical Properties

The compressive strength in cylindrical and cubes was determinate according Table 5. The specimens were made in Materials Testing Lab of University of Brasília (LEM/UnB). During the first 24 h after molding, were submerged in a lime and water solution inside to a chamber with controlled temperature (21 ± 2 °C) and humidity (50 ± 2%). After this time, the specimens were removed of water solution, deformed and returned to the humid chamber until the age of test.

Table 5. Summary mechanical properties tests.

Compressive strength	Age (days)	Dimensions (cm)	Number	Reference
Cylindrical specimens	28	5 × 10	6	NBR 5739:2018 [13]
Cubical specimens	28	4 × 4 × 4	12	NBR 13279:2005 [14]

3.2 Autogenous Shrinkage

The methodology for autogenous shrinkage was based on the propose of Terzawa with modifications by Silva [3]. For this test, three specimens were made for each studied mixture. The determination was performed from time zero (T_0), that was experimentally determined [8]. Readings were taken at intervals as shows in Fig. 2, until 28 days. The polystyrene sheet is to reduce friction and allow the movement. The samples were sealed using plastic film, silver tape and silicone glue, to avoid any water loss in order to analyze only the autogenous shrinkage of the material. After the molding, the set was taken to room with humidity and temperature controlled. The weight of the set was determined in the beginning and in the end to evaluate the loss of water.

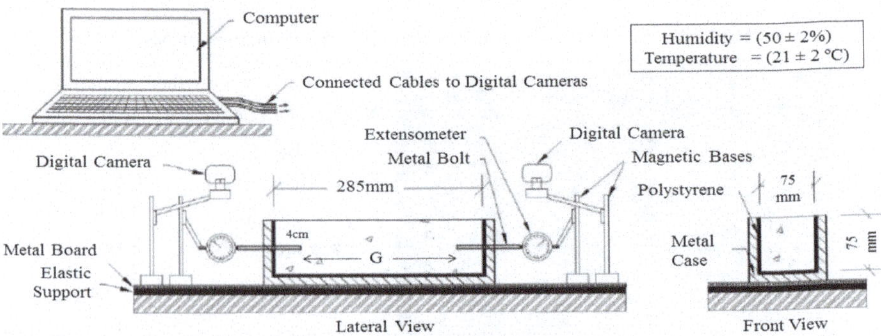

1) Insert the metal bolts in the specimens, leaving it 4cm inside;
2) The distance "G" will be used to calculate the autogenous shrinkage;
3) Attach the metal bolts to the extensometers, fixed to the metal board by magnetic bases to prevent movement;
4) Position the cameras in front of the extensometers to take the readings;
5) Set the cameras to take uninterrupted readings for 28 days in intervals:
   - 1 photo every 30 minutes during the first 3 days
   - 1 photo every 2 hours from 3 to 7 days
   - 1 photo every 4 hours from 7 to 28 days

Fig. 2. A general scheme of the autogenous shrinkage test [3].

4 Results and Discussion

4.1 Mechanical Properties

In Table 6 is verified that the SAP causes a decrease in the compressive strength. The mixture SAP02 showed a reduction of 16 and 7% in the compressive strength in cylinders and cubes. For the SAP03, this percentage reduction was 30% and 14%. Several studies report the loss of mechanical resistance caused by SAP, when compared to REF with the same w/c ratio (without curing water) [2, 15–17]. Due to this loss, some publications question whether it is appropriate to use SAP as an internal curing agent. There is yet no consensus among researchers as to whether additional water introduced (internal curing water) does or does not participate in the hydration [18]. If positive, a comparison of mixtures containing SAP with a reference whose total w/c ratio is similar to the total w/c is required. The loss of mechanical properties may be related due to the increase in the w/c ratio by the curing water, since the w/c total is understood as the w/c ratio plus internal curing water. This comparison is more appropriate and does not penalize the SAP. All deviations are according to establish by NBR 5739:2018 [13] and NBR 13279:2005 [14].

Table 6. Results of the mechanical properties at 28 days

Mixture	Compressive strength					
	Cylinders (MPa)	Deviation (MPa)	Percentage of REF (%)	Cubes (MPa)	Deviation (MPa)	Percentage of REF (%)
REF	100.5	4.3	100%	109.9	4.3	100
SAP02	84.5	3.2	84%	101.7	5.0	93%
SAP03	70.1	3.2	70%	94.4	3.1	86%
NS1	108.3	2.5	108%	117.5	1.6	107%
NS2	112.1	2.3	112%	119.0	3.8	108%
SAP02-NS1	85.3	3.6	85%	104.6	1.2	95%
SAP02-NS2	90.1	2.9	90%	106.3	2.0	97%
SAP03-NS1	84.0	5.0	84%	103.3	2.3	94%
SAP03-NS2	84.7	2.2	84%	106.5	1.3	97%

In concretes with NS, there was an increase in the compressive strength. In NS1, there was an increase of 8% and 7% in the compressive strengths in cylinders and cubes. For NS2, these values were 12% and 8%, respectively. The resistance provided by NS occurs due to the continuous hydration and the formation of large amounts of C-S-H and aluminosilicate. Another reason for this is the fact that NS behaves not only as a filler to improve the microstructure, but also as an activator to promote pozzolanic reactions with CH and leading to a higher precipitation of the hydrated products, making the homogeneous, dense and compact microstructure [19]. The results of the concrete containing both additions show that the NS caused an increase in compressive strength, when compared with the mixture containing only SAP. However, this increase was not sufficient to compensate for all the loss caused by SAP. The addition of NS provided a 2% and 7% increase for SAP02-NS1 and SAP02-NS2, relative to the SAP02, for cylinders, 3% and 4% in cubes. For SAP03, the behavior was similar. The compressive strength in cylinders and cubes obtained for SAP03 was 70.1 and 94.4 MPa (respectively, 70% and 86% of the REF). On the other hand, the mixtures SAP03-NS1 and SAP03-NS2 obtained values equal to 84.0 and 84.7 MPa in the test in cylinders, and 103.3 and 106.5 MPa for the cubic test specimens, showing an increase of the order of 20% and 11% with respect to the SAP03, for the cylinders and cubes. The best resistance gain result in the mixtures containing SAP occurred for SAP03-NS1. In this case, the compressive strength in cylinders was reduced by 16% in relation to the REF, and in the mixture containing only SAP in the same content, without NS (SAP03), the percentage of reduction of the resistance was twice (30%). Therefore, the addition of 1% NS was able to recover half of the lost. These results are consistent with Pourjavadi *et al.* [20], who verified that small additions of NS could compensate the negative effect of SAP.

4.2 Autogenous Shrinkage

Figure 3 shows the average results of autogenous shrinkage obtained for the 3 specimens. Both mixtures SAP02 and SAP03 were very efficient in reducing autogenous shrinkage; however, SAP03 content obtained a superior performance. While SAP02 reduced the autogenous shrinkage in 56% of the REF value at 3 days and 24% at 28 days, the SAP03 shows a decrease of 44% (3 days), and at 28 days this decrease was 70% of REF. It is interesting to observe that the effect of SAP was reduced over the time. This is important since stress development does not depend only on the absolute amplitude of the autogenous shrinkage, but also to the age of concrete, since the concrete is more prone to crack in the early ages, so the development of autogenous shrinkage in advanced ages is not so harmful. This behavior may be related to desorption kinetics of the curing water of the SAP. Much of the water absorbed by SAP was released at the earliest ages, making its effect on autogenous shrinkage more pronounced [17].

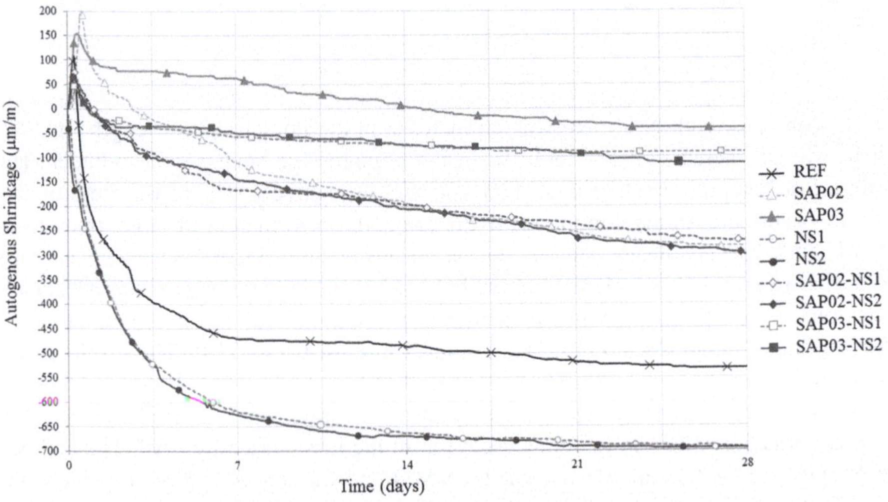

Fig. 3. Results of the autogenous shrinkage.

The NS mixtures did not presents peaks of expansion, indicating absence of initial expansion. This fact may be related to the high pozzolanic activity of NS, which react with the expansive CH, reducing the amount of this and producing more C-S-H. The NS mixtures presents lower porosity and therefore, have less space to accommodate the volumetric variations, being more sensitive to these variations [20]. Is important to note that the autogenous shrinkage observed at 28 days for mixtures with

NS reached values higher than the REF. The increase of the autogenous shrinkage in concretes with NS may be related to the elevation of the hydration temperature, since the NS present a high impact on the speed of the hydration reactions. The temperature development in concretes containing different types of NS rise is inversely proportional to the size of the NS particles [21]. This fact can accelerate the process of self-desiccation and the appearance of capillaries where the tensile stresses responsible for autogenous shrinkage will develop. In the NS mixtures, the result was lower than the REF but very close to each other, not justifying 2% NS instead of 1%. It is possible to note that there was a reduction of autogenous shrinkage in the concrete containing both additions. Compared to REF, SAP02-NS1 had a value of −272 μm/m, and the SAP02-NS2, −312 μm/m. The SAP03-NS1 mixture had a value of −91 μm/m, and the SAP03-NS2 mixture, −89 μm/m. This fact is due to the rather pronounced peak of expansion obtained in concrete containing only SAP. It is important to note that this potentiating effect of NS, when used with SAP, in reducing autogenous shrinkage was not expected, and is especially due to the absence of expansion peaks in NS mixtures, as opposed to non-negligible initial expansion values obtained with SAP. Some researchers, contrary to what happens in this research, adopt effective autogenous shrinkage only as the value measured from the moment when the curve of autogenous shrinkage versus time reaches the negative axis, thus disregarding the peaks of material expansion. In further work, this mechanism of the early age expansion can be evaluated.

5 Conclusions

The SAP has been an excellent strategy to mitigate autogenous shrinkage. The content that presented the best results was SAP03, with 70% reduction observed. The effectiveness of SAP in reducing autogenous shrinkage was not influenced by the addition of NS. The use of 2% of NS did not cause an additional effect in relation to 1%, thus not justifying the use of the higher content. When used without SAP, the NS increases autogenous shrinkage by 10%. There was a significant decrease in the mechanical properties due to the SAP. This behavior was expected and was reported by several researchers. SAP03 resulted in a reduction of 20% in relation to the REF. The ability of NS to increase the mechanical properties of cement based materials is also reported. In this work it was verified that 2% NS promoted increase of 10% in the compressive strength. However, the most interesting property in this work was to evaluate the ability of NS to compensate the losses of strength caused by SAP. In this regard, the NS is very promising, since all concretes containing both additions obtained higher values when compared to the concrete with only SAP. The SAP03-NS1, when compared to SAP03, showed a 20% increase. Although the values obtained did not completely compensate the loss, the results were satisfactory.

References

1. Aitcin, P.C.: High-Performance Concrete, 1st edn. E&FN Spon, London (1998)
2. Lura, P.: Autogenous deformation and internal curing of concrete. Doctoral thesis, Delft University of Technology, Delft (2003)
3. Silva, E.F.: Variações dimensionais em concretos de alto desempenho contendo aditivo redutor de retração. Doctoral thesis. Universidade Federal do Rio de Janeiro, Rio de Janeiro, Brasil (2007)
4. Jensen, O.M., Hansen, P.F.: Water-entrained cement-based materials I. Cem. Concr. Res. **31**, 647–654 (2001)
5. Friedrich, S.V.: Superabsorbent polymers. In: Mechtcherine, V. (ed.) RILEM Technical Committee 225-SAP. Springer, Dordrecht (2012)
6. Mechtcherine, V., et al.: Mitigating early age shrinkage of ultra-high performance concrete by using SAP. In: Tanabe, T., et al. (eds.) Shrinkage and Durability Mechanics of Concrete and Concrete Structures - CONCREEP-8, London (2009)
7. Igarashi, S., Watanabe, A.: Experimental study on autogenous deformation by Internal curing using SAP particles. In: Jensen, O.M., Lura, P., Kovler, K. (eds.) Pro Changes of Hardening Concrete, vol. 52. RILEM, France (2006)
8. Craeye, B., Schutter, G.: Experimental evaluation of autogenous shrinkage by means of a vertical dilatometer. In: Jensen, O.M., Lura, P., Kovler, K. (eds.) Pro Changes of Hardening Concrete, vol. 52. RILEM, France (2006)
9. Cunha, T.A., et al.: Determination of time zero in high strength concrete containing superabsorbent polymer and nano-silica. J. Build. Pathol. Rehabil. **1**, 18 (2016)
10. Gleize, P.J.: Nanotecnologia e materiais de construção. In: Isaia, G.C. (ed.) Materiais de Construção Civil e Princípios de Ciência e Engenharia de Materiais, vol. 2, 2nd edn. Ibracon, São Paulo (2007)
11. Sobolev, K., Sanchez, F.: Nanotechnology in concrete. Constr. Build. Mater. **24**, 2060–2071 (2010)
12. Mechtcherine, V., Reinhardt, H.W.: Application of Superabsorbent Polymer (SAP) in Concrete Construction: State of the Art Report Technical Committee TC-225-SAP. Springer, Dordrecht (2007)
13. Associação Brasileira de Normas Técnicas ABNT, NBR 5739: Concreto - Ensaio de compressão de corpos de prova cilíndricos, São Paulo, Brasil (2018)
14. Associação Brasileira de Normas Técnicas ABNT, NBR 13279: Argamassa para assentamento e revestimento de paredes e tetos - Determinação da resistência à tração na flexão e à compressão, São Paulo, Brasil (2005)
15. Dudziak, L., Mechtcherine, V.: Reducing the cracking potential of ultra-high performace concrete by using super absorbent polymers. In: Proceedings of the international conference on Advanced Concrete Materials, pp. 11–19 (2010)
16. Silva, E.F., et al.: Effects of SAP on the autogenous shrinkage and compressive strength on high strength concrete. In: International RILEM Conference on Application of SAP, Dresden, Germany: [s.n.], pp. 211–219 (2014)
17. Manzano, M.A.R.: Estudo experimental de microconcretos de alta resistência modificados com polímeros superabsorventes (PSA). Doctoral thesis, Universidade de Brasília, Brasília, Brasil (2016)
18. Hashoult, M.T., et al.: Can superabsorent polymers mitigate autogenous shrinkage of internally cured concrete without compromising the strength? Constr. Build. Mater. **31**, 226–230 (2012)

19. Aleen, S.A.E., Heikal, M., Morsi, W.M.: Hydration characteristic, thermal expansion and microstructure of cement containing nano-silica. Constr. Build. Mater. **59**, 151–160 (2014)
20. Pourjavadi, A., et al.: Improving the performance of cement-based composites containing superabsorbent polymers by utilization of nano-SiO2 particles. Mater. Des. **42**, 94–101 (2012)
21. Belkowitz, J.S., et al.: Impact of nano-silica size and surface area on concrete properties. ACI Mater. J. **112**, 419–428 (2015)

Cement Replacement and Improved Hydration in Ultra-High Performance Concrete Using Biochar

Anjaneya Dixit[1(✉)], Souradeep Gupta[2], Sze Dai Pang[1], and Harn Wei Kua[2]

[1] Department of Civil and Environmental Engineering, National University of Singapore, Singapore, Singapore
anjaneya.dixit@u.nus.edu, ceepsd@nus.edu.sg
[2] School of Design and Environment, National University of Singapore, Singapore, Singapore
gupta.souradeep@u.nus.edu, bdgkuahw@nus.edu.sg

Abstract. This study demonstrates the efficacy of biochar in a dual role of internal curing agent as well as a viable candidate for cement replacement in ultra-high performance concrete (UHPC). Bio-char (BC), a product of pyrolysis of mixed wood saw dust, was prepared by pyrolytic conversion at 500 °C. Biochar produced was manually grinded into micron-sized particles. It was then pre-soaked for 24 h to achieve saturation, which was then mixed to replace 2%, 5% and 8% of cement by wt. in UHPC. Isothermal calorimetry (ITC) tests showed that the presence of BC improved the hydration in BC-UHPC mix compared to reference. This is further confirmed from the bound-water measurements using thermo-gravimetric analysis (TGA), where a substantial improvement in BC-UHPC mix compared to reference was observed. Scanning electron microscope (SEM) images revealed that pores of BC serve as nucleation sites for hydration products. The drop in compressive strength in BC-UHPC mix were limited to 10% of the reference mix, and comparable strength was achieved at 5% replacement level. Overall, the results indicate that bio-char from wood waste can be a potential mineral admixture in UHPC, which might be effective to reduce cement demand and create novel avenue for waste valorisation.

Keywords: Ultra-high performance concrete (UHPC) · Biochar · Hydration · Internal curing

1 Introduction

Global trend of exponential growth in infrastructure projects has led to a commensurate intensification of humankind's impact on the environment. Concrete remains as the most widely used construction material due to its cost and availability. Cement, the most integral constituent in concrete production, has a global annual production standing at approximately 4000 Mt/year and expected to grow by 12–23% till 2050 (International Energy Agency 2018). Cement production, unfortunately, is also a bane to the

W. P. Boshoff et al. (Eds.): SAP 2019, RILEM Bookseries 24, pp. 222–229, 2020.
https://doi.org/10.1007/978-3-030-33342-3_24

environment, responsible for emitting about 1 ton of CO_2 for every ton of cement produced. The past few decades have seen an incredible surge in the production of cement, so has the carbon emissions associated with it. The enormous carbon footprint of the construction materials and its grave effects in the coming years has made it imperative to look for sustainable materials that can reduce demand of carbon intensive raw materials such as cement. Reducing cement usage by alternative sustainable materials is an effective way to curtail the carbon footprint given the significant greenhouse gas emissions associated with cement production. Replacement of cement with pozzolanic materials such as of fly ash (FA), rice husk ash (RHA), silica fume (SF) etc. has been extensively done in this field. Research on supplementary cementitious materials such as ground granulated blast furnace slag (GGBFS) have even resulted in development of codes dedicated to their usage in concrete (e.g. EN 15167-1).

A key shortcoming in conventional concrete compared to steel is the strength-to-weight ratio (i.e. the specific strength), and steel possess a substantially higher value than concrete. This leads to typical reinforced concrete (RC) structures having much larger dead loads compared to steel structures (although the latter may be much costlier). In the studies leading to development of stronger concretes, the invention of ultra-high strength concrete (UHSC) in the 1970s followed by reactive powder concrete (RPC) and ultimately the ultra-high performance concrete (UHPC) in 1990s (Richard and Cheyrezy 1995; Yoo and Banthia 2016) was a major milestone. Highly durable concrete with strength up to 250 MPa under compression and porosity below 2% was now achievable. But this came with a heavy price, figuratively as well as literally. One of the key factors for such remarkable strength in UHPC is attributed to high cement content of 800–1000 kg/m^3 per cum. of UHPC (Yoo and Banthia 2016) compared to conventional concrete. Although use of UHPC offers stronger and a more durable infrastructure, the high cement content in UHPC limits its environmental and economic sustainability for wide-scale application. It is, therefore, necessary to investigate alternative materials which can reduce cement demand in UHPC without significantly affecting its strength development.

In this endeavour, biochar (BC) has been used in this study as material candidate for partial cement replacement in UHPC mix. Biochar is the pyrolytic product of solid biomass like sawdust and food waste which is obtained on calcining biomass up to or above 300 °C. Pyrolysis changes the structure of biomass due to the decomposition of hemicellulose, cellulose and lignin at different temperature ranges (Weber and Quicker 2018). Depending on preparation conditions, BC offers high surface area with wide range of particle size, making it suitable as a fine additive for cementitious composites. Moreover, the pores of BC are active in absorbing water (Gupta et al. 2017, 2018b; Weber and Quicker 2018), which make it a favourable for internal curing application. Water-pre-soaked in BC particles prior to mixing can be desorbed during hardening stage (Gupta et al. 2018b), thus improving the microstructure and strength development of the composite. Choi et al. (2012) reported that BC from hardwood can be applied to replace up to 5% by wt. cement in mortar, resulting in higher strength development by 10–12% compared to reference. However, it is to be noted that high dosage of BC (beyond 5 by wt% of cement) led to drastic loss in workability due to excessive water absorption by the particles. Gupta and Kua (2019) in their study on mortar with biochar found an increase in strength of 30% under air-cured conditions. The study presented herein is an attempt to replace cement in UHPC using BC finer than 125 μm. BC in pre-soaked condition is used in UHPC to replace 2%, 5% and 8% cement by wt. The effect of BC on compressive strength, hydration kinetics and degree of hydration (bound water) have been presented in this paper.

2 Materials and Methodology

Biochar was prepared by following the methodology reported in Gupta et al. (2018b). Locally available mixed saw-dust waste was dried and subjected to pyrolysis at a temperature of 500 °C in a muffle furnace. The furnace is fitted with an escape vent for volatile and organic compounds to avoid its re-deposition on the bio-char surface. The heating rate was 10 °C/min, and the pyrolysis time was maintained for 1 h. The prepared BC was allowed to cooled down to room temperature and then stored in an air-tight container. Thereafter, it was crushed and sieved to obtain fraction finer than 125 μm. The ingredients for UHPC mix included ordinary Portland cement (C), silica fume (SF) from Elkem Materials, silica sand (SS) and quartz powder (QP) from SAC Corporation. The mixing proportion and mix nomenclature are presented in Table 1. The original cement content was replaced by pre-soaked biochar while maintaining the same water-to-(C+BC) ratio as that of the reference (R1). The replacement levels were 2%, 5% and 8% by wt. of cement and the superplasticizer-to-cement (SP/C) ratio was kept at 2.50%. It should be noted that the absorption of water by BC leads to a decrease in free water and hence, loss in workability of the mix (Gupta and Kua 2019). If this is compensated by using additional water, it would lead to an increase in the w/c ratio. A potential downside of higher w/c is the decrease in strength and increase in porosity of the hardened mix. Therefore, the total water content (kg per m^3 of UHPC) was kept same for all mixes. Additionally, it is hypothesized that the stored moisture in BC during mixing would later be released for internal curing and improve the hydration even when the total w/c ratio is kept the same as the reference.

Table 1. Mix proportion for UHPC with biochar.

Mix	Cement	Biochar	Quartz powder	Silica fume	Silica sand	Water
R1	1.00	0.00	0.35	0.25	1.1	0.265
F2	0.98	0.02				
F5	0.95	0.05				
F8	0.92	0.08				

Cube samples (50 mm) were prepared to test for compressive strength at 1, 7, and 28 days. The average value of three cubes have been reported in the next section. Curing conditions were maintained at ambient condition: 30 °C at 65% relative humidity. Isothermal calorimetry (ITC) test was performed on pastes samples to investigate the effect on hydration kinetics. Thermo-gravimetric analysis (TGA) were also done to ascertain the bound water in the pastes. For this purpose, approximately

35 mg of finely grounded hydrated paste was heated from room temperature to 950 °C at a rate of 10 °C/min in an inert N_2 environment. The details of the paste proportions are given in Table 2. SEM images of tested specimen were taken to study the suitability of BC to host hydration reactions on its surface and/or inside its pores.

Table 2. Mix proportion of pastes prepared for ITC and TGA tests.

Mix	Cement	Biochar	SP/Cement	Water
R1/R2[a]	1.00	0.00	0.005/0.015[a]	0.265
F2	0.98	0.02	0.005	
F5	0.95	0.05	0.005	
F8	0.92	0.08	0.015	

[a]The SP content had to be increased for paste F8 during ITC tests due to workability issues. A separate reference mix, R2 with the same increase in SP was tested for a judicious comparison with biochar.

3 Results and Discussions

Figure 1 shows the results of the ITC tests done on paste samples. The results indicate role of BC particles as internal curing agent and nucleation sites for cement hydration. The acceleration in hydration for samples with BC is evident in Fig. 1a, where the hydration peaks for biochar pastes show a remarkable leftward shift, especially for F8, where the acceleration was more than 12 h compared to R2. Furthermore, the results from the cumulative heat evolved shown in Fig. 1b suggest that the water absorbed by the BC pores assisted in additional hydration. The UHPC-BC pastes released up to 10% higher heat compared to their respective reference samples. The internal curing ability of BC, behaving as micro-reservoirs ensuring supply of moisture, is also evident by SEM images (Fig. 2), wherein deposition of hydration products in and around BC pores is observed.

The results from TGA are shown in Fig. 3 and Table 3. The differential TG (DTG) curves have been shown till 500 °C to focus on the loss of water due to dehydration of hydration products (105–400 °C) and decomposition of portlandite (400–500 °C), which are the primary markers to judge the degree of hydration. The DTG curves for samples with BC show a discernible shift compared to the reference paste, especially for F5. The TGA data for mass loss shown in Table 3 indicates a noticeable increase in the bound water, reflected in the dehydration occurring for the temperature range of 105–400 °C. This mass loss, attributed primarily to the dehydration of the C-S-H gels, ettringite and AFm phases (Huang et al. 2017). It can be, therefore, inferred that not only did BC improve the hydration in early age as suggested by ITC results, the effect was noticeable even in later ages till 28th day. This can be attributed to the internal curing action of BC through which the absorbed water in the BC pores facilitated in the formation of higher amount of CSH gels and other associated hydration products compared to the reference mix.

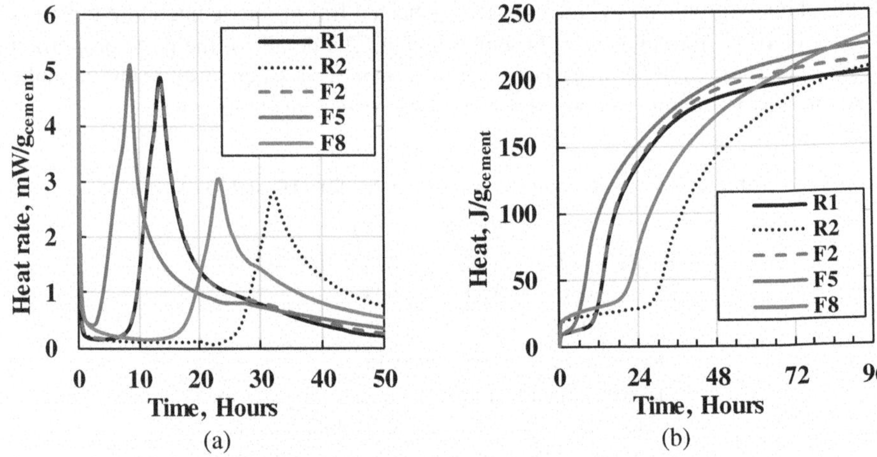

Fig. 1. Heat evolution curves from ITC tests showing (a) accelerated hydration and, (b) higher heat of hydration in samples with BC

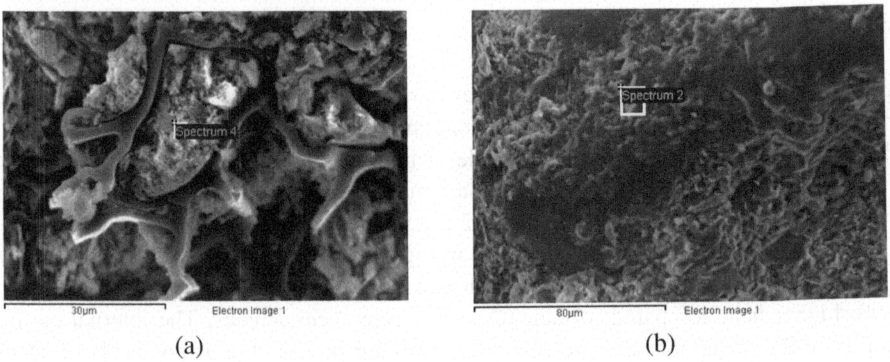

Fig. 2. SEM images showing deposition of hydration products (a) inside bio-char pores, and (b) on bio-char surface

The results from the compressive strength tests on cube samples are shown in Fig. 4, wherein a drop in strength was observed with the use of BC. The remarkable strength of UHPC is due to its dense and virtually defect-free microstructure. Since, BC is porous and relatively weaker than the UHPC matrix, it acts as micro-defect, affecting the strength. However, the improved hydration and generation of additional CSH gels and AFm phases due to the internal curing by BC could have negated the weakness induced by it, as reflected in the strength of mix F5 (144 MPa) compared to reference mix R1(150 MPa).

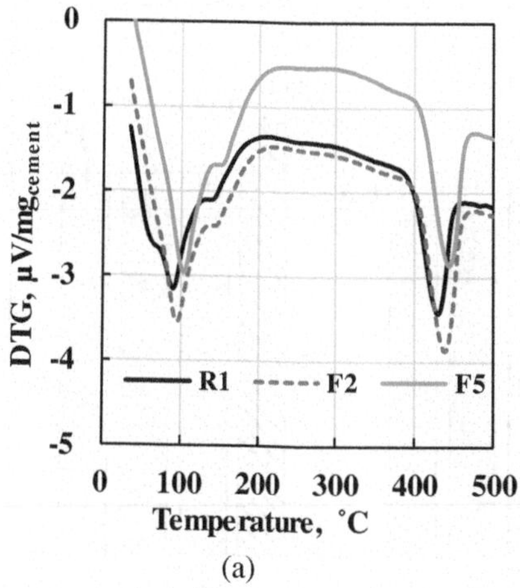

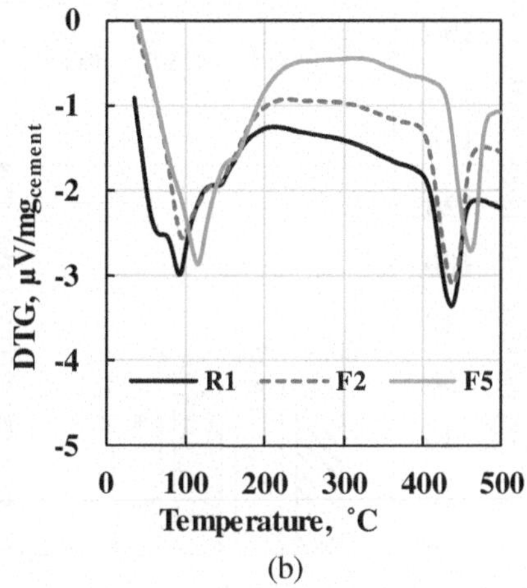

Fig. 3. DTG curves for samples at (a) 7 days and (b) 28-days. Samples with BC showed higher weight loss between 105–500 °C, indicating better degree of hydration

Table 3. Results of mass loss (% of sample wt.) from TGA tests

Temp	105–400 °C	400–500 °C	Total
R1, 7 days	5.88	2.05	7.93
F2, 7 days	7.19	2.94	10.13
F5, 7 days	7.91	3.14	11.05
R1, 28 days	5.52	2.15	7.67
F2, 28 days	7.11	3.43	10.54
F5, 28 days	8.30	3.08	11.38

The production of biochar has the potential of curtailing approx. 870 kg CO_2-equivalent of greenhouse emissions for every ton of dry feedstock pyrolyzed. In cementitious composites, a comparative analysis done by Gupta et al. (2018a) showed that adding 2% by wt. BC in normal mortar reduced the Global Warming Potential (GWP) by 15%. The carbon footprint of UHPC is much higher compared to normal concrete owing to the high cement content. Therefore, replacement of cement with BC would have high ecological benefits. In Singapore's context, land availability for waste disposal is limited, and use of BC as a concrete admixture can be an effective way of waste recycling and valorization. The results from this study indicate the potential of BC as an admixture in UHPC to curtail the cement demand and reduce the carbon footprint of UHPC.

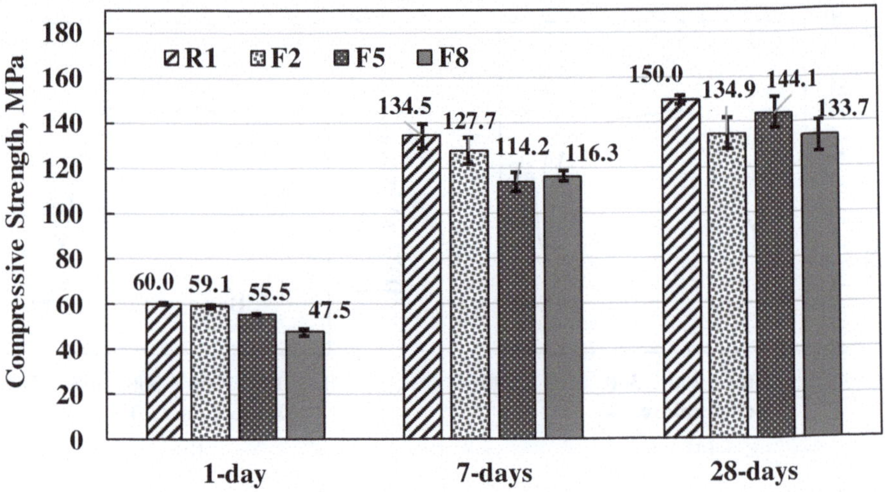

Fig. 4. Compressive strength of mixes at 1, 7 and 28-days, showing reduction in strength with BC addition

4 Conclusion

This study investigated the application of wood waste bio-char as a potential admixture for internal curing as well as reducing cement content in a conventional UHPC mix. The primary concern was the effect of bio-char on the early age hydration kinetics and strength development in UHPC. The results from the compressive strength tests showed that the strength was comparable at 28-days, with mix F5 displaying similar values as compared to the reference mix. The ITC results support the efficacy of BC in accelerating the hydration kinetics, while SEM images establish that BC particles act as sites for deposition of hydration products. The results from TGA indicate that BC improved the degree of hydration by formation of additional hydration products. Overall, this study indicates that biochar can be used to partially replace cement and generate internal curing action, resulting in enhancement of hydration degree and achieve similar compressive strength. The study concludes that bio-char from mixed wood waste can be a novel admixture for UHPC mix, which would enhance recycling of wood waste and reduce demand for landfilling area

Acknowledgment. This research was supported in the National University of Singapore by Singapore Ministry of Education Academic Research Fund Tier 1 Grant (No. R-302-000-183-114).

References

Choi, W.C., Yun, H.Do, Lee, J.Y.: Mechanical properties of mortar containing bio-char from pyrolysis. J. Korea Inst. Struct. Maint. Insp. **16**, 67–74 (2012)

Gupta, S., Kua, H.W., Tan Cynthia, S.Y.: Use of biochar-coated polypropylene fibers for carbon sequestration and physical improvement of mortar. Cem. Concr. Compos. **83**, 171–187 (2017). https://doi.org/10.1016/J.CEMCONCOMP.2017.07.012

Gupta, S., Kua, H.W., Low, C.Y.: Use of biochar as carbon sequestering additive in cement mortar. Cem. Concr. Compos. **87**, 110–129 (2018a). https://doi.org/10.1016/j.cemconcomp.2017.12.009

Gupta, S., Kua, H.W., Pang, S.D.: Biochar-mortar composite: Manufacturing, evaluation of physical properties and economic viability. Constr. Build. Mater. **167**, 874–889 (2018b). https://doi.org/10.1016/j.conbuildmat.2018.02.104

Gupta, S., Kua, H.W.: Carbonaceous micro-filler for cement: effect of particle size and dosage of biochar on fresh and hardened properties of cement mortar. Total. Environ., Sci (2019). https://doi.org/10.1016/J.SCITOTENV.2019.01.269

Huang, W., Kazemi-Kamyab, H., Sun, W., Scrivener, K.: Effect of replacement of silica fume with calcined clay on the hydration and microstructural development of eco-UHPFRC. Mater. Des. **121**, 36–46 (2017). https://doi.org/10.1016/j.matdes.2017.02.052

International Energy Agency: Technology Roadmap: Low-Carbon Transition in the Cement Industry (2018)

Richard, P., Cheyrezy, M.: Composition of reactive powder concretes. Cem. Concr. Res. **25**, 1501–1511 (1995). https://doi.org/10.1016/0008-8846(95)00144-2

Weber, K., Quicker, P.: Properties of biochar. Fuel **217**, 240–261 (2018). https://doi.org/10.1016/j.fuel.2017.12.054

Yoo, D., Banthia, N.: Mechanical properties of ultra-high-performance fiber-reinforced concrete: A review. Cem. Concr. Compos. **73**, 267–280 (2016). https://doi.org/10.1016/j.cemconcomp.2016.08.001

Solidification of Two-Component Grouts by the Use of Superabsorbent Polymers as Activator

Christoph Schulte-Schrepping(✉), David Ov, and Rolf Breitenbücher

Institute for Building Materials, Department of Civil and Environmental Engineering, Ruhr-University Bochum, Universitätsstraße 150, Bochum 44801, Germany
christoph.schulte-schrepping@rub.de

Abstract. In mechanized shield tunneling, the annular gap between the tunnel structure and the surrounding soil needs to be filled with an adequate grouting mortar to ensure a rapid and safe bedding of the segment rings and to minimize settlements on the surface above the tunnel lining. After mounting of the segment rings and filling of the annular gap, a rapid solidification of the used grout must prevent possible displacements or a floating of the tunnel. In the case of nearly impermeable soils, two-component grouts are necessary, which develop an adequate strength and stiffness in a short period of time by the use of powerful activators like water glass (component B). In addition to the commonly activated cementitious materials, it is feasible to ensure an immediate and sufficient bedding by physical effects. Therefore, the use of superabsorbent polymers (SAP) as component B has been investigated. Experimental studies have been carried out in a systematic way in order to determine the type of the SAP, which leads to a sufficient absorption rate in the alkaline pore water of a cementitious grout. After identification of a suitable "alkali-stable" polymer, tests were carried out in order to examine the necessary amount of SAPs in a slightly modified one-component grout (component A) to cause a sufficient solidification of the whole system within a short period of time. Next to this, the short and long term strength development like shear strength or compressive strength of the combined system (component A and B) were determined. Considering the state of the art of the structural design of the grouting technology on a tunnel boring machine and the generally used liquid activators, a permanent pre-suspension of the SAP was tested and also the strength development of the activated system was examined.

Keywords: Two-component grouts · Mechanized tunneling · Superabsorbent polymers

1 Introduction

A gap between the surrounding soil and the tunnel lining arises during mechanized tunnel driving. This annular gap needs to be filled with an adequate grout in order to ensure a safe-bedding of the tunnel structure as well as to minimize settlements above

W. P. Boshoff et al. (Eds.): SAP 2019, RILEM Bookseries 24, pp. 230–238, 2020.
https://doi.org/10.1007/978-3-030-33342-3_25

the tunnel. This becomes a very important topic when buildings or other underground structures are located above the tunnel construction. In mechanized tunneling, the grout has to fulfill contradictory requirements. Initially, a good workability (i.e. low viscosity) of the mortar is required in order to pump the material through pipes. Once the gap is filled completely, an immediate increase in shear strength and stiffness is necessary to prevent floating of the tunnel lining. While a high initial water content in the mixture ensures the workability, large amounts of the mixing water become redundant when the mortar reaches its final position, namely the annular gap. Dewatering of one-component grouts in this stage is usually done by a pressure-diffusion-driven consolidation process. However, the efficiency as well as the time span for the consolidation process depends strongly on the permeability of the surrounding soil. For drilling in nearly impermeable soils, the strength increase by consolidation is usually not sufficient. As a consequence, the tunnel lining is no longer appropriately supported by the grout, which results in high bending moments and possible damage to the tunnel structure. To overcome this problem, a two-component grout can be used instead, wherein the strength development is forced by the addition of a powerful activator like water glass. In this paper, the innovative concept of using superabsorbent polymers (SAP) as an physical activator in order to accelerate the solidification process is presented. From a chemical point of view, SAP are cross-linked polyelectrolytes, which contain a high amount of fixed charges attached to the polymer chains. When these polymers come in contact with an aqueous solution, they are able to absorb and bind large amounts of water, while the cross-links between the polymer chains prevent the gel from dissolving in the fluid (Fig. 1).

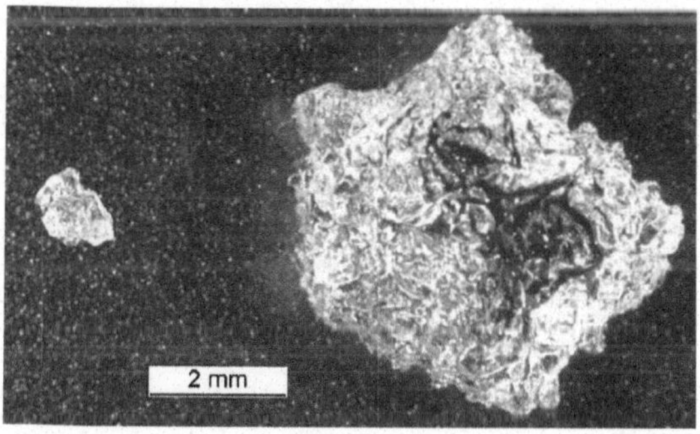

Fig. 1. SAP particle in dry conditions (left) and after water absorption (right).

By using an adequate type as well as the right amount of SAP, an immediate solidification of the grout in the annular gap can be achieved. Therefore, the water uptake and the absorption capacity has to be investigated against the background of the pore fluid chemistry. In addition, the strength development of such physical activated grouts within the first minutes and hours but also in long term has to be considered.

The here shown results are part of the Collaborative Research Center 837 "Interaction Modeling in Mechanized Tunneling" funded by the German Research Foundation (DFG), in which the Institute for Building Materials at the Ruhr University Bochum is developing innovative multi-component grouts.

2 Annular Gap Grouting

The complete and homogeneous filling of the annular gap ensures the position stability of the tunnel lining, prevents it from floating and reduces surface settlements. The development of a suitable grout currently is carried out on an empirical basis and the requirements of the fresh and solid grout properties are basically project-specific and vary in large ranges. This affects in particular the required strength and stiffness development. At the material-technological level, in the case of two-component grouts, these properties are primarily controlled by the content and composition of the binder and by the activator content. Herein the knowledge about the strength development, in particular in the young age of the grout is significant, to ensure a permanent position stability of the tunnel. Usually, the back filling of the annular gap (Fig. 2, left) takes place simultaneous to the advance of the tunnel boring machine through pilaster strips in the shield tail (Maidl et al. 2011). In the case of two-component grouts the component A and B are mixed together shortly before they enter the annular gap (TAC Corporation 2019).

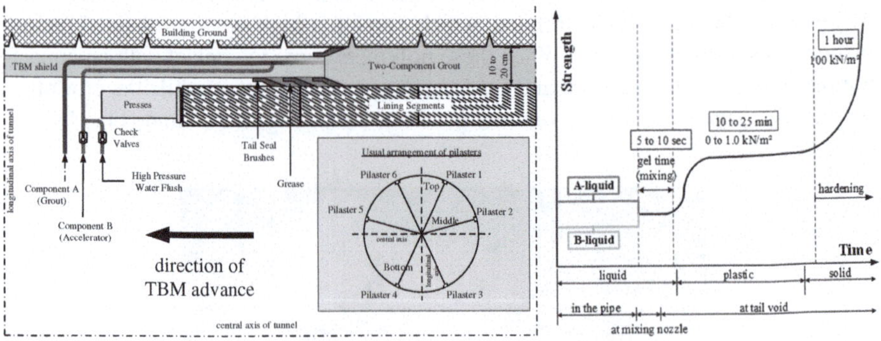

Fig. 2. Left: Schematic illustration of the annular gap grouting and typical pilaster strip arrangement; Right: Strength development of two-component grouts (Hashimoto et al. 2006).

In the case of typical activated two-component grouts, after the two components are mixed, the material gels within a few seconds and develops a mechanical strength within the first hours (Hashimoto et al. 2006) (Fig. 2, right). The gelling of the two-component grout is necessary in water-bearing soils to prevent excessive washing out of the grout (EFNARC 2005). The required strength development is project-specific and depends on the structural boundary conditions such as the advance speed and the ring construction time. Possible control variables to adapt the grout to varying boundary conditions as well as fundamental systematic factors influencing the strength development but also their interactions with the workability of component are not available by now.

3 Experimental Studies

In a first step, special attention has been paid to the absorption characteristics of SAP within the first minutes, since this is the time period in which the gelation of the activated grout should take place. To simulate realistic conditions, next to tap water, bleeding water was used, which was filtrated from component A of a typical two-component grout by pressure-induced drainage. This grout was produced with typical additives (retarding agent) as well as source materials comparable to well-established mixtures in the construction practice (Schulte-Schrepping et al. 2018). The SAP-types used are shown in Table 1.

Table 1. Tested superabsorbent polymers.

SAP-type		Particle size distribution
SAP 1	Cross-linked polyacrylic acid homopolymers, potassium neutralized	0.2–1.0 mm
SAP 2	Cross-linked sodium polyarcylate	–
SAP 3	Cross-linked sodium polyacrylate	0.1–0.5 mm

As a part of the experimental investigations in a first step, the water absorption capacity (Fig. 3) according to DIN 18132 of three different SAP with different particle distributions and chemical properties (Table 1) was determined.

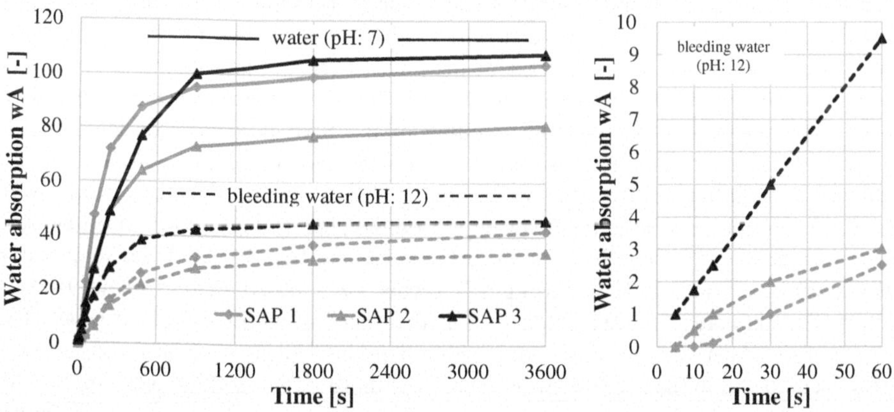

Fig. 3. Water absorption of different SAP in different solutions (pH: 7-water; and pH: 12-pore water/bleeding water of a cementitious grout).

When considering the absorption in water, it becomes clear that SAP 2 has a considerably lower maximum water absorption after one hour (w_A: 80 [-]). This SAP is obviously much coarser than SAP 1 and 3, which explains this behavior. SAP 1 and 3 behave similarly from 900 s and primarily differ in the initial water absorption speed. It can be seen when using bleeding water (pH: 12), that the maximum water absorption capacity and the water absorption speed of all SAP-types are significantly reduced in comparison to tap water. Possible reasons for this are the salts and alkalis present in the solution, a therefore a reduced osmotic pressure and a possible shielding effect of the anions by cations penetrating into the SAP. Looking at the time window up to 60 s (Fig. 3, right), it can be seen that SAP 3 is able to absorb the pore water significantly more strongly. According to the manufacturer, this SAP is declared "alkali-stable" and has the highest fineness of the tested SAP, which combined leads to this comparatively rapid fluid absorption in the bleeding water.

In addition, to describe the essential material characteristics for the use as an activator for annular gap grouts, the change in viscosity caused by different concentrations of SAP after addition to different solutions (tap water, pH: 7 and bleeding water, pH: 12) was investigated. The development of the torque before and after SAP addition within these tests was measured, in order to determine the potential of a gelation as required by a suitable annular gap grout.

By the viscosity measurements it can be seen from Fig. 4, left, that an increase of the amount of SAP leads to higher viscosity and that the absorption of the pore fluid (bleeding water, pH: 12) leads in general to a delayed and decreased water absorption, also when the SAP concentration is higher. SAP 3 is declared as an alkali-stable polymer and its water absorption starts relatively quickly after 60 s even under such unfavorable boundary conditions and reaches within approx. 600 s the maximum value. These rapid water absorption and therefore solidification is an essential characteristic of an appropriate physical activator. As seen in Fig. 4, right, the SAP 3 (6 kg/m^3) is able to cause a rapid solidification of the grout, in this case of a typical semi-active one-component grout (cement content: 60 kg/m^3). For one-component grouts, aggregates up to 8 mm are generally used. This leads after the absorption of the free water by the SAP to an increase of the internal friction and therefore to a shear strength development. In this case, the packing density of the solids is not increased after water absorption, because the swelling of the SAP is a more or less volume constant process. For determining the shear strength and compressive strength development, three different one-component grouts with different cement contents were used, because the cement is the major impact on the chemical composition of the pore fluid to be absorbed by the SAP. Herein an inert/cement free (I-0) and two semi-active (B-60: 60 kg/m^3 and B-120: 120 kg/m^3) grouts were tested (Thewes and Budach 2009). The untreated SAP were added in a "dry state" to the ready mixed grout. The results of the shear strength tests, done with a hand-held shear vane according to DIN 4094-4 compared to the average shear strength of those grouts without addition of SAP after dewatering for 30 min at 2.5 bar from (Youn 2016) are shown in Fig. 5 on the left. The compressive strength development at different ages after SAP addition are shown in Fig. 5 on the right.

As seen by the results of Fig. 5, left, the shear strength within 10 to 60 min after activation depends largely on the cement content of the grout. With a higher cement content, the shear strength is decreased, because of the unfavorable chemical composition of the pore water and therefore the reduced water absorption capacity of the SAP

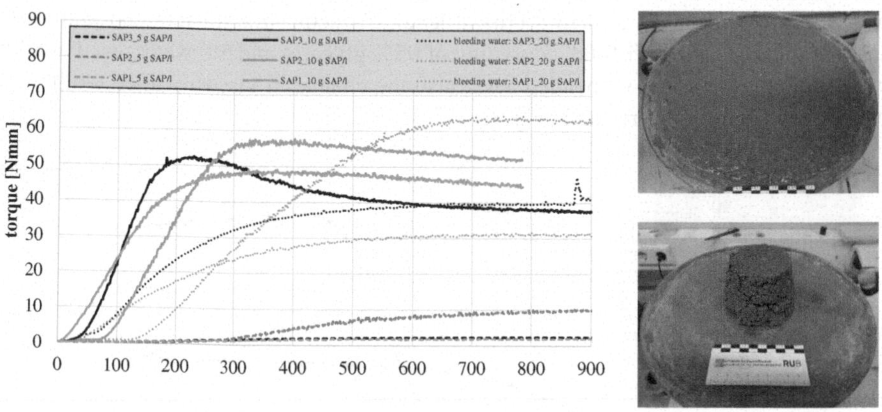

Fig. 4. Left: Change in viscosity due to SAP addition in different solutions (pH: 7 and 12). Right: Solidification of a typical one-component grout within 2 min after SAP addition.

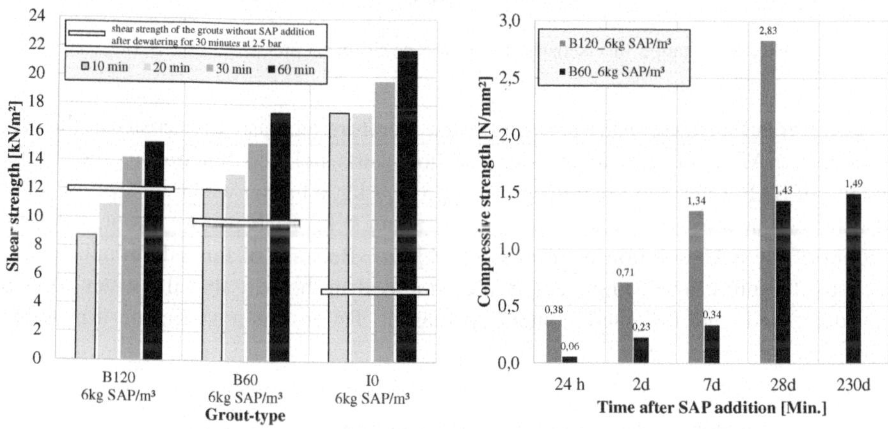

Fig. 5. Shear strength and compression strength development.

(see Fig. 3). This leads to a larger proportion of free water in the system, which leads to a lower internal friction. Compared to the shear strength of the non-activated grouts, the use of SAP leads in general to a higher shear strength after 30 min under typical boundary conditions. As a sufficient shear strength after 30 min, a minimum value of about 2.0 kN/m^2 can be assumed (Thewes and Budach 2009). All combinations fulfill this requirement. By the compressive strength tests, it can be seen, that a higher cement content leads to a higher compressive strength. When using the cement-free grout (I-0) no compressive strength could be determined.

Next to the dry addition of the SAP and under consideration of the current injection technology, a pre-suspension of the SAP in a suitable carrier fluid was investigated. Therefore different solutions with sodium but also nonpolar fluids like glycerin were tested. Only with glycerin as carrier fluid, a stable suspension could be achieved in which the SAP do not swell to a critical level. To achieve a suitable and pumpable suspension, 16.7 g carrier fluid, consisting of 95% glycerin and 5% water, per 1 g SAP were used. With the pre-suspended SAP, also the change in viscosity after addition to bleeding water was determined (Fig. 6).

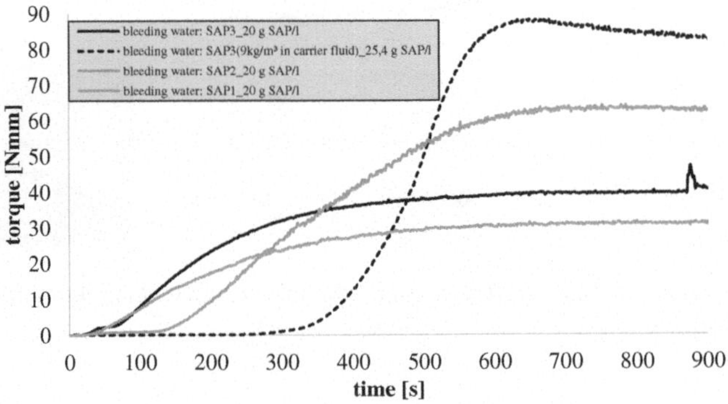

Fig. 6. Change in viscosity due to the addition of pre-suspended SAP.

As seen in Fig. 6 the water absorption and therefore increase in viscosity of the pre-suspended SAP compared to the dry addition starts delayed but due to the higher amount of SAP, referred to one m^3 grout approx. 9 kg, a higher water absorption could be achieved in the same time frame. When using 9 kg SAP per m^3 grout, the total amount of the SAP-solution is approx. 150 kg/m^3. Because of this additional fluid in the activated grout, the component A was modified in the way that the water content was decreased and the sand content increased. To ensure a sufficient flowability superplasticizer was used (Table 2).

Table 2. Base grout "B-60" and modified grout "mod. B-60".

grout/material [kg/m^3]	Cement	Water	Sand 0–1 mm	Sand 0–2 mm	Gravel 2–8 mm	Bentonite-suspension (conc. 6%)	Fly ash	Plasticizer
B-60	60	164	169	674	454	166	328	0
mod. B-60	60	68	272	735	598	166	328	2,4

With this pre-suspended SAP and the modified component A "mod. B-60", the shear strength was determined using different amounts of carrier fluid per gram SAP (Fig. 7, left). In Fig. 7, also a specimen of a physically activated grout is shown.

It can be seen, that with an SAP content of 9 kg/m^3 and an amount of carrier fluid of 16.7 g per gram SAP the requirements for the shear strength can be fulfilled after 30 min. Furthermore it gets obvious, that a reduction of the carrier fluid by 15 kg/m^3 (carrier fluid: 15 g sol./1 g SAP: 135 kg; 16.7 g sol./1 g SAP: 150 kg) leads to a significant increase in shear strength, immediately after addition. At the same time, it

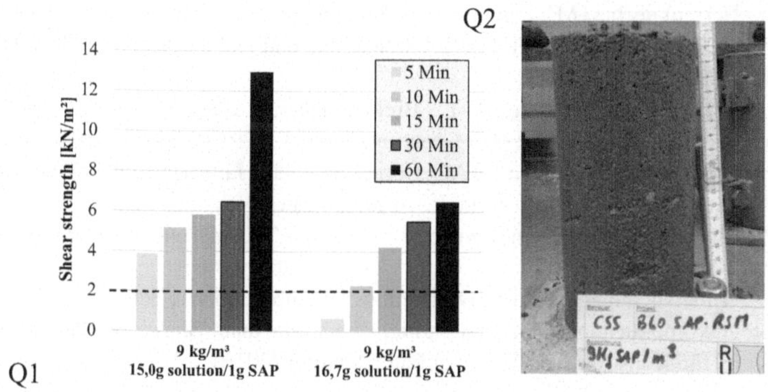

Fig. 7. Left: Shear strength development of the physically activated grout with pre-suspended SAP, Right: Specimen after 24 h (9 kg SAP/m^3).

was found that the reduction of the carrier fluid in component B resulted in a too rapid solidification after addition, so that the test specimens partially could only be produced with an insufficient homogeneity.

4 Conclusion

The use of superabsorbent polymers as component B for a rapid solidification of the annular gap grout proves to be a target-oriented and controllable instrument. Due to the enormous water absorption potential, a solidification and shear strength development occurs within a few minutes after addition, which meets the typical requirements for the early bedding of the tunnel lining. Clear correlations between the amount of SAP, the chemistry of component A and the content of carrier fluid can be identified. Higher shear strengths can be controlled by the cement content in component A and by the quantity of SAP or in the case of pre-suspended SAP by the quantity of the carrier fluid. From a process engineering point of view, a pre-suspension of SAP is preferable, since an insufficient mixing or an agglomeration of the SAP particles cannot be excluded when using a dry addition of SAP.

Acknowledgement. Financial support was provided by the German Science Foundation (DFG) in the framework of project B3 of the Collaborative Research Center SFB 837. This support is gratefully acknowledged.

References

EFNARC: Specification and Guidelines for the use of specialist products for Mechanised Tunnelling (TBM) in Soft Ground and Hard Rock. Farnham, UK (2005)

Hashimoto, T., Konda, T., Brinkman, J., Feddema, A., Kano, Y.: Simultaneous backfill grouting, pressure development in construction phase and in the long-term. In: Tunnelling. A Decade of Progress. GeoDelft 1995–2005. Taylor & Francis (2006)

Maidl, B., Herrenknecht, M., Maidl, U., Wehmeyer, G.: Maschineller Tunnelbau im Schildvortrieb. Ernst u. Sohn Verlag für Architektur und technische Wissenschaften GmbH, Berlin (2011)

Schulte-Schrepping, C., Youn-Cale, B.-Y., Breitenbücher, R.: Strength development of Two-Component Grouts for Annular Gap Grouting. Tunnel **3**(2018), 24–33 (2018)

TAC Corporation. www.tac-co.com. TAC Corporation. 11 January 2019. http://www.tac-co.com/enoutline/TAC%20Presentation.pdf. Accessed Jan 2017

Thewes, M., Budach, C.: Mörtel imTunnelbau - Stand der Technik und aktuelle Entwicklungen. BauPortal (2009)

Youn, B.-Y.: Untersuchungen zum Entwässerungsverhalten und zur Scherfestigkeits- entwicklung von einkomponentigen Ringspaltmörteln im Tunnelbau. Shaker Verlag, Schriftenreihe des Instituts für Konstruktiven Ingenieurbau, Bochum (2016)

Internal Curing Using Superabsorbent Polymers for Alkali Activated Slag-Fly Ash Mixtures

Ying Wang[1(✉)], Luca Montanari[2], W. Jason Weiss[3], and Prannoy Suraneni[1]

[1] Department of Civil, Architectural, and Environmental Engineering, University of Miami, Coral Gables, FL 33146, USA
{yxw1043, suranenip}@miami.edu

[2] SES Group and Associates LLC, Turner-Fairbank Highway Research Center, McLean, VA 22101, USA
luca.montanari.ctr@dot.gov

[3] School of Civil and Construction Engineering, Oregon State University, Corvallis, OR 97331, USA
jason.weiss@oregonstate.edu

Abstract. Increased shrinkage is often noted as a concern for alkali activated materials. In this study, two slag-fly ash paste and mortar mixtures with slag:fly ash ratios of 30:70 and 50:50 activated using 4M sodium hydroxide are formulated. The effects of two dosages of a commercial superabsorbent polymer (SAP) on the reaction heat, strength gain, autogenous shrinkage, drying shrinkage, and mass loss behavior are presented here. The SAP increases the heat of reaction of the alkali activated pastes, however, this increase is less than 5% at 7 days. The SAP slightly decreases the compressive strength of the alkali activated mortars, and this decrease is generally less than 10% at 1, 7, and 28 days. The SAP significantly reduces the ultimate autogenous shrinkage (by more than 50%) and reduces the drying shrinkage (by 15–30%) of the mortars. Mixtures with SAP have autogenous shrinkage between 50–300 $\mu\varepsilon$ and drying shrinkage between 600–700 $\mu\varepsilon$. When SAP is used, the mass loss in the mortars increases, however, the slope of the mass loss-drying shrinkage curve decreases. Shrinkage mitigation in the studied mixtures increases as the SAP dosage increases. Further studies on this system, and on other binders, activator combinations, and SAP types are currently ongoing.

Keywords: Superabsorbent polymers · Alkali activated materials · Fly ash · Slag

1 Introduction

Alkali activated materials and other alternative cementitious materials are gaining increasing acceptance due to their attractive properties including low CO_2 emissions, rapid early-age strength gain, and durability (Juenger et al. 2011). However, increased levels of shrinkage for such materials, specifically, for alkali activated slags and

W. P. Boshoff et al. (Eds.): SAP 2019, RILEM Bookseries 24, pp. 239–247, 2020.
https://doi.org/10.1007/978-3-030-33342-3_26

calcium aluminate cements, have been noted (Kraft et al. 2004; Atiş et al. 2009). In this study, the ability of superabsorbent polymer (SAP) to mitigate autogenous and drying shrinkage of alkali activated slag-fly ash mixtures is studied. In addition, the effect of partial SAP replacements on the shrinkage mitigation is also studied (Montanari et al. 2017). Prior research on partial replacement has revealed that the autogenous shrinkage mitigation increases as the partial replacement increases (Montanari et al. 2017). A limited amount of research has been performed on the use of SAP to reduce shrinkage in alkali activated materials. These works, one on slag (Song et al. 2016) and the other on slag:fly ash mixtures (with slag content less than 30%) (Tu et al. 2019) studied mixtures with significantly higher activator concentrations. Both showed autogenous shrinkage mitigation using SAP; drying shrinkage was not studied.

2 Materials and Methods

Commercially available Class F fly ash and slag conforming to standard specifications were used in this study. Their detailed characterization using X-ray fluorescence, X-ray diffraction, and laser diffraction is currently ongoing; some details for the fly ash are presented elsewhere (Khatibmasjedi et al. 2019). For alkali activation, a 4M NaOH solution is prepared using reagent grade material. Typical alkali activation for structural material requires higher activator concentrations and/or high temperature, therefore, it is expected that the materials that result here will not show considerable compressive strength. However, they are expected to be significantly more environmentally friendly that typical alkali activated materials. A commercially available angular crosslinked anionic polyacrylamide SAP was used; its detailed characterization and results with cementitious pastes are presented elsewhere (Montanari et al. 2017; Montanari et al. 2018). For the 4M NaOH, the SAP absorption is calculated to be 15 times, based on prior testing with the teabag method (Montanari et al. 2018). Alkali activated pastes and mortars were prepared using slag:fly ash ratios of 30:70 and 50:50. The mixture design was based on the results of preliminary testing – the liquid/solid ratio was 0.40; and the (standard) sand content was 57% of the total mixture by mass. The SAP dosages were 0.13% and 0.25% by mass of binder; these amounts showed significant autogenous shrinkage mitigation with portland cement mixtures (Montanari et al. 2017). Pastes and mortars were mixed according to standard procedures; details of such procedures are presented elsewhere (Khatibmasjedi et al. 2019). Isothermal calorimetry for 7 days was done on the pastes. The following tests were done on the mortars: compressive strength evolution (samples cured in a moist room till testing; tests run for 91 days), autogenous shrinkage (for 60 days), drying shrinkage (for 60 days), and mass loss measurements (for 60 days). These tests were run according to ASTM standard procedures, described in more detail elsewhere (Khatibmasjedi et al. 2019). Some of these tests are still ongoing and will be published in full elsewhere; the results presented here are somewhat preliminary.

3 Results and Discussion

Figure 1 shows the heat release normalized to binder mass (where binder is the sum of slag and fly ash mass) for 7 days. Repeat samples showed between 2–3% difference in heat release at 7 days. As expected (Tu et al. 2019), as the slag amount in the binder increases, the heat release increases significantly (by approximately 40%). The effect of SAP is subtle, but evident. As the amount of SAP increases, the heat release increases. With the larger SAP dosage, the heat release increases by approximately 5%. These results suggest an enhancement in the extent of reaction, consistent with results showing 5–10% increase in degree of hydration of cement when using similar amounts of (the same) SAP (Montanari et al. 2017).

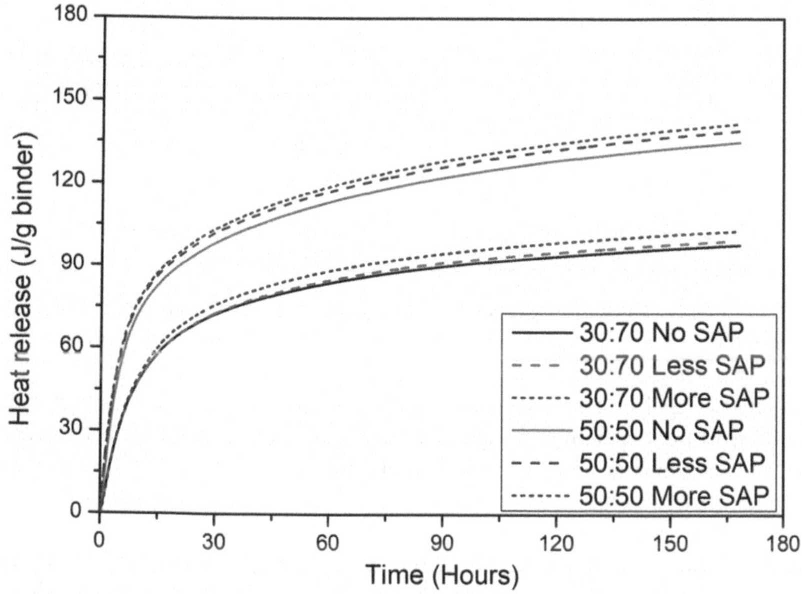

Fig. 1. Cumulative heat evolution of alkali activated paste mixtures.

Figure 2 shows the compressive strength evolution of the alkali activated mortars (average of three tested samples). Coefficient of variation (CV) of three tested samples at all ages was generally between 5 to 10%. Compressive strengths are low, with values at 28 days below 25 MPa. The compressive strength increases as the amount of slag increases; however, this difference is not obvious at later ages. The SAP generally decreases the compressive strength. This decrease is not always seen, with an average reduction lower than 10% among the considered mortars. This statistical significance of this value may be low considering the CV values (5–10%). Nevertheless, these results are consistent with other studies in literature (Song et al. 2016; Tu et al. 2019).

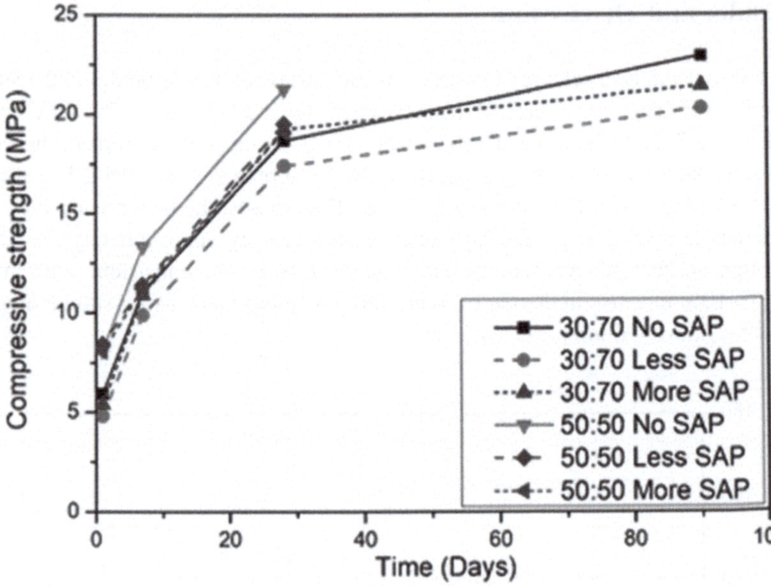

Fig. 2. Compressive strength evolution of alkali activated paste mixtures.

Figures 3 and 4 show the autogenous shrinkage of the 30:70 and 50:50 slag-fly ash mortar mixtures, respectively. As the slag amount increases, the autogenous shrinkage increases (by 100–250 microstrains (με)), qualitatively consistent with findings elsewhere (Tu et al. 2019). The SAP substantially decreases the autogenous shrinkage by about 100–250 με, depending on age, SAP dosage, and slag amount. The decrease is more than 50%, and in agreement with literature (Tu et al. 2019). The autogenous shrinkage decreases as the SAP amount increases, consistent with literature (Montanari et al. 2017).

Figures 5 and 6 show the drying shrinkage of the 30:70 and 50:50 slag-fly ash mortar mixtures, respectively. As the slag amount increases, the drying shrinkage increases (by 100–300 με). Increase in drying shrinkage with increasing slag amount is generally expected, however, drying shrinkage in alkali activated materials is notoriously complex (Thomas et al. 2017). The SAP reduces the drying shrinkage (by 100–250 με; 15–30%), with increased reduction as the amount of SAP increases. The reductions in drying shrinkage with SAP are almost exactly the same as the reductions in autogenous shrinkage, suggesting that the SAP is only reducing the autogenous part of the drying shrinkage; but further investigation is required (De la Varga et al. 2018).

Figures 7 and 8 show the mass loss-drying shrinkage behavior of the 30:70 and 50:50 slag:fly ash mortar mixtures, respectively. The behavior is *generally* linear, although deviations clearly exist, and the linearity is less readily apparent when compared to portland cement mixtures (Thomas et al. 2017; Khatibmasjedi et al. 2019). As the slag amount increases, the mass loss decreases but the slope of the mass loss-drying shrinkage curve increases approximately 2 times. As the SAP amount increases, the mass loss slightly increases, but the slope of the mass loss-drying shrinkage curve

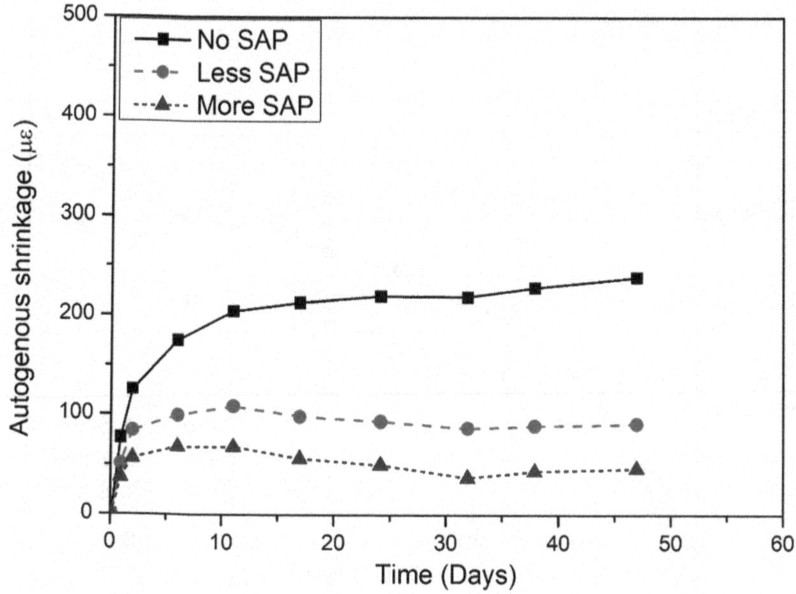

Fig. 3. Autogenous shrinkage in 30:70 slag-fly ash alkali activated mortars.

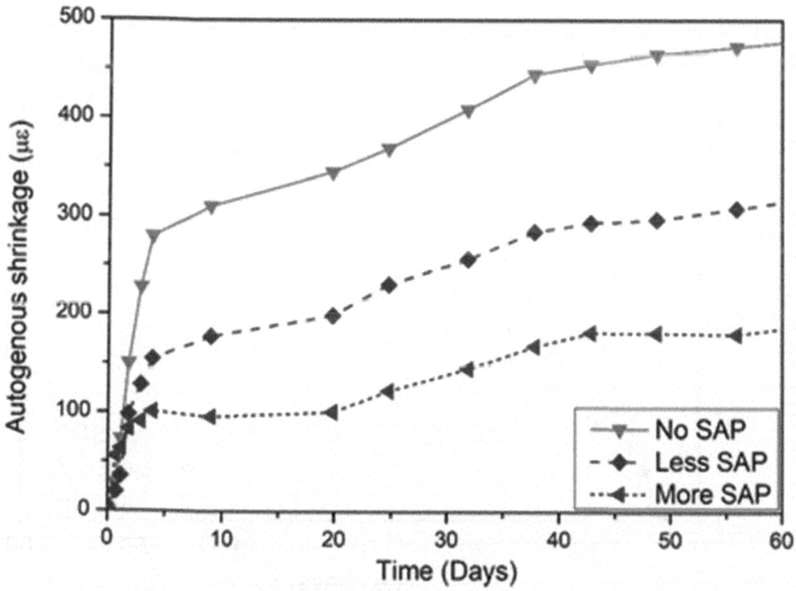

Fig. 4. Autogenous shrinkage in 50:50 slag-fly ash alkali activated mortars.

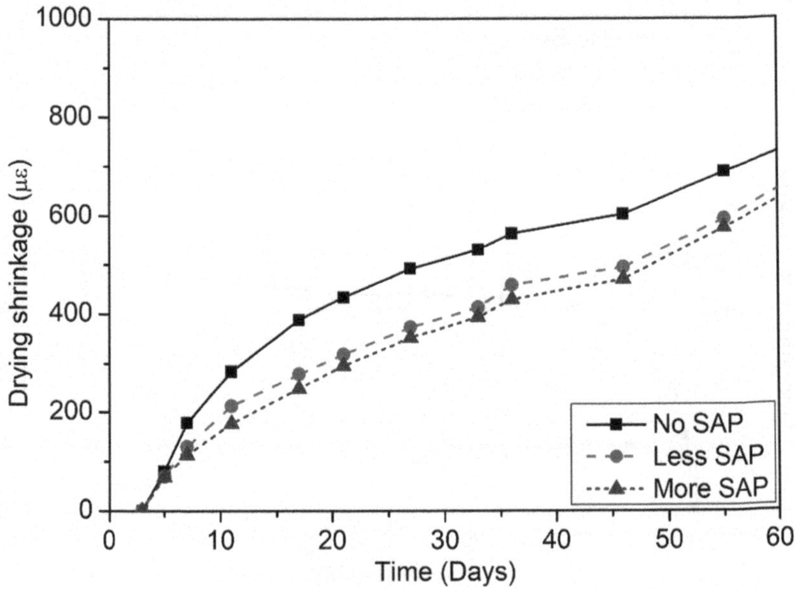

Fig. 5. Drying shrinkage in 30:70 slag-fly ash alkali activated mortars.

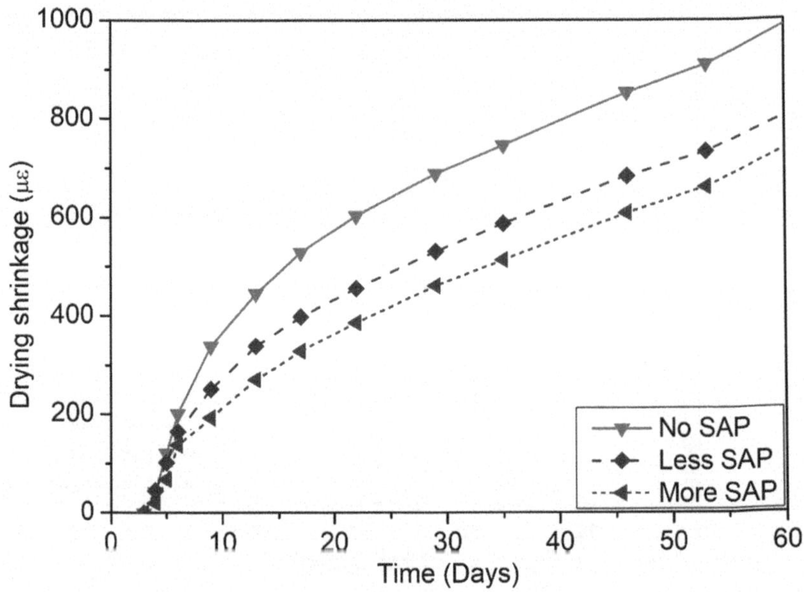

Fig. 6. Drying shrinkage in 50:50 slag-fly ash alkali activated mortars.

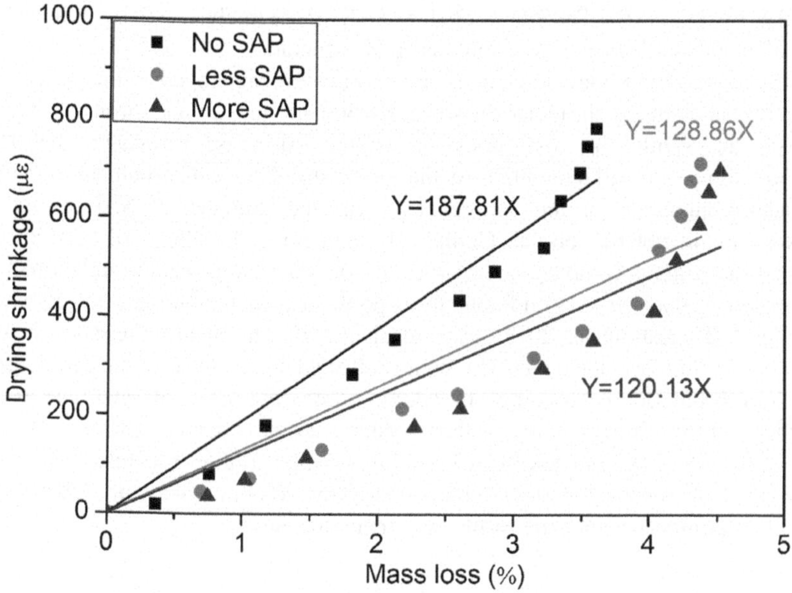

Fig. 7. Relationship between drying shrinkage and mass loss in 30:70 slag-fly ash alkali activated mortars.

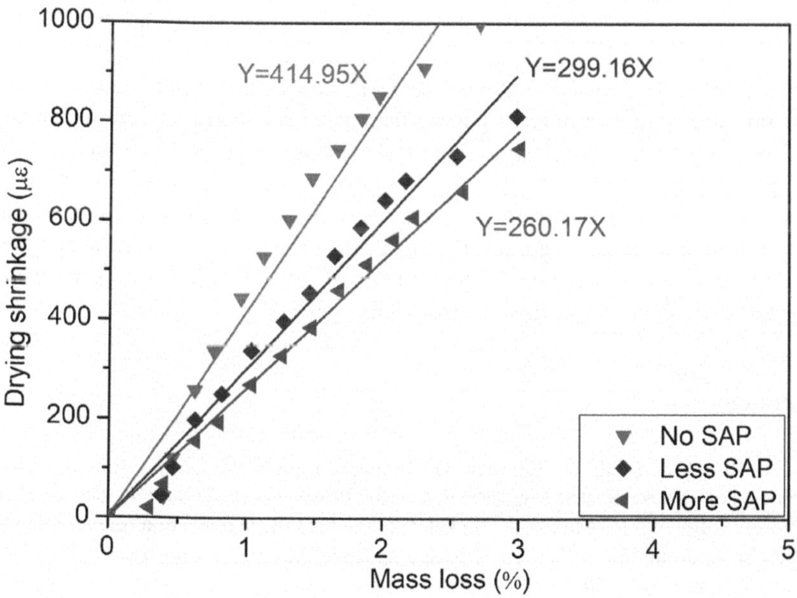

Fig. 8. Relationship between drying shrinkage and mass loss in 50:50 slag-fly ash alkali activated mortars.

decreases by up to 60%. The effects of slag and SAP are likely caused by differences in water availability, degree of reaction, and pore structure.

When considering both autogenous and drying shrinkage together, the mortars with SAP have autogenous shrinkage between 50–300 με and drying shrinkage between 600–700 με. While the SAP does substantially decrease shrinkage, the drying shrinkage values are still quite high. As the drying shrinkage values in these mortars are significantly higher than the autogenous shrinkage, the use of SAP to mitigate shrinkage might not be optimal. Completely explaining the effects of SAP in alkali activated materials is complex as the processes of autogenous and drying shrinkage in such materials may be very different from portland cements and are not completely understood (Thomas et al. 2017; Fang et al. 2019). The results presented here are somewhat preliminary; the data is still being collected and analysis of the shrinkage and mass loss behaviors is ongoing. In addition, measurements of later age relative humidity and the states of water in these systems are also being performed. As properties of fly ash, slag, and SAP can be quite variable, further research will focus on extending these results for other binders, activator combinations, and SAP types in order to determine whether the results are generalizable.

4 Conclusions

The effects of two dosages of a commercial SAP on the properties on slag-fly ash mixtures with slag:fly ash ratios of 30:70 and 50:50 activated using 4M sodium hydroxide were studied. The SAP slightly increases the heat of reaction (5% increase at 7 days) in pastes and slightly decreases the compressive strength in mortars (generally less than 10% at 1, 7, and 28 days). The SAP reduces the shrinkage substantially - autogenous shrinkage is reduced by more than 50% and the drying shrinkage by 15–30%. The mortars with SAP have autogenous shrinkage between 50–300 με and drying shrinkage between 600–700 με. The SAP slightly increases the mass loss, however, it decreases the slope of the mass loss-drying shrinkage curve. As the drying shrinkage values in these mortars are significantly higher than the autogenous shrinkage, the use of SAP to mitigate shrinkage might not be optimal. The observed results are generally similar to those seen with portland cement mixtures.

References

Atiş, C.D., Bilim, C., Çelik, Ö., Karahan, O.: Influence of activator on the strength and drying shrinkage of alkali-activated slag mortar. Constr. Build. Mater. **23**(1), 548–555 (2009)

De la Varga, I., Spragg, R., Muñoz, J., Nickel, C., Graybeal, B.: Application of internal curing in cementitious grouts for prefabricated bridge concrete elements connections. Adv. Civ. Eng. Mater. **7**(4), 628–643 (2018)

Fang, G., Bahrami, H., Zhang, M.: Mechanisms of autogenous shrinkage of alkali-activated fly ash-slag pastes cured at ambient temperature within 24 h. Constr. Build. Mater. **171**, 377–387 (2019)

Juenger, M.C.G., Winnefeld, F., Provis, J.L., Ideker, J.H.: Advances in alternative cementitious binders. Cem. Concr. Res. **41**(12), 1232–1243 (2011)

Khatibmasjedi, M., Ramanathan, S., Suraneni, P., Nanni, A.: Shrinkage behavior of cementitious mortars mixed with seawater. Adv. Civ. Eng. Mater. **8**(2), ACEM20180110 (2019)

Kraft, L., Engqvist, H., Hermansson, L.: Early-age deformation, drying shrinkage and thermal dilation in a new type of dental restorative material based on calcium aluminate cement. Cem. Concr. Res. **34**(3), 439–446 (2004)

Montanari, L., Suraneni, P., Weiss, W.J.: Accounting for water stored in superabsorbent polymers in increasing the degree of hydration and reducing the shrinkage of internally cured cementitious mixtures. Adv. Civ. Eng. Mater. **6**(1), 583–599 (2017)

Montanari, L., Suraneni, P., Tsui-Chang, M., Villani, C., Weiss, J.: Absorption and desorption of superabsorbent polymers for use in internally cured concrete. Adv. Civ. Eng. Mater. **7**(4), 547–566 (2018)

Song, C., Choi, Y.C., Choi, S.: Effect of internal curing by superabsorbent polymers – internal relative humidity and autogenous shrinkage of alkali-activated slag mortars. Constr. Build. Mater. **123**, 198–206 (2016)

Thomas, R.J., Lezama, D., Peethamparan, S.: On drying shrinkage in alkali-activated concrete: improving dimensional stability by aging or heat-curing. Cem. Concr. Res. **91**, 13–23 (2017)

Tu, W., Zhu, T., Fang, G., Wang, X., Zhang, M.: Internal curing of alkali-activated fly ash-slag pastes using superabsorbent polymer. Cem. Concr. Res. **116**, 179–190 (2019)

Effect of High Plasticiser Dosage on Ultra High-Performance Fibre Reinforced Concrete (UHPFRC)

Megan Weyers[1] and Elsabe P. Kearsley[2(✉)]

[1] Department of Civil Engineering, University of Pretoria, Lynnwood Road, Pretoria, South Africa
meganweyers@icloud.com

[2] Department of Civil Engineering, University of Pretoria, Lynnwood Road, Private Box X20, Pretoria, South Africa
elsabe.kearsley@up.ac.za

Abstract. The overall objective of this study was to develop an optimised UHPFRC matrix based on the modified Andreasen and Andersen optimum particle packing model by using available South African materials. The focus of this study was to determine the optimum combined fibre and superplasticiser content for UHPC by using a response surface design. The UHPFRC was appropriately designed, produced and tested. The flowability, density and mechanical properties of the designed UHPFRC were measured and analysed. It is clear from the results, that both the fibre and superplasticiser content play a significant role in the flowability of the fresh concrete. The addition of fibres significantly improved both the compressive and indirect tensile strength of the UHPFRC. However, the addition of superplasticiser slightly decreased both the compressive and indirect tensile strength of the UHPFRC. Both the fibre and superplasticiser content had an insignificant effect on the modulus of elasticity. The results show that the superplasticiser content can be increased if a more workable mix is required without decreasing the strength significantly. The study demonstrate that it is possible to efficiently produce a dense and workable UHPFRC with relatively low binder amount and low fibre content. This can result in more cost-effective UHPFRC, thus improving the practical application thereof.

Keywords: Ultra-High-Performance Fibre Reinforced Concrete (UHPFRC) · Modified Andreasen and Andersen particle packing model · Response surface design · Superplasticiser content · Fibre content

1 Introduction

Ultra-High-Performance Concrete (UHPC) is described as concrete with a minimum compressive strength of 150 MPa (Habel et al. 2008). When fibres are added, the term Ultra-High-Performance Fibre Reinforced Concrete (UHPFRC) is used. Abbas et al. (2016) state that UHPC is a construction material that demonstrates enhanced durability and mechanical properties. Economic construction can be promoted through decreasing

W. P. Boshoff et al. (Eds.): SAP 2019, RILEM Bookseries 24, pp. 248–255, 2020.
https://doi.org/10.1007/978-3-030-33342-3_27

the cross-section dimensions of members, thereby reducing materials used and lowering the installation costs as well as reducing the maintenance and increasing the lifespan (Richard and Cheyrezy 1995).

Sustainable development is currently a critical global issue and construction industries need to save energy and lower the environmental impact. The use of UHPFRC in the construction industry has been limited due to its high initial material cost, perceived CO_2 emission and energy consumption (Habert et al. 2013). Weina et al. (2017) state that supplementary cementitious materials can be used to reduce the cost and environmental impact of UHPFRC. These materials include silica fume (SF), ground granulated blast-furnace slag (GGBS) and fly ash (FA). According to Yu et al. (2014), the published mix compositions for UHPFRC are mostly given without any detailed explanation. To date questions regarding optimising the concrete mix composition for UHPFRC and using binders efficiently remains unanswered.

According to Neville (1995), the hydrated cement fraction is only about 40% of the cement used in UHPFRC. Thus, most of the cement particles just act as filler materials. An optimal particle packing of granular ingredients of concrete is the key to a durable and sustainable concrete (Hüsken 2010). A reduction in the cement content should be possible by using filler materials. These filler materials must be included in the entire grading of the mix together with the binding materials, in order to achieve the densest possible packing. The optimised particle packing of granular ingredients in concrete can result in a denser granular structure, with enhanced mechanical properties and improved porosity.

Beside the design of the concrete matrix, the efficient use of steel fibres is also crucial (Kim et al. 2008). The steel fibres are added into the mix, without considering the efficiency of the amount of fibres used. Kahanji et al. (2017), recommends reducing the unit cost of UHPFRC by reducing the steel fibre content. According to Graybeal (2007), tensile properties are greatly enhanced by the addition of steel fibres, however, the steel fibres significant decrease the flowability of the fresh concrete. According to Abbas et al. (2016), the effective addition of superplasticiser can improve the workability of UHPFRC. Various studies (Fehling et al. 2008) used superplasticiser dosages ranging between 1% and 8% by cement weight for enhancing the workability of UHPC mixtures. An optimum fibre content is required to balance the workability and mechanical performance (Wille et al. 2011). According to Li et al. (2016), it is necessary to investigate the effect of steel fibres on the flowability of UHPC before their utilisation.

Consequently, the objectives of this study were to efficiently develop an optimised concrete matrix based on the modified Andreasen and Andersen optimum particle packing model by using available South African materials. The focus of this study was to determine the optimum combined fibre and superplasticiser content for UHPC by using a response surface design.

2 Experimental Programme

A Class I 52.5R Portland cement with a relative density of 3.14, was used with Undensified Silica Fume (USF) and Unclassified Fly Ash (FA) as cementitious materials. Silica sand with a maximum size of 1.50 mm together with dolomite sand with a maximum size of 4.75 mm were chosen as the fine aggregate and dolomite stone with a maximum size of 6.70 mm was chosen as the coarse aggregate. Straight micro-steel fibres with a length of 13.0 mm and a diameter of 0.2 mm were used in all mix designs. The tensile strength of these fibres is 2500 MPa. Polycarboxylate ether-based superplasticiser was added to improve the workability of the mixes.

Fuller and Thompson (1907) showed that the properties of concrete are affected by the packing of concrete aggregates. They concluded that a continuous grading of the aggregates in the concrete mixture can help to significantly improve the concrete properties. However, the minimum particle size was not incorporated. Funk and Dinger (1994) proposed a modified model based on the Andreasen and Andersen Equation. In this study, the modified Andreasen and Andersen particle packing model, as shown in Eq. (1), was utilised to optimise the concrete mixture composition for all the granular materials:

$$P(D) = \frac{D^q - D^q_{min}}{D^q_{max} - D^q_{min}} \tag{1}$$

where D is the particle size (μm), P(D) is a fraction of the total solids being smaller than size D, D_{max} is the maximum particle size (μm), D_{min} is the minimum particle size (μm) and q is the distribution modulus (Yu et al. 2014). It is currently understood that by applying different q values, different types of concrete can be designed by using Eq. (1), since q determine the proportions (%) of particles between the fine and coarse particles in the mixture (Brouwers and Radix 2005). Based on the recommendation by Hunger (2010), the value of q is fixed at 0.23 in this study, since numerous fine particles are used to produce the UHPFRC matrix. According to Yu et al. (2014), the mass proportions (%) of each individual material in the concrete mix are adjusted until the deviation between the target curve and the composed mix, expressed as the sum of squares of the residuals (RSS) at defined particle sizes, is minimised (see Eq. (2)). The composition of the concrete is then considered optimal.

$$RSS = \frac{\sum_{i=1}^{n} \left(P_{mix}\left(D_i^{i+1}\right) - P_{tar}\left(D_i^{i+1}\right)\right)^2}{n} \tag{2}$$

where P_{mix} is the composed mix, the P_{tar} is the target grading calculated from Eq. (1), and n is the number of points (between D_{min} and D_{max}) used to calculate the deviation. The grading curve for the composed mix is shown in Fig. 1 (RSS = 15.9; R^2 = 0.97), together with the target curve obtained from the modified Andreasen and Andersen particle packing model as well as the Fuller curve.

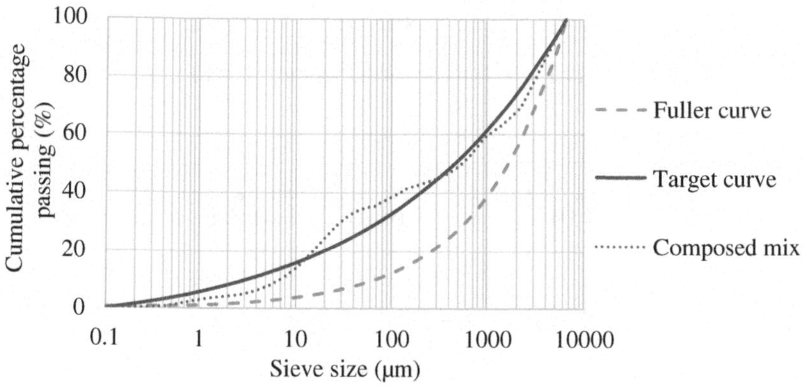

Fig. 1. Particle size distribution of the Fuller curve, target curve and the resulting grading curve of the composed mix.

According to Montgomery (2001), response surface methodology (RSM) is a set of statistical and mathematical methods that assists in the analyses of responses that are influenced by several variables. The objective of RSM is the optimisation of the response. RSM has a variety of response surface designs that can be used. Central Composite Design (CCD) was used in this study. CCD is an augmented version of the factorial design with centre and axial points (Montgomery 2001). CCD is useful in response surface modelling because it makes it possible to efficiently develop first-order and second-order models. To model curvature, a second-order model is required. Equation (3) is the general form of the second order polynomials.

$$y = \beta_0 + \sum_{j=1}^{k} \beta_j x_j + \sum_{i<j} \sum_{i<j} \beta_{ji} x_i x_j + \sum_{j=1}^{k} \beta_{jj} x_j^2 + \in \quad (3)$$

where $\sum_{j=1}^{k} \beta_{jj} x_j^2$ is the quadratic effect of a single variable, $\sum_{i<j} \sum_{i<j} \beta_{ji} x_i x_j$ is the interaction effect between two variables, β_0, β_i, β_{ii}, β_{ij} is the regression coefficients, x_i, x_j is the investigated factors, k is the number factors and $\in$ is the observed noise error. CCD consists of 2k axial points, 2^k factorial points and n number of centre runs. The number of independent variables is represented by k. The three different types of points that define the region of interest for a two-factor design are illustrated in Fig. 2. The factorial points are positioned on the corners of the square, the axial points are positioned a distance alpha away from the centre point in the positive and negative sides of each axis and the centre points are positioned at the intersection of the two axes. Alpha and the number of centre points specifies whether a response surface design is rotatable. The response surface design has an improved quality of prediction when it is rotatable ($\alpha = \pm 1.414$).

The mix designs obtained from the central composite design is tabulated in Table 1. The central mix was cast three times to test the repeatability of the experiment. The mixes were cast in a random order.

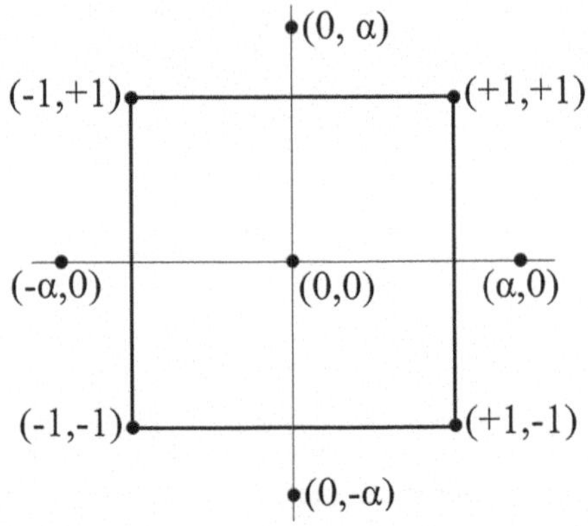

Fig. 2. Central composite design for two variables.

Table 1. Mix compositions from the central composite design (kg/m^3).

No.	C	USF	FA	W	D-A	D-S	S-S	SP	SMF
1	547	147	147	147	253	1014	218	15.0	157
2	542	146	146	146	251	1005	216	17.2	213
3	543	146	146	146	252	1007	217	22.5	157
4	538	145	145	145	249	997	214	22.5	236
5	543	146	146	146	252	1007	217	22.5	157
6	543	146	146	146	252	1007	217	22.5	157
7	540	145	145	145	250	1000	215	30.0	157
8	550	148	148	148	255	1019	219	17.2	102
9	549	148	148	147	254	1017	219	22.5	78.5
10	537	145	145	144	249	995	214	27.8	213
11	544	147	147	146	252	1009	217	27.8	102
RD	3.14	2.45	2.22	1.00	2.85	2.91	2.67	1.06	7.85

C: cement, USF: undensified silica fume, FA: fly ash, W: water, D-A: dolomite aggregate, D-S: dolomite sand, S-S: silica sand, SP: superplasticiser, SMF: short micro fibre, RD: relative density.

All the dry materials, except for USF, were mixed together for approximately one minute. The USF was excluded from the dry mixing since it was undensified and would not mix with the other materials, but instead becomes airborne. Thereafter, the water and admixture were added to the mix, after which the USF was added. After approximately five minutes, the fibres were evenly distributed by hand. After the fibres were added, the concrete was mixed for another four minutes. A flow table test was conducted to measure the flowability of the UHPFRC mixes. To ensure proper compaction, all the casting was done on a vibrating table and the specimens were vibrated for approximately one minute. All the specimens were then left to cure under a curing blanket in a temperature-controlled room (25 °C) for one day. After one day, all the specimens were demoulded and placed in a 25 °C water bath until the day of testing (28 days).

3 Experimental Results and Discussion

The surface plot for the flow of the fresh UHPFRC mixtures is shown in Fig. 3(a). The data illustrates the variation of the flow of UHPFRC with different short micro fibre and superplasticiser contents. The flow varied between 180 mm and 430 mm. A linear model was used and the R^2-value for the surface plot was 0.84. It is clear from the surface plot, that both the fibre and superplasticiser content play a significant role in the flowability of the fresh concrete. Although increased fibre content or decreased superplasticiser content reduced the workability, all mixtures were sufficiently workable as to allow the facile placement and compaction of concrete specimens.

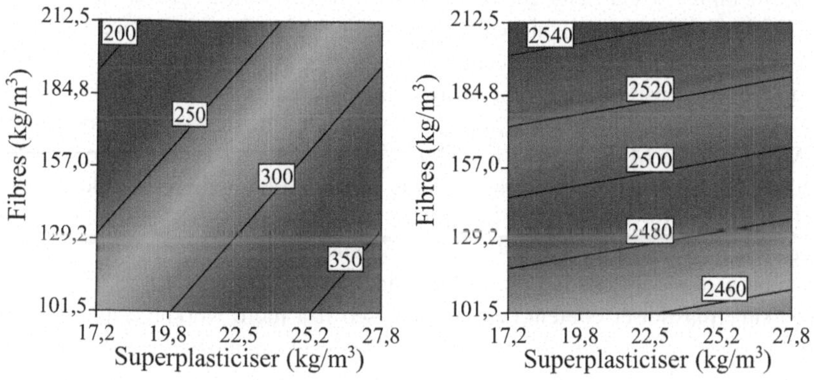

Fig. 3. Surface plot for (a) flow in mm and (b) density in kg/m³.

The density obtained from the UHPFRC cubes is illustrated in Fig. 3(b). The data shows the variation of the density of the UHPFRC with different short micro fibre and superplasticiser contents. The density varied between 2430 kg/m³ and 2550 kg/m³. A linear model was used and the R^2-value for the surface plot was 0.96. It is important to note that the density decreased with the addition of superplasticiser and increased with the addition of fibres. The densest mixes obtained were the ones with high fibre amounts and low superplasticiser contents.

The cube compressive strength of the UHPFRC mixes is illustrated on the surface plot in Fig. 4(a). The compressive strength varied between 144 MPa and 168 MPa. A linear model was used and the R^2-value for the surface plot was 0.83. The addition of fibres significantly improved the compressive strength of the UHPFRC. However, the addition of superplasticiser slightly decreased the compressive strength of the UHPFRC.

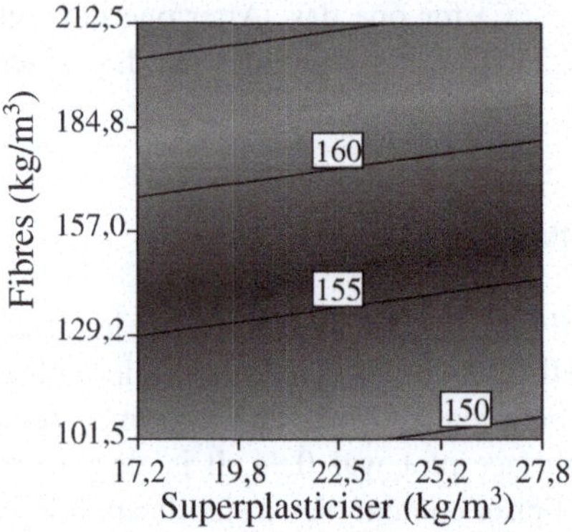

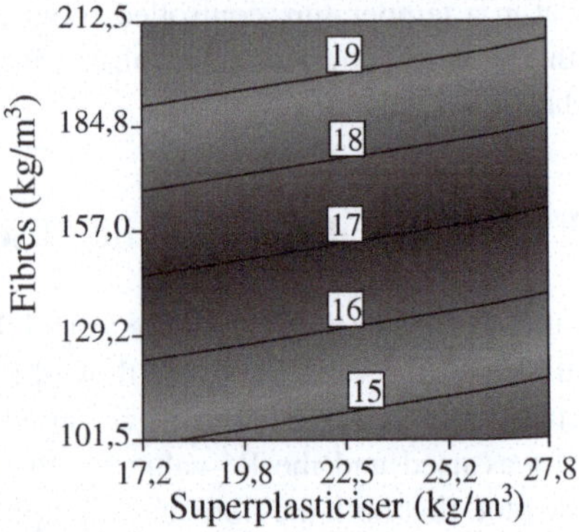

Fig. 4. Surface plot for (a) the compressive strength in MPa and (b) the indirect tensile strength in MPa.

The modulus of elasticity for the UHPFRC mixes varies between 49.5 GPa and 53.0 GPa and the results obtained a R^2-values of 0.96. The difference between the mixes was deemed to be insignificant.

The indirect tensile strength (splitting cylinder strength) of the UHPFRC mixes is shown on the surface plot in Fig. 4(b). The indirect tensile strength varied between 13.6 MPa and 20.5 MPa. A linear model was used and the R^2-value for the surface plot was 0.89. The addition of fibres significantly improved the indirect tensile strength of the UHPFRC. On the other hand, the addition of superplasticiser slightly decreased the indirect tensile strength of UHPFRC. The indirect tensile strength results follow the same trend as the compressive strength results.

The superplasticiser content can be increased if a more workable mix is required without decreasing the strength significantly. The results demonstrate that it is possible to produce workable UHPFRC with relatively low binder and fibre amounts.

4 Conclusion

The study presents a method to effectively develop optimised UHPFRC by using available South African materials. Working towards an efficient application of binders, superplasticiser and fibres in UHPFRC, the modified Andreasen and Andersen particle packing model and surface response design methodology were utilised. The UHPFRC was appropriately designed, produced and tested. Using the modified Andreasen and Andersen particle packing model and surface response design methodology, it is possible to efficiently produce a dense UHPFRC with a relatively low binder amount, low fibre content and good workability. This can make UHPFRC more cost-effective and improve the practical application thereof.

References

Abbas, S., Nehdi, M.L., Saleem, M.A.: Ultra-high-performance concrete: mechanical performance, durability, sustainability and implementation challenges. Int. J. Concr. Struct. Mater. **10**, 271–295 (2016)

Brouwers, H.J.H., Radix, H.J.: Self compacting concrete: theoretical and experimental study. Cem. Concr. Res. **35**, 2116–2136 (2005)

Fehling, E., Schmidt, M., Stuerwald, S.: Proceedings of the 2nd International Symposium on Ultra High Performance Concrete, Kassel, Germany, p. 902 (2008)

Fuller, W.B., Thompson, S.E.: The laws of proportioning concrete. Trans. Am. Soc. Civ. Eng. **33**, 222–298 (1907)

Funk, J.E., Dinger, D.R.: Predictive Process Control of Crowded Particulate Suspensions, Applied to Ceramic Manufacturing. Kluwer Academic Publishers, Boston (1994)

Graybeal, B.: Compressive behaviour of ultra-high-performance fibre-reinorced concrete. ACI Mater. J. **104**, 146–152 (2007)

Habel, K., Charron, J.P., Braike, B., Hooton, R.D., Gauvreau, P., Massicotte, B.: Ultra-high-performance fibre reinforced concrete mix design in central Canada. Can. J. Civ. Eng. **35**, 217–224 (2008)

Habert, G., Denarié, E., Šanja, A., Rossi, P.: Lowering the global warming impact of ridge rehabilitation by using Ultra High-Performance Fibre Reinforced Concretes. Cem. Concr. Compos. **38**, 1–11 (2013)

Hunger, M.: An integral design concept for ecological self-compacting concrete. Ph.D. thesis, Eindhoven University of Technology, Eindhoven, the Netherlands (2010)

Hüsken, G.: A multifunctional design approach for sustainable concrete with application to concrete mass products. Ph.D. thesis, Eindhoven University of Technology, Eindhoven, the Netherlands (2010)

Kahanji, C., Ali, F., Nadjai, A.: Structural performance of ultra-high-performance fibre reinforced concrete beams. Struct. Concr. **18**, 249–258 (2017)

Kim, S., Park, J., Kang, S., Ryu, G.: Effect of filling method on fibre orientation and dispersion and mechanical properties of UHPC. In: Proceedings of the 2nd International Symposium on Ultra High Performance Concrete, Kassel, Germany, pp. 185–192 (2008)

Li, P.P., Yu, Q.L., Brouwers, H.J.H., Yu, R.: Fresh behaviour of ultra-high performance concrete. In: Proceedings of the 9th International Concrete Conference, Environment, Efficiency and Economic Challenges for Concrete, Dundee, Scotland, UK, pp. 635–644 (2016)

Montgomery, D.C.: Design and Analysis of Experiments, 5th edn. Wiley, New York (2001)

Neville, A.M.: Properties of Concrete. Longman House, Burnt Mil, Harlow (1995)

Prem, P.R., Bharatkumar, B.H., Iyer, N.R.: Influence of curing regimes on compressive strength of ultra-high-performance concrete. Sādhanā **38**(6), 1421–1431 (2013)

Richard, P., Cheyrezy, M.: Composition of reactive powder concretes. Cem. Concr. Res. **25**, 1501–1511 (1995)

Weina, M., Valipour, M., Khayat, K.H.: Optimization and performance of cost-effective ultra-high performance concrete. Mater. Struct. **50**, 29 (2017)

Wille, K., Naaman, A.E., El-Tawil, S.: Optimizing ultra-high-performance fibre-reinforced concrete. Concrete Int. **33**, 35–41 (2011)

Yu, R., Spiesz, P., Brouwers, H.J.H.: Mix design and properties assessment of ultra-high-performance fibre reinforced concrete. Cem. Concr. Res. **56**, 29–39 (2014)

Author Index

W. P. Boshoff et al. (Eds.): SAP 2019, RILEM Bookseries 24, pp. 257–258, 2020.
https://doi.org/10.1007/978-3-030-33342-3

The manufacturer's authorised representative in the EU is Springer
Nature Customer Service Centre GmbH, Europaplatz 3, 69115 Heidelberg,
Germany. If you have any concerns regarding our products, please
contact ProductSafety@springernature.com

Printed and bound by CPI Group (UK) Ltd, Croydon, CR0 4YY
25/09/2026
02222869-0012